Universitext

Universitext

Universitext is a series of textbooks that presents material from a wide variety of mathematical disciplines at master's level and beyond. The books, often well class-tested by their author, may have an informal, personal even experimental approach to their subject matter. Some of the most successful and established books in the series have evolved through several editions, always following the evolution of teaching curricula, to very polished texts.

Thus as research topics trickle down into graduate-level teaching, first textbooks written for new, cutting-edge courses may make their way into *Universitext*.

More information about this series at http://www.springer.com/series/223

Stephen Bruce Sontz

Principal Bundles

The Classical Case

 Springer

Stephen Bruce Sontz
Centro de Investigación en Matemáticas, A.C.
Guanajuato, Mexico

ISSN 0172-5939   ISSN 2191-6675   (electronic)
Universitext
ISBN 978-3-319-14764-2   ISBN 978-3-319-14765-9   (eBook)
DOI 10.1007/978-3-319-14765-9

Library of Congress Control Number: 2015934141

Mathematics Subject Classification (2010): 14D21, 53C05, 55R10, 58A05, 78A25, 81T13

Springer Cham Heidelberg New York Dordrecht London

Printed on acid-free paper

Springer International Publishing AG Switzerland is part of Springer Science+Business Media (www.springer.com)

A Lilia con muchísimo amor

Preface

Yet another text on differential geometry! But *why*? The answer is because this book is focused on one particular topic in differential geometry, that is, principal bundles. And the aim is to get the reader to an understanding of that topic as efficiently as possible without oversimplifying its foundations in differential geometry. So I aim for an early arrival at the applications in physics that give this topic so much of its flavor and vitality. But this is just the view from the classical perspective, which is why we speak of classical differential geometry (sometimes simply classical geometry) and refer to the topic of this volume as classical principal bundles.

There is a saying that all is prologue. As with any saying, it has a limited range of applicability. But here it is relevant since the topic of this volume serves as preparation for the corresponding quantum (or noncommutative) case, which will be the topic of the companion volume [44]. That is usually known as noncommutative geometry, and the corresponding bundles are known as quantum principal bundles. That is a newer field still being studied and refined by contemporary researchers. While none of this quantum theory has reached a stage of general consensus on "what it is all about" or on what is the "best" approach, we do know enough about the various approaches to be sure that certain common elements will dominate future research. And a lot of the intuition and motivation for the quantum theory comes from the classical theory presented in this volume, which serves as prologue.

Those are the goals of the two volumes. And as far as I am aware, these two texts together comprise the only published books devoted exclusively to the exposition of principal bundles in these two settings. That's *why*.

For *whom* are these books intended? The intended audience for these two volumes are folks who have some interest in learning topics of current interest in geometry and their relation with mathematical physics. I have basically two groups in mind.

The first group comprises mathematicians who have not seen the applications of principal bundles in physics. For them, the first part of this book should be more accessible since concepts are defined and theorems are proved, all according to the modern criteria of mathematical rigor. The second part of this book might require more effort for them since some amount of physical intuition is always helpful for understanding the applications.

The second group is physicists with some familiarity with the standard topics in physics (such as classical electrodynamics) but who have not seen how the mathematics used for those topics works in detail as an aspect of geometry. For this group, the first part of the book will likely require more work. So I have tried to include a lot of motivation and intuition to make that part more accessible for them. Consequently, my more mathematically inclined readers may at times find too many details for their level of expertise. Please be patient and understand why this is happening. In short, this book is meant to help bridge the communication gap between the two communities of physicists and mathematicians. In brief, that's for *whom*.

And *how*? I have tried to make the presentation throughout the book as intuitive as I possibly could. Ideas and intuitions are emphasized as being important elements in formulating the theory. I even think that intuition and ideas are more important than rigor, though they should not exclude rigor but rather precede it to give motivation. This is the "tricky bit," as the saying goes. But intuition is not a sufficient condition for getting things right, nor is rigor. Intuition in physics has led even Nobel laureates to arrive at ideas in contradiction to experiment. As an example, there were those who rejected the initial conjecture of parity being violated in the weak interactions. And mathematicians, even armed with rigor, have also fallen into error. The original "proof" of the four-color theorem comes to mind. Nobody is perfect!

Also, there are plenty of exercises to keep the reader active in the creation process that is essential to understanding, as Feynman so neatly puts it in the quote at the start of Chapter 1. The intention is that the exercises should give the reader hands-on experience with the ideas and intuitions. Some exercises are rather routine, while others are meant to challenge the reader. And I do not even tell you which ones are the routine exercises and which ones are the tough nuts to crack. All of that recognition is part of your own learning process. However, there is an appendix with further discussion of the exercises, including hints. I recommend that you hold off on looking at this (as well as the multitude of other texts on these topics) for as long as possible. If not longer. Even so, some problems may remain quite difficult. But this manner of presentation is intentional and is meant to help the reader in the long run.

I have been accused of asking the reader to write major parts of the book by doing difficult exercises with no hints or suggestions in the text itself. Well, that is true. The more of this book that you, my kind reader, can write, the better off you will be. And sometimes I even give an exercise before acquainting you with the usual tools required for solving it. So gratification is not always immediate. Welcome to the real world of science! This is meant to be a difficult book, much more so than the usual introductory technical texts on the market. That's *how*.

I have avoided a strictly historical approach since the way we have arrived at this theory can obscure its logical structure. For example, the Dirac monopole appeared in the 1930s but is presented here near the end of the book as a special case of the theory developed decades later.

Mathematicians should be aware that physics is a discipline with its own logical structure, though that is not always clear to mathematicians. Ideas flow into other ideas. Results that involve very particular ideas (such as the Yang–Mills theory) then motivate generalizations, which in turn illuminate the original particular case (among other examples) in new ways. This is not the paradigm of definition, theorem, proof. Nonetheless, it is a logical structure in its own right. In the sections on applications to physics, I have tried to include a lot more than the usual amount of motivation based on physics ideas in the hope of helping my mathematically inclined readers. The physicists among my readers might find this boring, and so I kindly request their patience and understanding. But they might be more challenged by the translation of these ideas into a mathematically rigorous geometric formulation.

These are texts meant for learning the material, for either actual students or others who want to learn about principal bundles. The pace is designed for use in a course or for self-study. My experience is that there is enough material here for a one-year graduate course although undergraduates with an adequate background and motivation could profit from such a course.

These are not definitive treatises meant only for the purpose of giving experts a place to look up all the variants of theorems. The experts should not expect too much from either of these volumes, except perhaps as a way of organizing these topics for their own courses. I have not included a multitude of fascinating topics, both in geometry and in physics. These introductory texts should serve to motivate the reader to carry on with study and research in what is *not* presented as well as in what is presented.

The prerequisites for this volume consist of a bit of many things, such as the basic vocabulary of group theory (not to be confused with group therapy); a smattering of linear algebra, including tensor products (even though this will be briefly reviewed); multivariable calculus of real variables, including vector calculus notation; at least a vague appreciation of what a nonlinear differential equation is; some talking points from elementary topology (such as compact, open, closed, Hausdorff, continuous, etc.); and something about categories as a system of notation and diagrams, but not as a theory. And a bit of physics for the chapters with such applications may be useful though I have tried to keep that material as self-contained as possible.

More than anything else, the present volume is my own personal take on classical differential geometry, the role that principal bundles play in it, and how all this relates to physics. Since the goal of the book, as its title reveals, is the exposition of the theory of principal bundles in the classical case, not all of the quite fascinating topics of differential geometry will be presented in the first chapters, but more than enough to get us to that goal. But to understand what principal bundles are "good for" requires meaningful examples as well. And since my personal motivation comes from physics, I devote the remainder of the book to several chapters on examples taken from physics. These chapters make this book more than just another introduction to principal bundles. I tried to make these as self-contained in terms of the physics content as I could, but I rely heavily on the mathematical theory developed earlier in the book. I have always felt that learning physics with little more than a solid mathematics background is far easier than the other way around. So I beseech my physicist readers not to despair during the first part of this book. The trip may be tougher than you'd like, but the payoff should be well worth it.

Nothing here is original. I can merely hope that my way of presenting this material has an appealing style that makes things intuitive and accessible. This book is my own personal response to the challenge Richard Feynman posed to himself, and to no one else, in the famous last blackboard quote cited in Chapter 1.

This is also the place to give thanks to all those who helped me create my own understanding of these topics. It turns out to be the sum of many contributions, some small, some quite large. And all over a long period of time. Many are people whom I have never met but only know through their publications. Names like Courant and Robbins come to mind since they were the first to show me how to think about mathematics in their book [5]. Others have spoken directly to me, but about topics that seem to be worlds away from differential geometry. A name like Larry Thomas comes to mind since he was the first to show me how to think about mathematical research. Also, when I was a graduate student at Virginia, David Brydges gave a very pretty course about gauge theory that helped me put a lot of details that I vaguely knew into sharper focus. The list goes on and on. Please, my friends and colleagues, do not be offended that you are not explicitly mentioned here. The list is way too long.

Also, I gratefully thank all those at Springer who produced a book out of a manuscript. Among those, special thanks go to Donna Chernyk, my editor. The comments of the anonymous referees also helped me improve the book. For those not aware of the details of this conversion process, let me say that it is neither a function nor a functor as far as I can figure out, but it is certainly a lot of work.

This volume is based on several introductory graduate courses that I have taught over the years. Among the participants in those courses, very special thanks are due to Claudio Pita for his helpful comments and continual insistence on ever-clearer explanations.

But there is one person who ignited the spark that made differential geometry quite clear from the very start. And that is Arunas Liulevicius. I most graciously thank him. He taught the first courses I ever took on differential geometry. His intelligent, rigorous discourse, blended with an authentic, bemused humor, made a great impact on me as a model of how to approach mathematics in general, not just these topics. And I never had to un-learn anything—which is a sort of litmus test in itself too! In terms of the specific topics in the beginning of this book, I am more indebted to him than to anyone else.

As for errors, omissions, and all other sorts of academic misdemeanors, there is no one to blame but myself. Please believe me that though I be guilty of whatever shortcoming, I am innocent of any malicious intent. In particular, omissions in the references are mere reflections of my limited knowledge. I request my kind reader to help me out with a message to set me right. I will be most appreciative.

Guanajuato, Mexico   Stephen Bruce Sontz
September 2014

Contents

1   Introduction 1

2   Basics of Manifolds 7
    2.1   Charts 7
    2.2   The Objects - Differential Manifolds 9
    2.3   The Morphisms: Smooth Functions 11
    2.4   The Derivative 12
    2.5   The Tangent Bundle 16
    2.6 T as a Functor 23
    2.7   Submanifolds 24
    2.8   Appendix: C^∞-functions 26

3   Vector Bundles 31
    3.1   Transition Functions for the Tangent Bundle 31
    3.2   Cocycles 33
    3.3   Smooth Vector Bundles 34
    3.4   Bundles Associated to the Tangent Bundle 36
    3.5   Sections 40
    3.6   Some Abstract Nonsense 41
    3.7   Appendix: Tensor Products 42

4   Vectors, Covectors, and All That 49
    4.1   Jacobi Matrix 49
    4.2   Curves 51
    4.3   The Dual of Curves 52
    4.4   The Duality 54

5   Exterior Algebra and Differential Forms 59
    5.1   Exterior Algebra 59
    5.2   Relation to Tensors 64
    5.3   Hodge Star 65
    5.4   Differential Forms 71

        5.5   Exterior Differential Calculus 73
        5.6   The Classical Example 76

6   **Lie Derivatives** 81
        6.1   Integral Curves 81
        6.2   Examples 85
        6.3   Lie Derivatives 87

7   **Lie Groups**.. 93
        7.1   Basics 93
        7.2   Examples 96
        7.3   One-Parameter Subgroups 98
        7.4   The Exponential Map 100
        7.5   Examples: Continuation 101
        7.6   Concluding Comment 103

8   **The Frobenius Theorem** 105
        8.1   Definitions 105
        8.2   The Integrability Condition 109
        8.3   Theorem and Comments 110

9   **Principal Bundles**.. 111
        9.1   Definitions 111
        9.2   Sections 117
        9.3   Vertical Vectors 118

10   **Connections on Principal Bundles** 121
        10.1   Connection as Horizontal Vectors 121
        10.2   Ehresmann Connections 124
        10.3   Examples 129
        10.4   Horizontal Lifts... 134
        10.5   Koszul Connections 138
        10.6   Covariant Derivatives 141
        10.7   Dual Connections as Vertical Covectors 143

11   **Curvature of a Connection** 145
        11.1   Definition 145
        11.2   The Structure Theorem... 146
        11.3   Structure Equation for a Lie Group.................................. 149
        11.4   Bianchi's Identity... 149

12   **Classical Electromagnetism** 151
        12.1   Maxwell's Equations in Vector Calculus............................. 151
        12.2   Maxwell's Equations in Differential Forms.......................... 156
        12.3   Gauge Theory of Maxwell's Equations 163
        12.4   Gauge Transformations and Quantum Mechanics 166

13 Yang–Mills Theory 173
    13.1   Motivation 173
    13.2   Basics 177
    13.3   Yang–Mills Equations 187
    13.4   Mass of the Mesons 191

14 Gauge Theory 195
    14.1   Gauge Theory in Physics 196
    14.2   Gauge Fields in Geometry 200
    14.3   Matter Fields in Geometry 221

15 The Dirac Monopole 225
    15.1   The Monopole Magnetic Field 225
    15.2   The Monopole Vector Potential 227
    15.3   The Monopole Principal Bundle 230

16 Instantons 233
    16.1   Quaternions 233
    16.2   Compact Symplectic Groups $Sp(n)$ 235
    16.3   The Ehresmann Connection on the Hopf Bundle 238
    16.4   BPST Instantons 241

17 What Next? 247

Correction to: Principal Bundles C1

A   Discussion of the Exercises 249

Bibliography 271

Index 273

Abbreviations

ASD   anti-self-dual
BPST   Belavin, Polyakov, Schwartz, Tyupkin
EL   Euler–Lagrange
CR   Cauchy–Riemann
GR   general relativity
LIVF   left invariant vector field
ODE   ordinary differential equation
QCD   quantum chromodynamics
SD   self-dual
SI   Système International d'Unités

The original version of this book was revised. The correction to this book is available at
https://doi.org/10.1007/978-3-319-14765-9_18

Chapter 1
Introduction

What I cannot create I do not understand.

— *Richard Feynman in*
"Feynman's Office: The Last Blackboards" [14]

This volume started out as notes in an attempt to help my students in a course on classical differential geometry. Not feeling constrained to use one particular text, I went off on my merry way introducing some of the basic structures of classical differential geometry (standard reference: [30]) that are used in physics. When the students requested specific references to texts, I would say that any one from a standard list of quite excellent texts would be fine. But my approach was not to be found in any *one* of them. Rather, the students had to search here and there in the literature and then try to piece it all together. So my notes were just that: a piecing together of things well known. However, in the spirit of the famous saying of Feynman (see Ref. [14]) noted above, some considerable part of the development of the subject is left for the reader to do in the exercises. Of course, in that same spirit, the reader should create explicitly all of the material presented here.

The main points of our approach to classical differential geometry follow:

- **Avoidance of explicit coordinates**
  This is a point of view advocated by Lang in [32]. We prefer to put the emphasis on what the coordinate functions really represent, namely, charts that are homeomorphisms onto open sets in a model space. This also eliminates an inundation of indices and so results in formulas that are much easier to understand. However, when we feel that a coordinate-free presentation is too cumbersome, we will use local coordinates.
- **The use of diagrams in lieu of formulas**
  This may seem to contradict the previous point but is meant to complement it. The point is that every formula can be expressed in words, but it is much preferable to use a formula and, among all possible formulas, the clearest of them all. (This is a sort of optimization problem whose study should be taken seriously!) But often the clearest formula can be rewritten as an even clearer diagram. And so we propose to do just that. An example of this is our definition of an Ehresmann connection, where we present the usual formulas but *also* their equivalent diagrams, which we find to be more transparent.

© Springer International Publishing Switzerland 2015
S.B. Sontz, *Principal Bundles*, Universitext, DOI 10.1007/978-3-319-14765-9_1

- **The definition of differential forms**
  We use a definition that explicitly displays 1-forms as dual maps (with values in the real line) to vector fields. Consequently, vector-valued differential forms are defined as maps with the same domains but with values in a given vector space.
- **The affine operation**
  We present the affine operation (also known as the translation function) as a basic structure defined on every Lie group G as well as on the fibers of every principal bundle with structure group G.
- **Early use of cocycles**
  We introduce at an early stage the concept of cocycle as the structure behind the construction of the tangent bundle and consequently in the definitions of vector bundle and principal bundle.
- **The importance of the fiber product**
  This is a well-known construction, but for whatever reason, it does not seem to be very popular in introductions to the subject matter. We find it to be extremely useful in our definition of differential forms k-forms for $k \geq 2$ and in the discussion of the affine operation.
- **The Maurer–Cartan form as an Ehresmann connection**
  The point here is that a seemingly trivial example of an Ehresmann connection gives us the definition of the Maurer–Cartan form of a Lie group. This leads to the natural problem of the classification of all Ehresmann connections on a Lie group considered as the total space of the principal bundle over a one-point space. This rather easy, yet rather important problem is rarely mentioned in introductory texts. But the answer illuminates the discussion of gauge fields and so deserves to be mentioned. In particular, proofs based on extended calculations are replaced by arguments based on the underlying structure of the theory, thereby revealing just why the calculations had to work out right.
- **The use of categories**
  We use category theory as an efficient language for describing many aspects of the theory. While this language is not absolutely necessary for understanding this book, this is a good time to learn it since it is essential language for the companion volume [44]. For reasons totally unclear to me, there still are some mathematicians who are allergic to this language and prefer to avoid it at any cost. As far as I am concerned, one might as well eliminate vector spaces as unnecessarily abstract constructs.
- **One definition of smooth**
  Paraphrasing Gertrude Stein, one of our maxims will be smooth is smooth is smooth. This means that we introduce one definition of a smooth manifold and one definition of a smooth map between smooth manifolds. While this seems to be a completely straightforward and uncontroversial application of the more general *categorical imperative*, it is too often violated in practice. For example, smooth distributions and smooth sections are sometimes given independent definitions as if they were not particular cases of the concept of a smooth map. We consider such an approach to be distracting, to say the least.

- **Not all concepts are defined**

  This might well be deemed heresy! Sometimes no definition is given at all since a lot of basic vocabulary is readily available in the 21st century on the Internet. Some terms that one would expect to be defined in 20th-century (or earlier) books are simply not defined. Just a few examples are linear, quotient space, and disjoint union.

  Of course, rigorous definitions are the essential first step of the basic *logical* process: definition, theorem, proof. However, the *mathematical* process begins with ideas, not definitions. Only then is the logical process followed. So some effort is made to explain clearly the ideas behind those definitions that are presented. But sometimes only an idea, even just a vague idea, is given and the definition, theorem, proof sequence is left as an exercise. This is part and parcel of the philosophy behind Feynman's maxim. This philosophy will be practiced in moderation. For example, it is used mostly in the discussion of results from category theory, which the reader need not dominate on a first reading. This will give the reader the opportunity to practice transforming ideas into rigorous definitions, which can then be used to prove theorems. The next step is to "learn" how to have ideas, but we leave that for another book.

- **Tangent vectors are underemphasized**

  If this had not been explicitly mentioned, the reader might never have noticed. Sure, the tangent vectors and tangent fields are here. So is the tangent bundle. But I deliberately only sketch the theory of vector fields and their integral curves, a standard topic! I am saying that these structures are not as central as they were once thought to be. (Heresy!) The important structures are their duals: cotangent vectors, 1-forms, and the cotangent bundle. But from our traditional, historically limited perspective, we do not understand how to grasp the dual structures directly as things in themselves. In this regard, we note that a generally accepted folklore of noncommutative spaces is that their "correct" infinitesimal structure is modeled in classical spaces by the de Rham theory, which is based on differential forms. In the companion volume [44], there will be little mention of tangents and their analogs in the noncommutative setting.

The reader may already be aware of the (whimsical!) definition of classical differential geometry as the study of properties invariant under change of notation. As with any pleasantry, there is a ring of truth to this. Our presentation suffers (as most others do) from a dependence on one system of notation. The novice should realize that further studies in this subject matter will necessarily include an inordinate amount of time spent in learning how to express things already learned in terms of another notation. Other introductions to these topics with their own notations and points of view exist in abundance, but I find some of them to be directed only at physicists wishing to learn some mathematics (just half of my point of view), some of them to be monographs rather than texts, and some to be unreadable for the typical beginner, whether physicist or mathematician. Of course, the encyclopedic two volumes [30] by S. Kobayashi and K. Nozimu are an excellent reference for differential geometry for those who have become more expert in the field.

So far I have mostly spoken about classical differential geometry. In [44], the sequel to this volume, I go on to present the more recent noncommutative theory. The word "classical" used to describe this volume is meant to contrast it with the theory in [44] of noncommutative geometry (also known as quantum geometry). This latter is again not original material, but the sources are different. Much of the motivation for noncommutative geometry is to be found in the classical theory even though the latter has an importance and an interest that extend further than its being a motivating factor for the newer noncommutative geometry. And this is why I start off with classical differential geometry and continue on to noncommutative geometry. However, a continuing challenge for research in noncommutative geometry is to find new concepts and applications that are more than mere imitations of the classical theory presented here.

We now present a brief outline of the contents of this book. Chapters 2 to 5 are essentially a translation of multivariable calculus into the modern language of differential geometry. In this sense, the last section of Chapter 5 is the denouement of that development. Of course, vector bundles, and more specifically the tangent bundle, are not much in evidence in the standard courses on multivariable calculus, but the point here is that they provide a concise language for discussing that material. A consequence is that there are no "real" theorems in these chapters. Rather, all the results are more or less consequences of the definitions. So the definitions are the nontrivial material in these chapters! An exception is the appendix to Chapter 3 on tensor products. This material is covered in many standard algebra courses, but in my experience, mathematics students often only know how to "add" vector spaces (direct sum) and not how to "multiply" them (tensor product). So I decided to include this material in an appendix for those who need to see it developed thoroughly. I imagine that many physics students will have to read this carefully too. Chapter 2 also has an appendix for those who need a more solid understanding of C^∞-functions, but proofs are not given.

Chapter 6 is an introduction to integral curves of vector fields and some of their applications. The existence and uniqueness of integral curves are shown to be equivalent to the same properties for an associated ordinary differential equation. This provides enough formalism to discuss the flows of electric and magnetic fields, which gets us into a bit of physics. The chapter concludes with the Lie derivative, very much a basic structure in differential geometry.

Chapter 7 is a frighteningly brief glimpse at the immense field of Lie group theory. It is hoped to be enough to get us through the theory of principal bundles, which is the main interest.

Next, Chapter 8 is mostly a cultural excursion into mathematically nontrivial territory, the Frobenius theorem. But we do motivate the definition of a distribution of subspaces in a tangent bundle. And this will be used later in the theory of connections. However, we do not prove the Frobenius theorem. Understanding what it says is plenty for now.

Chapters 2 to 8 provide the background material for launching into the topic of this book, principal bundles, which is dealt with in the remaining chapters. My intent has been to do this as quickly as is reasonably possible without skipping over the ever-important motivating intuitions. My more mathematically inclined readers may have seen some of that background already. Everyone is invited to skip sections or chapters that are already known, or mostly known. One can always refer back to skipped parts as need be. The rest of the book is meant to be a challenge for all of my potential readers. In some sense, the book really begins with Chapter 9. That is where the meat of the matter is begun to be presented. For my readers who do not have the prerequisites for understanding the prerequisite material before Chapter 9, please do not despair! There are more leisurely texts to get you up to speed. My personal favorite is Lee's text [33]. Actually, Lee proves a lot of the material that I present without proof. My view is that proofs can be important, but that understanding what is going on is even more important and can be sufficient, especially on a first pass through new material—but with the caveat that one has to really understand what is going on. Later on one can return to clean up the split infinitives that litter the way taken.

Chapters 9 to 11 provide a lot of material on principal bundles, their Ehresmann connections, and the curvature of those connections. These three chapters are the mathematical and geometrical core of the book.

Then Chapters 12 to 16 concern various applications of the general theory of principal bundles in physics. A lot of the exposition is aimed at providing physical motivation for people with backgrounds in mathematics. But there is also a lot of detailed description of how the physics is translated into the geometrical language of principal bundles for those with a physics background. So these chapters are not only aimed at all of my audience, but they also are meant to serve as an attempt to encourage more, and clearer, communication between these two communities. So it is the style of these chapters, even more than their well-known content, that gives this book whatever characteristic flavor it may have.

The text itself ends with Chapter 17, a one-page farewell to my kind readers, with some suggestions as to what they might want to do next with this knowledge. Of course, opening up new avenues of research not mentioned there would also be most welcome.

There is finally an appendix with discussions of almost all of the exercises. It is best to avoid looking at any of that, except after a long, hard consideration of an exercise. Remember Feynman's challenge!

Chapter 2
Basics of Manifolds

2.1 Charts

We start off with the idea of a chart. This is sometimes called a system of coordinates, but we feel, as does Lang (see [32]), that this both obscures the basic idea and impairs the recognition of an immediate generalization to Banach manifolds by introducing scads of unnecessary notation. The idea that Banach spaces provide an appropriate scenario for doing differential calculus goes back at least to the treatise [8] of Dieudonné. An extremely well-written text on smooth manifolds is Lee's book [33].

The setup is a Hausdorff topological space M.

Definition 2.1 A chart *in M with values in the vector space $\mathbb{R}^n$ (resp., in the Banach space B) is a pair (U, ϕ) such that U is an open subset of M in the topology of M and $\phi : U \to \phi(U) \subset \mathbb{R}^n$ is a homeomorphism onto an open set $\phi(U)$ in $\mathbb{R}^n$ [resp., $\phi : U \to \phi(U) \subset B$ is a homeomorphism onto an open set $\phi(U)$ in B]. Here $n \geq 0$ is an integer.*

Usually, we think about the case $n \geq 1$, though it is important to note that everything works (in a rather trivial way to be sure) when $n = 0$. We remind the reader that $\mathbb{R}^0 = \{0\}$, the trivial (but *not* empty) vector space.

Definition 2.2 An atlas *for M is a set of charts, say $\{(U_\alpha, \phi_\alpha)\}_{\alpha \in A}$, such that $\bigcup_\alpha U_\alpha = M$ and there is one integer n such that $\phi_\alpha(U_\alpha) \subset \mathbb{R}^n$ for all $\alpha \in A$ [resp., there is one Banach space B such that $\phi_\alpha(U_\alpha) \subset B$ for all $\alpha \in A$].*

From now on, we will consider the case where the *model space* is $\mathbb{R}^n$ rather than a general Banach space. However, the reader should bear in mind that a lot of this goes through *mutatis mutandis* with a Banach space B in place of $\mathbb{R}^n$. We remind the reader that there are (not finite-dimensional) Banach spaces that have no basis. So the generalization to manifolds with such a Banach space as the model space must not use coordinate notation. But somehow this fact is beside the point, which is simply to eliminate confusing, unneeded notation.

© Springer International Publishing Switzerland 2015
S.B. Sontz, *Principal Bundles*, Universitext, DOI 10.1007/978-3-319-14765-9_2

So the situation of interest for us is a topological space M such that every point has a neighborhood that looks like an open set in some (fixed) Euclidean space. One says: M is *locally Euclidean*. The existence of an atlas already greatly restricts the sort of topological spaces under consideration. However, one can impose more structure on the atlas and, consequently, further restrict the class of topological spaces being studied.

Definition 2.3 *We say that an atlas for M is a C^k-atlas (or that it is of class C^k) provided that for each pair of charts, say (U_α, ϕ_α) and (U_β, ϕ_β), in the atlas we have that the map*

$$\phi_\beta \circ \phi_\alpha^{-1} : \phi_\alpha(U_\alpha \cap U_\beta) \to \phi_\beta(U_\alpha \cap U_\beta) \tag{2.1}$$

is a function of class C^k. Furthermore, we say that $\phi_\beta \circ \phi_\alpha^{-1}$ is the transition function (or the change of coordinates or the change of charts) from the chart (U_α, ϕ_α) to the chart (U_β, ϕ_β).

Several comments are in order here.

- First, let us note that both $\phi_\alpha(U_\alpha \cap U_\beta)$ and $\phi_\beta(U_\alpha \cap U_\beta)$ are open subsets of $\mathbb{R}^n$. So when we say that $\phi_\beta \circ \phi_\alpha^{-1}$ is a function of class C^k here, we mean that in the sense of multivariable calculus. See the appendix to this chapter for more details on the class C^k.
- Second, in spite of the usual insistence on using different notation for a function and for its restriction to a subdomain, we are using ϕ_α both for the function in the chart (U_α, ϕ_α), that is,

$$\phi_\alpha : U_\alpha \to \phi_\alpha(U_\alpha), \tag{2.2}$$

as well as for its restriction

$$\phi_\alpha : U_\alpha \cap U_\beta \to \phi_\alpha(U_\alpha \cap U_\beta), \tag{2.3}$$

which, by the way, makes all the sense in the world even when $U_\alpha \cap U_\beta$ is empty. Each of these versions of ϕ_α (namely, as given in (2.2) and in (2.3)) is a homeomorphism between its domain and its codomain. Analogous comments hold, of course, for our usage of the notation ϕ_β. This is how to interpret these maps in (2.1):

$$\phi_\alpha(U_\alpha \cap U_\beta) \xrightarrow{\phi_\alpha^{-1}} U_\alpha \cap U_\beta \xrightarrow{\phi_\beta} \phi_\beta(U_\alpha \cap U_\beta).$$

- Finally, we note that $k \geq 1$ is an integer or $k = \infty$. Most results hold for $k \geq 1$, though some only work for $k \geq 2$. However, we will only discuss the case $k = \infty$ here unless otherwise indicated. So we will drop the prefix C^k and simply say *atlas* when $k = \infty$.
  We use *smooth* as a synonym for class C^∞.

Definition 2.4 *We say a chart* (U, ϕ) *in M is* compatible *with a C^k-atlas* $\{(U_\alpha, \phi_\alpha)\}_{\alpha \in A}$ *for M if the union,*

$$\{(U_\alpha, \phi_\alpha)\}_{\alpha \in A} \cup (U, \phi),$$

is again a C^k-atlas on M.

So we throw one more chart into the atlas to get a new set of charts. Then we demand that this new set of charts be a C^k-atlas on M. In particular, $\phi(U)$ will be an open set in the same model space that is being used for the original atlas.

Exercise 2.1 *This definition is equivalent to demanding the condition that $\phi \circ \phi_\alpha^{-1}$ and $\phi_\alpha \circ \phi^{-1}$ be of class C^k for every $\alpha \in A$.*

We can introduce a partial order on atlases in the following way.

Definition 2.5 *Suppose that* $\mathcal{U} = \{(U_\alpha, \phi_\alpha)\}_{\alpha \in A}$ *and* $\mathcal{V} = \{(V_\gamma, \psi_\gamma)\}_{\gamma \in \Gamma}$ *are atlases on M. Then we write* $\mathcal{U} \ll \mathcal{V}$ *if each chart* (U_α, ϕ_α) *in* $\mathcal{U}$ *is also a chart in* $\mathcal{V}$, *that is, if there exists a* $\gamma \in \Gamma$ *such that* $(U_\alpha, \phi_\alpha) = (V_\gamma, \psi_\gamma)$. *Here equality means equality; that is,* $U_\alpha = V_\gamma$ *(the* same *open subset of M) and* $\phi_\alpha = \psi_\gamma$ *(the same function with the same domain and the same codomain).*

As a preliminary to the definition of differential manifold, we will use the following.

Definition 2.6 *A* differential structure *on a Hausdorff topological space M is a maximal smooth atlas, that is, a C^∞-atlas that is maximal among all such atlases with respect to the partial order $\ll$.*

Theorem 2.1 *Every smooth atlas $\mathcal{U}$ on M determines a unique differential structure on M containing it.*

Notice that we are *not* saying that a given Hausdorff topological space that has an atlas actually has a *smooth* atlas. (Indeed, there do exist Hausdorff topological spaces that have an atlas but do not have any smooth atlases. See [29] and [43]. We note in passing that a Hausdorff topological space that has an atlas is called a *topological manifold.*) We are saying that if a Hausdorff topological space does have a smooth atlas, then it has an associated differential structure.

Exercise 2.2 *Prove Theorem 2.1.*

2.2   The Objects - Differential Manifolds

Next, we define one of the central concepts of this theory.

Definition 2.7 *A* differential manifold *is a Hausdorff topological space M together with a given differential structure. This is also called a* smooth manifold. *Also, we define the* dimension *of M to be n, where* $\mathbb{R}^n$ *is the model space for M. The notation for this is* dim $M = n$.

Thus, Theorem 2.1 says that any smooth atlas on M uniquely makes M become a differential manifold.

When we speak of a manifold, we often refer to the topological space M and leave the differential structure out of the notation. This is truly sloppy, since it is well known that a given topological space can have inequivalent differential structures. But it is a quite common abuse of notation. (We will define later what equivalence of differential structures means.)

Examples:

- Any open set $U \subset \mathbb{R}^n$ with atlas $\{U, \iota_U\}$, where $\iota_U : U \to \mathbb{R}^n$ is the inclusion map, is a differential manifold of dimension n.
- The unit n-sphere $S^n := \{x \in \mathbb{R}^{n+1} \mid ||x|| = 1\}$ can be made into a differential manifold of dimension n. Here, $||\cdot||$ is the Euclidean norm on $\mathbb{R}^{n+1}$. Some care must be given here since an atlas also has to be specified. One way is to use the $2n + 2$ open hemispheres

$$U_{+,k} := \{x = (x_1, x_2, \ldots, x_{n+1}) \mid x_k > 0\}$$

$$\text{and} \quad U_{-,k} := \{x = (x_1, x_2, \ldots, x_{n+1}) \mid x_k < 0\}$$

for $k = 1, 2, \ldots, n + 1$, together with the maps $\phi_{+,k} : U_{+,k} \to \mathbb{R}^n$ and $\phi_{-,k} : U_{-,k} \to \mathbb{R}^n$ given by

$$\phi_{+,k}(x_1, \ldots, x_{n+1}) := (x_1, \ldots, x_{k-1}, x_{k+1}, \ldots, x_{n+1})$$

$$\phi_{-,k}(x_1, \ldots, x_{n+1}) := (x_1, \ldots, x_{k-1}, x_{k+1}, \ldots, x_{n+1})$$

for $k = 1, 2, \ldots, n + 1$. Note that $\phi_{+,k}$ and $\phi_{-,k}$ are different functions since their domains are different. Actually, their domains are disjoint. The reader is advised to draw a picture of this situation for the cases $n = 1$ and $n = 2$.
Another atlas for S^n is given by only two charts, each of which is a stereographic projection, one from the "north pole" $(0, 0, \ldots, 0, 1)$ and the other from the "south pole" $(0, 0, \ldots, 0, -1)$. We leave the details of this, including a proof that the same differential structure is obtained, as an exercise for the reader.
- The real projective space $\mathbb{R}P^n := S^n/\mathbb{Z}_2 = \mathbb{R}^{n+1} \setminus \{0\}/\mathbb{R} \setminus \{0\}$ can be made into a differential manifold of dimension n.
- The complex projective space $\mathbb{C}P^n := S^{2n+1}/S^1 = \mathbb{C}^{n+1} \setminus \{0\}/\mathbb{C} \setminus \{0\}$ can be made into a differential manifold of dimension $2n$.
- A Riemann surface is a differential manifold of dimension 2.

Exercise 2.3 *Fill in all the details for these examples.*

In the jargon of category theory, the differential manifolds are the objects in a category. In general, a category is a collection of objects with some sort of common "structure" together with a set of morphisms for each pair of objects. The morphisms can be thought of (and for our examples, will be) functions that preserve the "structure." The mysterious "structure" resides in the relation between the objects and the morphisms and need not be explicitly specified beyond that! This is one of those minor details that may be difficult to grasp at first. Examples help. We continue in the next section with the definition of the morphisms in the category we are setting up.

2.3 The Morphisms: Smooth Functions

We now have a class of "objects" of interest, namely, the class of differential manifolds. The viewpoint of category theory is that there should also be morphisms, also known as "arrows" between pairs of these objects and that these arrows should "preserve" the structure of the objects. These will be the smooth functions, which we will now define. Since this is a local concept for functions in Euclidean space, we will use the fact that manifolds are locally Euclidean to define this concept for functions between manifolds. This will be the first occurrence of a leitmotif that we will see repeatedly.

Definition 2.8 *Suppose that $f : M \to N$ is a continuous function and that M and N are smooth manifolds of dimension m and n, respectively. Then we say that f is smooth if, for every $x \in M$, we can find charts (U, ϕ) in M and (V, ψ) in N with $x \in U$ and $f(x) \in V$ and $f(U) \subset V$ such that $\psi \circ f \circ \phi^{-1} : \phi(U) \to \psi(V)$ is a C^∞-function from the open set $\phi(U) \subset \mathbb{R}^m$ to the open set $\psi(V) \subset \mathbb{R}^n$.*

Diagrammatically, we have

$$\mathbb{R}^m \supset \phi(U) \xrightarrow{\phi^{-1}} U \xrightarrow{f} V \xrightarrow{\psi} \psi(V) \subset \mathbb{R}^n.$$

Notice that the function $f : U \to V$ in this diagram is not really the original function $f : M \to N$ but rather its restriction to U, which we sometimes denote by $f|_U$. However, we will usually not use the correct notation since it leads to a lot of cumbersome formulas.

Exercise 2.4 *Suppose that $f : M \to N$ is a continuous function between differential manifolds. Show that f is smooth if and only if, for every choice of charts (U, ϕ) in M and (V, ψ) in N satisfying $f(U) \subset V$, we have that $\psi \circ f \circ \phi^{-1} : \phi(U) \to \psi(V)$ is a C^∞-function.*

Basic Properties of Smooth Functions

- The identity map $1_M : M \to M$ is smooth.
  (For any set S, we denote the identity function $S \to S$ by 1_S.)
- If $f : M \to N$ and $g : N \to P$ are smooth, then $g \circ f : M \to P$ is smooth.
- If $f : M \to N$ is smooth, then $f \circ 1_M = f$ and $1_N \circ f = f$.
- $f(gh) = (fg)h$ provided the domains and codomains hook up correctly.

One summarizes these properties by saying that we have a *category* with *objects* given by the smooth manifolds and with *morphisms* (or *arrows*) given by the smooth functions between smooth manifolds. This is called the *category of differential manifolds* or the *category of smooth manifolds*.

As in any category, we can define an *isomorphism* to be a morphism $f : M \to N$ that has an inverse morphism $g : N \to M$, namely, $f \circ g = 1_N$ and $g \circ f = 1_M$. However, by tradition, in some categories an isomorphism is given another name. In the category of differential manifolds, an isomorphism is commonly called a *diffeomorphism*.

Given a Hausdorff topological space M with differential structures $\mathcal{A}_1$ and $\mathcal{A}_2$ (these being maximal smooth atlases), we say that $\mathcal{A}_1$ and $\mathcal{A}_2$ are *equivalent* if there is a diffeomorphism $\phi : (M, \mathcal{A}_1) \to (M, \mathcal{A}_2)$ from M with the first differential structure to M with the second differential structure. Note that ϕ need not be the identity function.

Exercise 2.5 *The real line $\mathbb{R}$ can be provided with the usual differential structure as well as the differential structure determined by the atlas with one chart, $\{(\mathbb{R}, t)\}$, where $t : \mathbb{R} \to \mathbb{R}$ is given by $t(x) := x^3$. Prove that these are equivalent differential structures but that the identity is not a diffeomorphism between them.*

I believe it was Milnor in [37] who first showed by giving an example that a Hausdorff topological space can have several inequivalent differential structures. In any case, his example in that paper was S^7. Differential structures on S^n that are not equivalent to the standard differential structure are called *exotic spheres*. Later, in the 1980s, it was shown that $\mathbb{R}^n$ has exactly one differential structure (up to equivalence) if and only if $n \neq 4$. Moreover, in that time period it was shown that $\mathbb{R}^4$ has infinitely many inequivalent differential structures. Those differential structures on $\mathbb{R}^4$ that are not equivalent to the standard differential structure are called *fake* $\mathbb{R}^4$s. Terms such as "exotic" and even more so "fake" tend to have emotional connotations that may not necessarily be beneficial for scientific discussions, particularly when one of the interlocutors does not have much background in scientific matters.

Exercise 2.6 *With respect to the standard differential structures, prove that*

- *S^m is diffeomorphic to S^n if and only if $m = n$.*
- *$\mathbb{R}^m$ is diffeomorphic to $\mathbb{R}^n$ if and only if $m = n$.*
- *S^m and $\mathbb{R}^n$ are not diffeomorphic.*

Convention: We often drop the word "smooth," so that we say "atlas" instead of "smooth atlas," "function" instead of "smooth function," and "manifold" instead of "smooth manifold," and so on.

2.4 The Derivative

We have arrived at a rather strange stage in the development of this theory. We now know what it means for a function to be smooth, that is, C^∞, but we still do not have a definition for the derivative of a smooth function. We will now address this crucial issue. Following our general philosophy, we start with the Euclidean case. So consider

$$\mathbb{R}^m \supset U \xrightarrow{f} V \subset \mathbb{R}^n,$$

where U and V are open and f is differentiable in U. Then the derivative of f at a point $x \in U$ is a linear map, denoted $Df(x) : \mathbb{R}^m \to \mathbb{R}^n$. Actually, it is the best linear approximation to f at the point $x \in U$ in the sense that

$$f(x + h) = f(x) + (Df(x))h + o(||h||)$$

for $h \in \mathbb{R}^m$ as $h \to 0$. Note that the left-hand side is well defined for small $||h||$. When there is no ambiguity, we simply write $Df(x)h$. The remainder term is of order $o(||h||)$ as a function with values in $\mathbb{R}^n$. (This notation is called *Landau's little o*.) Explicitly, this means that

$$\frac{||f(x + h) - f(x) - Df(x)h||_{\mathbb{R}^n}}{||h||_{\mathbb{R}^m}} \to 0$$

when $||h|| \equiv ||h||_{\mathbb{R}^m} \to 0$. It is shown in a calculus course that this best linear approximation is unique. Of course, the map $Df : U \to \mathcal{L}(\mathbb{R}^m, \mathbb{R}^n)$, defined by $x \mapsto Df(x)$, is in general nonlinear. (In fact, since U in general is not a vector space, it is ridiculous even to expect Df to be linear.) Here

$$\mathcal{L}(\mathbb{R}^m, \mathbb{R}^n) := \{T : \mathbb{R}^m \to \mathbb{R}^n \mid T \text{ is linear}\}$$

is the vector space of all linear maps from $\mathbb{R}^m$ to $\mathbb{R}^n$. It has dimension mn.

Since the Euclidean spaces $\mathbb{R}^m$ and $\mathbb{R}^n$ have canonical ordered bases (which also give us canonical coordinates), we can write $Df(x)$ as a matrix in terms of those bases. This is called the *Jacobi matrix* $J(f)$ of f. It is an $n \times m$ matrix-valued function of $x \in U$. Since we prefer to avoid coordinates, we will not put much emphasis on this matrix for now. Unfortunately, in some presentations of the differential calculus, the Jacobi matrix is given a prominent role, while its underlying linear map is often only barely mentioned or is even completely ignored.

However, we do note that the entries of a Jacobi matrix are the partial derivatives of the components $f(x) = (f_1(x), f_2(x), \ldots, f_n(x))$ with respect to the coordinates $x = (x_1, x_2, \ldots, x_m)$. So

$$J(f) = \left(\frac{\partial f_i}{\partial x_j} \right) \quad \text{for } 1 \leq i \leq n, \ 1 \leq j \leq m.$$

We will come back to this point of view in more detail in Chapter 4.

Exercise 2.7 *To expand on a previous comment, the reader should understand what this theory of differentiation means when*

(i) $m = 0, n \neq 0$.
(ii) $m \neq 0, n = 0$.
(iii) $m = n = 0$.

Still within this Euclidean context, we will now write the derivative in yet another way. We define $Tf : U \times \mathbb{R}^m \to V \times \mathbb{R}^n$ by

$$Tf(x, v) := (f(x), Df(x)v) \tag{2.4}$$

for $x \in U$ and $v \in \mathbb{R}^m$. Also, $\tau_U : U \times \mathbb{R}^m \to U$ is defined by

$$\tau_U(x, v) := x, \tag{2.5}$$

and similarly for τ_V. Notice by considering the Jacobi matrix $J(f)$ that Tf is of class C^{k-1} if f is of class C^k with $k \geq 1$ an integer. (Class C^0 simply means continuous.) However, if f is of class C^∞, then Tf is again of class C^∞. This is one of the reasons that we prefer to work with C^∞-functions.

The definition in (2.4) seems a bit strange since the derivative only appears in the second entry, while the first entry seems to be there for "bookkeeping" purposes only. Actually, it an extremely clever definition, as the next exercise shows.

Exercise 2.8 *Suppose $f : U \to V$ and $g : V \to W$ are smooth functions, where U, V, and W are open subsets of the Euclidean spaces $\mathbb{R}^m$, $\mathbb{R}^n$, and $\mathbb{R}^p$, respectively. Prove that $T(g \circ f) = Tg \circ Tf$. After reading about diagrams below, restate this result in terms of a commuting diagram.*

Also, prove that $T(1_U) = 1_{U \times \mathbb{R}^m}$.

We now have our first commutative diagram:

$$
\begin{array}{ccc}
{}^{\bullet}U \times \mathbb{R}^m & \xrightarrow{Tf} & V \times \mathbb{R}^n \\
\downarrow{\tau_U} & & \downarrow{\tau_V} \\
U & \xrightarrow{f} & V.
\end{array}
\tag{2.6}
$$

A *diagram* is just any collection of objects and arrows. It could simply be one arrow between two objects, for example. It does not even need to be connected. A diagram is said to be *commutative* if, for every ordered pair of objects in the diagram, the result of following arrows from the first object in the pair to the second object in the pair is independent of the route taken. Of course, often there is no way to go from the first object to the second. Or there is exactly one route. In either of these two cases, the commutativity condition is trivially satisfied. In the above diagram, there is only one ordered pair for which there are two (or more) possible routes. Namely, starting from $U \times \mathbb{R}^m$, there are two routes that end at V. So we have to check that

$\tau_V \circ Tf = f \circ \tau_U$. Written in this algebraic manner, we see the advantage of the diagram. The latter is a better notation since it exhibits more directly what needs to be proved. It is also easier to remember since each arrow carries its own domain and codomain. Then it is straightforward to connect the arrows to form a diagram.

One method of proving that a diagram is commutative is the *diagram chase*. This consists in taking a generic element in an appropriate object and seeing where the arrows take it down the various routes available. (We say that we *chase* the element around the diagram.) We now do this for the earlier diagram (2.6). So we start with $(x, v) \in U \times \mathbb{R}^m$. Then go across and then down:

$$(x, v) \xrightarrow{\ Tf\ } (f(x), Df(x)v)$$
$$\downarrow \tau_V$$
$$f(x).$$

Then starting with the same element but going down and then across, we get

$$(x, v)$$
$$\downarrow \tau_U$$
$$x \xrightarrow{\ f\ } f(x).$$

So we see that the same result is found for the two routes. This finishes the diagram chase argument and shows that the diagram is, in fact, commutative.

Definition 2.9 *Suppose $U \subset \mathbb{R}^m$ is open. Then we define*

$$TU = T(U) := U \times \mathbb{R}^m. \tag{2.7}$$

This is the tangent bundle *of U. We say that TU is the* total space *and that U is the* base space. *The map $\tau_U : TU \to U$ is called the* tangent bundle map. *Also, for every $x \in U$, we define the* fiber above x *to be the set $\tau_U^{-1}(x)$, that is, the complete inverse image of x under the map τ_U.*

Notice that the commutativity of the above diagram (2.6) implies (is actually equivalent to) the assertion that

$$Tf\,|_{\tau_U^{-1}(x)} \colon \tau_U^{-1}(x) \to \tau_V^{-1}(f(x)); \tag{2.8}$$

that is, Tf restricted to the fiber above x maps that fiber to the fiber above $f(x)$. We introduce a special notation for this restriction: $T_x f$. We also use the notation $T_x U := \iota_U^{-1}(x)$ for the fiber above $x \in U$. Hence, the diagram (2.8) becomes

$$T_x f : T_x U \to T_{f(x)} V.$$

We will shortly have more to say about this map.

It follows immediately that $T_x U = \tau_U^{-1}(x) = \{x\} \times \mathbb{R}^m$. In particular, this allows us to define a vector space structure on each fiber $T_x U$ in a most natural way:

$$(x, v_1) + (x, v_2) := (x, v_1 + v_2),$$
$$r(x, v) := (x, rv)$$

in the obvious notation. Thus, $T_x U$ is isomorphic to $\mathbb{R}^m$ and, moreover, by a canonical isomorphism, $T_x U \ni (x, v) \mapsto v \in \mathbb{R}^m$. Now it is easy to verify that

$$T_x f : T_x U \to T_{f(x)} V$$

is a linear map with respect to these vector space structures on the fibers. So not only does $T_x f$ map fibers to fibers as noted in the previous paragraph, but it maps fibers to fibers *linearly*.

2.5 The Tangent Bundle

Having produced all these spaces, maps, structures, and so forth for the Euclidean case, our next goal is to do the same for the manifold case. Again, since a manifold is locally Euclidean, we know what to do locally. The trick is to assemble (or bundle!) it all up into one consistent package. This will be the tangent bundle of a manifold together with all its concomitant structures.

So we fix a manifold M of dimension m and consider charts (U_α, ϕ_α) and (U_β, ϕ_β) on M. Since $\phi_\alpha(U_\alpha)$ and $\phi_\beta(U_\beta)$ are open sets in the Euclidean space $\mathbb{R}^m$, we know how to construct their tangent bundles:

$$T(\phi_\alpha(U_\alpha)) = \phi_\alpha(U_\alpha) \times \mathbb{R}^m \quad T(\phi_\beta(U_\beta)) = \phi_\beta(U_\beta) \times \mathbb{R}^m$$
$$\downarrow \qquad\qquad\qquad\qquad \downarrow$$
$$\phi_\alpha(U_\alpha) \qquad\qquad\qquad\qquad \phi_\beta(U_\beta).$$

Note that we often do not label arrows provided that the context makes it obvious what they are. Also, to save space (and make a more "functorial" diagram), we will mostly use the T notation for tangent bundles from now on. Of course, these charts on M are related by the transition function $\phi_\beta \circ \phi_\alpha^{-1}$. So these two tangent bundles should also be related over the overlap:

$$T(\phi_\alpha(U_\alpha \cap U_\beta)) \xdashrightarrow{?} T(\phi_\beta(U_\beta \cap U_\alpha))$$
$$\downarrow \qquad\qquad\qquad\qquad\qquad \downarrow$$
$$\phi_\alpha(U_\alpha \cap U_\beta) \xrightarrow{\phi_\beta \circ \phi_\alpha^{-1}} \phi_\beta(U_\beta \cap U_\alpha).$$

Thus, we have filled in the diagram using the transition function (2.1) on the base spaces, but we have not yet specified the map (indicated with a dashed arrow and a question mark) between the total spaces. Clearly the thing to do is put $T(\phi_\beta \circ \phi_\alpha^{-1})$ there! Before doing this, we want to introduce this notation for the transition functions: $\psi_{\beta\alpha} := \phi_\beta \circ \phi_\alpha^{-1}$. So we arrive at this commutative diagram:

$$
T(\phi_\alpha(U_\alpha)) \supset T(\phi_\alpha(U_\alpha \cap U_\beta)) \xrightarrow{T\psi_{\beta\alpha}} T(\phi_\beta(U_\beta \cap U_\alpha)) \subset T(\phi_\beta(U_\beta))
$$
$$
\downarrow \qquad\qquad \downarrow \qquad\qquad\qquad \downarrow \qquad\qquad\qquad \downarrow \qquad\qquad (2.9)
$$
$$
\phi_\alpha(U_\alpha) \supset \phi_\alpha(U_\alpha \cap U_\beta) \xrightarrow{\psi_{\beta\alpha}} \phi_\beta(U_\beta \cap U_\alpha) \subset \phi_\beta(U_\beta).
$$

By the way, the inclusion symbols $\supset$ and $\subset$ are considered to be arrows in this diagram; that is, they represent inclusion maps. Now the idea is that the bottom row of this diagram contains all of the "ingredients" for construction of the given manifold M, and so the top row should have all the "ingredients" for constructing a new manifold, the tangent bundle TM.

So let's examine the bottom row of (2.9) in greater detail. First, we define $V_\alpha := \phi_\alpha(U_\alpha)$ and $V_{\beta\alpha} := \phi_\alpha(U_\alpha \cap U_\beta)$ for $\alpha, \beta \in A$, the index set for the atlas on M that we are using. So we have two families of open sets in $\mathbb{R}^m$, namely, $\{V_\alpha\}_{\alpha \in A}$ and $\{V_{\beta\alpha}\}_{\beta, \alpha \in A}$, such that $V_{\beta\alpha} \subset V_\alpha$. We also have a family of smooth functions $\psi_{\beta\alpha} : V_{\beta\alpha} \to V_{\alpha\beta}$, the transition functions. Now the bottom row of (2.9) looks like this:

$$
V_\alpha \supset V_{\beta\alpha} \xrightarrow{\psi_{\beta\alpha}} V_{\alpha\beta} \subset V_\beta. \qquad\qquad (2.10)
$$

Now we would like to construct a manifold from these data: the V_α, the $V_{\beta\alpha}$, and the transition functions $\psi_{\beta\alpha}$. And the way to do this is to "paste" the sets V_α together, using the $\psi_{\beta\alpha}$ to identify certain points in V_α with certain points in V_β.

So we define a topological space X by taking the disjoint union of the V_αs with the disjoint union topology and quotienting out by $\sim$ (a still-undefined equivalence relation) with the quotient topology:

$$
X := \left(\bigsqcup_{\alpha \in A} V_\alpha \right) / \sim .
$$

And the equivalence relation $\sim$ is given by the transition functions; namely, each $x \in V_{\beta\alpha} \subset V_\alpha$ is defined to be equivalent to $y \in V_{\alpha\beta} \subset V_\beta$, denoted $x \sim y$, provided that $y = \psi_{\beta\alpha}(x)$. Of course, we have to verify that this is indeed an equivalence relation. To achieve this, we first record three key facts about the two families of open sets and the transition functions:

- (Fact R): $V_{\alpha\alpha} = V_\alpha$ and diagram (2.10) in the case $\alpha = \beta$ becomes

$$V_\alpha = V_{\alpha\alpha} \xrightarrow{\psi_{\alpha\alpha}} V_{\alpha\alpha} = V_\alpha$$

with $\psi_{\alpha\alpha} = 1_{V_{\alpha\alpha}} = 1_{V_\alpha}$.

- (Fact S): The functions $V_{\beta\alpha} \xrightarrow{\psi_{\beta\alpha}} V_{\alpha\beta}$ and $V_{\alpha\beta} \xrightarrow{\psi_{\alpha\beta}} V_{\beta\alpha}$ are inverse to each other.
- (Fact T): $V_{\beta\alpha} \cap \psi_{\beta\alpha}^{-1}(V_{\gamma\beta}) \subset V_{\gamma\alpha}$ and $\psi_{\gamma\beta} \circ \psi_{\beta\alpha}(x) = \psi_{\gamma\alpha}(x)$ for all $x \in V_{\beta\alpha} \cap \psi_{\beta\alpha}^{-1}(V_{\gamma\beta})$.

Exercise 2.9 *Prove these three facts and then use them to prove that $\sim$ is an equivalence relation.*

Notice that for $x \in V_\alpha$ given, there is at most *one* $y \in V_\beta$ with $x \sim y$. In particular, if $x, x' \in V_\alpha$ with $x \sim x'$, then $x = x'$. So when we divide out by $\sim$, we "paste" x to at most one element in V_β for $\beta \neq \alpha$ and to no other element of V_β.

Finally, we endow the topological space X with a differential structure and claim that it is canonically diffeomorphic to the originally given manifold M. To do this, we first note that the function $V_\alpha \to X$ given by $x \mapsto [x]$ (where $[x]$ is the $\sim$ equivalence class in X determined by $x \in V_\alpha$) is a homeomorphism onto its image, which we define to be W_α. Then we note that W_α is open in X and denote the inverse to the map just defined as κ_α. Therefore,

$$\kappa_\alpha : W_\alpha \to V_\alpha \subset \mathbb{R}^m$$

is also a homeomorphism. Next, we claim that $\{(W_\alpha, \kappa_\alpha)\}$ is an atlas on X, thereby making X into a differential manifold. Also, the transition functions $\kappa_\beta \circ \kappa_\alpha^{-1} : V_{\beta\alpha} \to V_{\alpha\beta}$ of this atlas are equal to $\psi_{\beta\alpha}$.

Exercise 2.10 *Fill in the details in the preceding paragraph. Also, show that there is a canonical diffeomorphism from M onto X.*

Now we return to the first row of the commutative diagram (2.9):

$$T(\phi_\alpha(U_\alpha)) \supset T(\phi_\alpha(U_\alpha \cap U_\beta)) \xrightarrow{T\psi_{\beta\alpha}} T(\phi_\beta(U_\beta \cap U_\alpha)) \subset T(\phi_\beta(U_\beta)). \qquad (2.11)$$

This is very similar to the bottom row of (2.9), except now we do not have that these spaces and maps come from a previously given manifold. What we propose to do is to *construct* a differential manifold from these data, in an analogous way to the reconstruction of M from the bottom row of (2.9).

We first note that (2.11) gives us two families of open sets, but now in $\mathbb{R}^{2m}$ instead of $\mathbb{R}^m$. They are

$$\tilde{V}_\alpha := T(\phi_\alpha(U_\alpha)) = \phi_\alpha(U_\alpha) \times \mathbb{R}^m \subset \mathbb{R}^m \times \mathbb{R}^m \cong \mathbb{R}^{2m}$$

and

$$\tilde{V}_{\beta\alpha} := T(\phi_\alpha(U_\alpha \cap U_\beta)) = \phi_\alpha(U_\alpha \cap U_\beta) \times \mathbb{R}^m \subset \tilde{V}_\alpha \subset \mathbb{R}^{2m}.$$

We also have a family of smooth functions from (2.11), namely,

$$\tilde{\psi}_{\beta\alpha} := T\psi_{\beta\alpha} : \tilde{V}_{\beta\alpha} \to \tilde{V}_{\alpha\beta},$$

which are candidates for being transition functions for the manifold we wish to define.

Exercise 2.11 *Show that the open sets $\tilde{V}_\alpha$ and $\tilde{V}_{\beta\alpha}$ and the functions $\tilde{\psi}_{\beta\alpha}$ satisfy Facts R, S, and T. In analogy with the definition of the relation $\sim$, which used the transition functions $\psi_{\beta\alpha}$, define a relation $\approx$ on $\bigsqcup_{\alpha\in A} \tilde{V}_\alpha$ using the functions $\tilde{\psi}_{\beta\alpha}$. Show that $\approx$ is an equivalence relation.*

So now we have all of the "ingredients" for constructing a differential manifold, though now these "ingredients" did not arise from a previously given differential manifold. But we carry on as before, only now this *defines* a new manifold:

$$TM := \left(\bigsqcup_{\alpha\in A} \tilde{V}_\alpha\right) / \approx = \left(\bigsqcup_{\alpha\in A} T(\phi_\alpha(U_\alpha))\right) / \approx,$$

where $\approx$ is the equivalence relation defined by the functions $\tilde{\psi}_{\beta\alpha}$.

Let's describe in more detail what this equivalence relation $\approx$ is all about. It says that

$$(\phi_\alpha(x), v) \in \phi_\alpha(U_\alpha) \times \mathbb{R}^m$$

is in the same $\approx$ equivalence class as

$$\begin{aligned}
\tilde{\psi}_{\beta\alpha}(\phi_\alpha(x), v) &= T\psi_{\beta\alpha}(\phi_\alpha(x), v) \\
&= (\psi_{\beta\alpha}(\phi_\alpha(x)), D\psi_{\beta\alpha}(\phi_\alpha(x))v) \\
&= (\phi_\beta(x), D(\phi_\beta \circ \phi_\alpha^{-1})(\phi_\alpha(x))v) \\
&\in \phi_\beta(U_\beta) \times \mathbb{R}^m
\end{aligned} \tag{2.12}$$

for all $x \in U_\alpha \cap U_\beta$. Here we used $\psi_{\beta\alpha} = \phi_\beta \circ \phi_\alpha^{-1}$. In words:

v is tangent to $\phi_\alpha(x)$ in chart α

if and only if

$D(\phi_\beta \circ \phi_\alpha^{-1})(\phi_\alpha(x))v$ is tangent to $\phi_\beta(x)$ in chart β

for all $x \in U_\alpha \cap U_\beta$. In physics, this property is often expressed by saying,

> "A vector is something that transforms like a vector."

To explain this jargon, we introduce some of our own jargon. We say that a point $x \in M$ *is contained in the chart* α if $x \in U_\alpha$. Then we can rephrase the above as follows: "A tangent vector at a point $x \in M$ is an equivalence class whose representatives (there being exactly one representative for each chart α containing the point x) are related by the transformation rule

$$(\phi_\alpha(x), v) \longrightarrow (\phi_\beta(x), D(\phi_\beta \circ \phi_\alpha^{-1})(\phi_\alpha(x))v)$$

for identifying the representative associated with the chart α containing x to the representative associated with the chart β containing x." But physicists have an intuition that mathematicians lack and thus find this rephrasing to be unnecessary and annoying pedantry.

By the way, this transformation is frequently described by indicating its effect on the second entry only. This transformation from chart α to chart β, namely, the linear map $v \mapsto D(\phi_\beta \circ \phi_\alpha^{-1})(\phi_\alpha(x))v$ of $\mathbb{R}^m$ to itself, is typically written in coordinate notation. But then the point x [or equivalently, $\phi_\alpha(x)$] is a rather mysterious object. This shows the clear advantage of including two factors in the definition of TU in equation (2.7).

The reader should note that "chart α" in the above could refer to the chart (U_α, ϕ_α) in M or to the corresponding *natural chart* $(\tilde{U}_\alpha, \tilde{\phi}_\alpha)$ in TM, which we now define much as we defined the charts $(W_\alpha, \kappa_\alpha)$ on X.

Exercise 2.12 *For each (U_α, ϕ_α) in the atlas for M, define a map*

$$\phi_\alpha(U_\alpha) \times \mathbb{R}^m \to TM$$

by sending each element $(\phi_\alpha(x), v) \in \phi_\alpha(U_\alpha) \times \mathbb{R}^m$ to its $\approx$ equivalence class in TM. Show that this map is a homeomorphism onto its image. Then define $\tilde{U}_\alpha \subset TM$ to be the image of this map. Next, define

$$\tilde{\phi}_\alpha : \tilde{U}_\alpha \to \phi_\alpha(U_\alpha) \times \mathbb{R}^m$$

to be the inverse map, again a homeomorphism. Show that these natural charts $\{(\tilde{U}_\alpha, \tilde{\phi}_\alpha)\}_{\alpha \in A}$ form a C^∞ atlas on TM whose corresponding transition functions satisfy

$$\tilde{\phi}_\beta \circ \tilde{\phi}_\alpha^{-1} = \tilde{\psi}_{\beta\alpha} = T\psi_{\beta\alpha}.$$

Show that TM thereby becomes a differential manifold with

$$\dim TM = 2 \dim M = 2m.$$

Moreover, the projections $\phi_\alpha(U_\alpha) \times \mathbb{R}^m \to \phi_\alpha(U_\alpha)$ *in the Euclidean space paste together to give a smooth function, called the* tangent bundle map

$$\tau_M : TM \to M.$$

These fit together to give the following commutative diagram for every $\alpha \in A$:

$$
\begin{array}{ccc}
TM \supset \tilde{U}_\alpha & \xrightarrow{\tilde{\phi}_\alpha} & \phi_\alpha(U_\alpha) \times \mathbb{R}^m \\
\Big\downarrow{\tau_M} \quad \Big\downarrow{\tau_M} & & \Big\downarrow \\
M \supset U_\alpha & \xrightarrow{\phi_\alpha} & \phi_\alpha(U_\alpha).
\end{array}
$$

We also have the identifications $\tilde{U}_\alpha = \tau_M^{-1}(U_\alpha) = TU_\alpha$.

Definition 2.10 *Let* M *be a smooth manifold. Then* TM *together with the map* $\tau_M : TM \to M$ *is called the* tangent bundle *of* M. *The manifold* TM *is called the* total space *of the tangent bundle, while the manifold* M *is called the* base space *of the tangent bundle. For each point* $x \in M$, *the* fiber *above* x *is defined to be* $\tau_M^{-1}(x)$. *One writes* $T_x M := \tau_M^{-1}(x)$ *and calls this the* tangent space *of* M *at* x.

Exercise 2.13 *Using the natural charts on* TM, *define a vector space structure on each tangent space. Show that each tangent space is isomorphic to* $\mathbb{R}^m$, *though not in general in a canonical way.*

As we shall see later, the tangent bundle of a manifold is a motivating example for the definition of a vector bundle. For the moment, let's simply note that even though TM is a manifold, we never want to use general charts on TM, but rather the natural charts defined above. This is because the latter respect both the vector space structure of the fibers and the tangent bundle maps, whereas general charts do not.

There are a multitude of (eventually equivalent) ways for defining the tangent bundle and the tangent spaces of a differential manifold. We have chosen the way presented here since it is congenial to a subsequent treatment of vector bundles and also general fiber bundles, of which principal bundles will be a major interest for us. For a reference with our point of view of the tangent bundle, see the text [27] of Husemoller. Also, Lang [32] uses almost the same approach, though he defines the equivalence relation $\approx$ pointwise [i.e., separately at each point in $x \in M$ and so obtains the tangent space $T_x(M)$ at x] and then takes the disjoint union of the spaces $T_x(M)$ to define the tangent bundle. We take a disjoint union first and then divide out by the globally defined $\approx$. This gives us the tangent bundle in one fell swoop; the tangent spaces are then defined as fibers in it. The upshot is that these two operations commute. Finally, for a superb treatment of fiber bundles, have a look in Chapter I of Hirzebruch's classic text [23]. As they say: Read the masters.

The concept of tangent space should be more intuitive for the reader than that of tangent bundle. After all, we all are familiar with the tangent line to a curve at a given point on the curve. And, similarly, we can picture the tangent plane to a surface at a given point on the surface. Actually, these concepts are literally ancient since they can be found in the works of Archimedes, Euclid, and others. Tangent space is simply the generalization of these concepts to a dimension higher than 2. On the other hand, when we bundle up all of the tangent spaces into one object, the tangent bundle, it is easy to lose perspective about the resulting structure. For example, it is clear what a tangent line to a circle is. It may not be quite so clear what the tangent bundle of a circle is. After all, when we draw the circle in the Euclidean plane, it turns out that the tangent lines for distinct points have (in general) a nonempty intersection. And their (not disjoint) union is the complement of the open disk inside the circle. But the tangent spaces for distinct points sit inside the tangent bundle as disjoint subspaces. To "see" the tangent bundle of the circle, one takes each tangent line and rotates it in Euclidean 3-space by 90 degrees so that it now passes perpendicularly through the plane of the circle but still passes through the same point of tangency. However, the tangent line no longer looks like a tangent! Nonetheless, we have now that tangent lines for distinct points are disjoint subspaces in the space that is so obtained. And that space is a cylinder whose base is the given circle. In particular, the tangent bundle of a circle is trivial (this being a technical term). But, again, the concept of tangent bundle is nontrivial (this now being a colloquial and subjective way of speaking) and seems to be the result of research work done in the 20th century. In particular, the idea that the tangent bundle has a naturally equipped topology and differential structure is not trivial. And when we say that the tangent bundle of a circle is a cylinder, we mean as a differential manifold. So the reader should feel no embarrassment if it takes some time to assimilate these concepts.

Exercise 2.14 *This is an exercise in the history of mathematics, so it will take more than pencil and paper to do it. The point is that the tangent space at a given point of a geometrical object is a geometrically intuitive notion, known by the ancient Greek mathematicians (such as Archimedes and Apollonius) for the cases of curves and surfaces. And the tangent space was clearly recognized as a geometric object in its own right. But the idea that the tangent bundle, the collection of all the tangent spaces, is also a geometric object is a rather more recent development. As the reader will probably agree, it is not an entirely obvious notion. So the exercise consists in finding out who was the first person to introduce the tangent bundle into the mathematical literature. I have some wild guesses as to who it might have been, but they do not even qualify as conjectures.*

Now at this conjuncture, the reader should feel somewhat concerned about two aspects of this theory. First, one should wonder what effect the choice of charts for the base manifold has on the definition of the tangent bundle. Second, there is the question of backward compatibility since we defined the tangent bundle first for open subsets of a Euclidean space and second for general manifolds. So does the second definition agree, as it should, with the first definition in the special case when the general manifold is an open subset of a Euclidean space? These are not totally unrelated concerns, but we leave them for the ruminations of the reader. But don't worry! Everything works out as one hopes. If you get frustrated with these concerns, try looking at [23] or [27].

Exercise 2.15 *As a warning to the reader about the limits of this method for constructing manifolds by pasting together open subsets of a Euclidean space, we present the following for your careful consideration. We take the index set A to be* $\{1, 2\}$.

Let $V_1 = V_2 = \mathbb{R}$ *and let* $V_{12} = V_{21} = \mathbb{R} \setminus \{0\}$. *(Of course,* $V_{11} = V_{22} = \mathbb{R}$ *and* $\phi_{11} = 1_{\mathbb{R}}$ *and* $\phi_{22} = 1_{\mathbb{R}}$ *are forced on us.) Then the only map left to specify is* $\phi_{12} : V_{12} \to V_{21}$ *since* $\phi_{21} = \phi_{12}^{-1}$. *We consider two cases:*

(i) $\phi_{12} : \mathbb{R} \setminus \{0\} \to \mathbb{R} \setminus \{0\}$ *is the identity.*
(ii) $\phi_{12} : \mathbb{R} \setminus \{0\} \to \mathbb{R} \setminus \{0\}$ *is given by* $\phi_{12}(x) = 1/x$ *for all* $x \in \mathbb{R} \setminus \{0\}$.

In each case, identify $(V_1 \sqcup V_2)/ \sim$. *Is this space a differential manifold?*

2.6 *T* as a Functor

In case the reader has forgotten, this discussion of the tangent bundle was prompted by our observation that we had not yet defined the derivative of a smooth $f : M \to N$. Actually, we did this in the development of this theory, but only in the Euclidean case [see the diagram (2.6)]. But having the definition locally, we can now proceed to the global definition. We pick charts (U, ϕ) in M and (V, ψ) in N with $f(U) \subset V$. So $\psi \circ f \circ \phi^{-1} : \phi(U) \to \psi(V)$ is a C^∞-function between open subsets of Euclidean spaces. This gives us the commutative diagram

$$
\begin{array}{ccc}
\phi(U) \times \mathbb{R}^m & \xrightarrow{T(\psi \circ f \circ \phi^{-1})} & \psi(V) \times \mathbb{R}^n \\
\downarrow{\scriptstyle \tau_{\phi(U)}} & & \downarrow{\scriptstyle \tau_{\psi(V)}} \\
\phi(U) & \xrightarrow{\psi \circ f \circ \phi^{-1}} & \psi(V).
\end{array}
$$

By passing to $\sim$ equivalence classes in the base space, we recover the given smooth function f. But on passing to the $\approx$ equivalence classes in the total space, we get a new function. Notation: $Tf : TM \to TN$.

Exercise 2.16 *Show that the definition of* Tf *does pass to the* $\approx$ *equivalence classes, that it is a smooth function, that we have a commutative diagram*

$$
\begin{array}{ccc}
TM & \xrightarrow{Tf} & TN \\
\downarrow{\scriptstyle \tau_M} & & \downarrow{\scriptstyle \tau_N} \\
M & \xrightarrow{f} & N,
\end{array}
$$

and that Tf *is a linear map on fibers; namely,* $T_x f : T_x M \to T_{f(x)} N$ *is linear for all* $x \in M$.

Theorem 2.2 *We have these two properties:*

- $T(1_M) = 1_{TM}$ *for all differentiable manifolds* M.
- *If* $f : M \to N$ *and* $g : N \to P$ *are smooth functions, where* M, N, *and* P *are differentiable manifolds, then* $T(g \circ f) = Tg \circ Tf$.

Exercise 2.17 *Prove this theorem.*

For those who speak a bit of the language called *category theory*, this theorem can be understood as saying that the assignment $M \mapsto TM$ for differential manifolds M together with the assignment $f \mapsto Tf$ for smooth functions f between differential manifolds combine to give us a *(covariant) functor* from the category of differential manifolds to itself. Later on, we will be able to interpret this result as saying that this is a functor from the category of differential manifolds to the category of (smooth) vector bundles.

This is a good point for the reader to pause, collect his or her thoughts, and reconsider this material carefully since this chapter is the foundation on which the rest of the theory is built. Besides mastering all the technical details (which is important!), the reader should also consider why we call this subject differential geometry. Roughly speaking, this comes from an idea known since the very first introduction to calculus: The derivative describes a geometric object, namely, the tangent. It may be helpful to keep this in mind when reviewing this chapter.

2.7 Submanifolds

So far we have only discussed a single manifold M. These are objects in a category. As with many categories, we can define subobjects, which in this setting are called submanifolds. Some things are clear about this. We should have that N is a subset of M and that N is a manifold in its own right. Furthermore, the inclusion map $N \to M$ should be smooth. But it has been found that this is not enough for a "good" definition. There is an extra, seemingly highly technical, condition that goes into the standard definition.

To take some of the mystery out of this extra condition, let's think about the smooth map $\iota : \mathbb{R}^k \to \mathbb{R}^m$, where $0 \le k \le m$ and

$$\iota(a_1, \ldots, a_k) := (a_1, \ldots, a_k, 0, \ldots, 0),$$

with $m - k$ occurrences of 0 and $a_j \in \mathbb{R}$ for $j = 1, \ldots, k$. We use this map to identify $\mathbb{R}^k$ with its image $\iota(\mathbb{R}^k)$ in $\mathbb{R}^m$. So we have $\mathbb{R}^k \subset \mathbb{R}^m$ in a standard way, which is called the *(standard) k-slice* in $\mathbb{R}^m$. (All the other k-planes in $\mathbb{R}^m$ are also called k-slices, but we will not need them here.)

The k-slice $\mathbb{R}^k \subset \mathbb{R}^m$ is now taken as the model for what a submanifold (here $\mathbb{R}^k$) should look like as a subset of a manifold (here $\mathbb{R}^m$). This global picture gets adapted to the local situation in every chart in the following definition.

Definition 2.11 *Let M be a manifold with $\dim M = m$. Suppose that N is a subset of M with the property that there exists some atlas of M, denoted $\{(U_\alpha, \phi_\alpha)\}$, such that there exists an integer k with $0 \le k \le m$ and*

$$\phi_\alpha : U_\alpha \cap N \to \phi_\alpha(U_\alpha) \cap \mathbb{R}^k$$

is a homeomorphism for every α such that $U_\alpha \cap N \ne \emptyset$.

Then we say that N is a submanifold *of M with $\dim N = k$. We also say that N has* codimension *$m - k$ in M.*

This definition says that the *part* of N in a chart corresponds to the *part* of the k-slice that lies in the image of the chart.

But the definition is nothing but a technical condition! Whatever has happened with the initial intuition? First, let's note that N has the subspace topology induced from M, and this makes N into a Hausdorff topological space.

Proposition 2.1 *Suppose that N is a submanifold of M with the above notations. Then $\{(U_\alpha \cap N, \phi_\alpha)\}$ is an atlas on the Hausdorff topological space N, which turns N into a differential manifold with $\dim N = k$.*

Moreover, the inclusion map $N \to M$ is smooth with respect to this smooth structure on N and the original smooth structure on M.

The straightforward proof of this proposition is left to the reader.

Exercise 2.18 *Many, but not all, of the inclusion maps into a manifold M give rise to a submanifold. Here are two canonical examples.*

- *The unit sphere S^k is a submanifold of codimension 1 in $\mathbb{R}^{k+1}$.*
- *The two-dimensional torus $\mathbb{T}^2$ can be defined as*

$$\mathbb{T}^2 := S^1 \times S^1 \subset \mathbb{R}^2 \times \mathbb{R}^2 \cong \mathbb{R}^4.$$

But one can also realize $\mathbb{T}^2$ as the quotient space $\mathbb{R}^2/\mathbb{Z}^2$, where $\mathbb{Z}^2$ is the integer lattice in $\mathbb{R}^2$, that is, all points (j, k), where j and k are integers. Then for every real number r, one first looks at the line in $\mathbb{R}^2$ given by the points (t, rt) for all $t \in \mathbb{R}$. Next, one looks at the image N_r of this line in the quotient space $\mathbb{T}^2$. Prove that N_r is a submanifold of $\mathbb{T}^2$ if and only if r is a rational number.

Submanifolds will not be seen much in this book, though Lie subgroups of a Lie group are a major, important exception. The following theorem, whose proof is an adaption of the implicit function theorem to this context, is the usual way one proves that a subset of a manifold is actually a submanifold.

Theorem 2.3 *Suppose $f : M_1 \to M_2$ is a smooth map of smooth manifolds and $y \in M_2$ is a regular value of f; that is, $Tf : T_x M_1 \to T_y M_2$ is a surjection for all $x \in f^{-1}(y)$. Then $N := f^{-1}(y)$ is a submanifold of M_1 whose codimension is $\dim M_2$.*

This theorem is proved in many texts, for example, in [33]. One often takes M_2 to be a Euclidean space. In particular, by taking $M_2 = \mathbb{R}$, we have that the submanifold $N = f^{-1}(y)$ has codimension 1; that is, it is a hypersurface.

As a cultural aside, let me note that a *critical value* of a smooth function is any value of it that is not a regular value. This important concept is not used in this book.

2.8   Appendix: C^∞-functions

It has been my sad experience that almost all students of mathematics have no idea what the properties of C^∞-functions are. At best, they know the definition, but little else. At worst, they "know" a totally false version of Taylor's theorem. The situation with students of physics can hardly be expected to be better. This brief appendix is meant to set the record straight by putting the facts down clearly and concisely. At this point, proofs are not the essential ingredient. For a wonderfully clear and detailed exposition of these facts with readable proofs, I recommend Strichartz's book [47].

Be warned that the results in this section will be needed to solve some of the exercises in later chapters.

First, here is the definition just to make sure we are all on the same page.

Definition 2.12 *Let U be a nonempty open subset of $\mathbb{R}^m$ and $f : U \to \mathbb{R}$ be a function. Let $k \geq 1$ be an integer. Then we say that f is of class C^k if all the partial derivatives of f of every order j satisfying $0 \leq j \leq k$ exist and are continuous. Next, we say that f is of class C^∞ or is a C^∞-function if f is of class C^k for all $k \geq 1$. Notation: $C^\infty(U)$ is the set (actually, vector space) of all such C^∞-functions.*

Taking $j = 0$ in the first part of Definition 2.12, we see that f itself is continuous. We can also develop this theory for complex-valued functions with the corresponding definition. If f takes values in $\mathbb{R}^n$, then we write $f = (f_1, \ldots, f_n)$ and require that all the entries f_j are of class C^k (resp., class C^∞) for f to be of class C^k (resp., class C^∞).

Also, $f : U \to \mathbb{R}^n$ is of class C^k for an integer $k \geq 1$ if and only if

$$Df : U \to \mathcal{L}(\mathbb{R}^m, \mathbb{R}^n)$$

exists, is continuous, and is of class C^{k-1}. This property can be used to give an equivalent recursive definition of class C^k for integer $k \geq 1$ and then, as above, a definition of class C^∞. This relates the abstract derivative Df with the more explicit Definition 2.12 using partial derivatives. This is done quite elegantly in Chapter 1 of [26] in the Banach space setting.

For simplicity, in the following we will always take $n = 1$ since all the ideas are the same as in the general case. We will take $U = (a, b)$ to be an open interval in $\mathbb{R}$. Also, we will only be interested in C^∞-functions.

The key fact underlying the following definition is that we have all the derivatives of f available to us for any point in its domain.

Definition 2.13 *Take $f \in C^\infty((a,b))$ and $c \in (a,b)$. Then for every integer $n \geq 0$, we define the* Taylor polynomial T_n *of f centered at c for all $x \in \mathbb{R}$ by*

$$T_n(x) = T_n(x;c) := \sum_{k=0}^{n} \frac{1}{k!} f^{(k)}(c)(x-c)^k,$$

where $f^{(k)}$ denotes the kth derivative of f. [In particular, $f^{(0)} = f$.]
   The Taylor series *of f centered at c is defined by*

$$T(x) = T(x;c) := \sum_{k=0}^{\infty} \frac{1}{k!} f^{(k)}(c)(x-c)^k$$

for all $x \in \mathbb{R}$ for which this infinite series converges.

Notice that T_n is indeed a polynomial with degree at most n. As with any polynomial, its domain can be taken to be the real line $\mathbb{R}$ as we have done here. The notation omits the function f and, when convenient, the center point c as well.

The Taylor series $T(x;c)$ always converges for $x = c$ and its value for that choice of x is $f(c)$. (For this we use the standard convention in analysis that $0^0 = 1$.) In general, we cannot say more than that about the convergence of the Taylor series. The Taylor series could converge at points $x \notin (a,b)$. The Taylor series can be understood differently as a formal power series in the formal variable $x - c$. We will not follow that approach.

For all points $x \in \mathbb{R}$ for which the Taylor series converges, we have by the very definition of convergence of an infinite series that

$$\lim_{n \to \infty} T_n(x) = T(x). \qquad (2.13)$$

Notice that the point x here has nothing to do with the domain of f. More importantly, this statement says nothing about the relation of the function f with this limit.

But then confusion arises over the relation of the Taylor polynomials and the Taylor series with the function f. Let's first note these unpleasant facts about the Taylor series:

- The Taylor series $T(x;c)$ could converge in some open interval $I \subset \mathbb{R}$ centered at c, but $T(x;c) \neq f(x)$ for all $x \in I \cap (a,b) \setminus \{c\}$.
- The Taylor series could diverge for all points $x \neq c$.

The first unpleasant fact is shown by explicitly finding a C^∞-function $f : \mathbb{R} \to \mathbb{R}$ such that $f(x) > 0$ for all $x \neq 0$ and $f^{(k)}(0) = 0$ for all $k \geq 0$. Let's think about what the graph of f looks like near $x = 0$ since it might conflict

with your intuition. That graph looks something like the graph of $f_n(x) = x^{2n}$
for n extremely large, except that it is even flatter at 0 for *every* $n \geq 1$ since
$f_n^{(2n)}(0) = (2n)! \neq 0$. Anyway, the Taylor series of f centered at $x = 0$ has all of
its coefficients equal to zero and so it converges at all $x \in \mathbb{R}$. And it converges to 0.
And that is not equal to $f(x)$ for $x \neq 0$.

Exercise 2.19 *Define* $g : \mathbb{R} \to \mathbb{R}$ *by* $g(x) := e^{1/|x|}$ *for* $x \neq 0$ *and* $g(0) := 0$.
Prove that $g^{(k)}(0) = 0$ *for all* $k \geq 0$ *and that this implies* g *is a* C^∞-*function. So
this function* g, *defined by a relatively simple expression, satisfies the properties of
the function* f *described in the previous paragraph.*

One of the major theorems on C^∞-functions is the next highly nontrivial result.
It clearly implies the second of these two unpleasant facts. It also says in some
intuitive sense that there are "many" C^∞-functions whose Taylor series diverges
for all $x \neq c$.

Theorem 2.4 *(Borel) Let* $\{a_k \mid k \geq 0\}$ *be any sequence of real numbers. Let* $\epsilon > 0$
and $c \in \mathbb{R}$ *be given. Then there exists a* C^∞-*function*

$$f : (c - \epsilon, c + \epsilon) \to \mathbb{R}$$

such that $f^{(k)}(c) = a_k$ *for all* $k \geq 0$.

Actually, the proof of this theorem consists in the construction of "many" such
functions f. This theorem might surprise you. You might think that it is yet another
unpleasant fact. But eventually these unpleasant facts change into well-known and
natural properties. Be patient!

The complementary "nice" property of the Taylor series motivates this definition:

Definition 2.14 *With the above notation and hypotheses, we say that* f *is real
analytic near* c *if*

- *the Taylor series* $T(x; c)$ *of* f *centered at* c *converges in some open nonempty
  interval* I *centered at* c *and*
- $T(x; c) = f(x)$ *for all* $x \in I \cap (a, b)$.

Notice that the property that a function can be expanded in its Taylor series is the
defining property of a set of functions and is not a theorem about a previously given
set of functions.

You may have a vague memory from some course that Taylor's theorem says
something about C^∞-functions. But given the above facts, what could it possibly
say? Certainly, there can be *no* assertion that the Taylor series converges in some
open interval, let alone that it converges to the function f. But if not that, what?
Well, here is one version of the result:

Theorem 2.5 *(Taylor) With the above notation and hypotheses, we have for every integer $n \geq 0$ that*

$$\lim_{x \to c} \left(\frac{f(x) - T_n(x;c)}{(x - c)^n} \right) = 0.$$

By the very definition of limit, this is a statement concerning the behavior of the function of x in the parentheses near but not equal to c. Described in a few words, Taylor's theorem states something about an infinite number of limits of certain functions, which are indexed by the integer $n \geq 0$, as $x \to c$. The difference in the numerator here is known as the *nth remainder term*, $R_n(x;c) := f(x) - T_n(x;c)$. So Taylor's theorem says that as $x \to c$, the remainder term $R_n(x;c) \to 0$ faster than the denominator $(x - c)^n \to 0$.

The convergence of the Taylor series at some given $x \in \mathbb{R}$ is equivalent to the existence of the *one* limit in (2.13) as $n \to \infty$. Taylor's theorem says absolutely nothing one way or the other about this particular limit.

An amazing property of C^∞-functions is that they can be nonzero and yet have compact support. This is not the case for the rather "small" class of real analytic functions. Here is one useful result in that direction.

Proposition 2.2 *Let $a < a' < b' < b$ be real numbers. Then there exists a C^∞-function $f : \mathbb{R} \to \mathbb{R}$ such that*

- $f(x) = 1$ *for all $x \in (a', b')$.*
- $f(x) = 0$ *for all $x \in \mathbb{R} \setminus (a, b)$.*
- $0 \leq f(x) \leq 1$ *for all $x \in \mathbb{R}$.*

Such a function f is called a *smooth approximate characteristic function* for the interval (a', b'). Such functions are not unique. In fact, they are plentiful, a statement that we leave for the reader's further consideration.

Exercise 2.20 *Let f be a smooth approximate characteristic function with the above notation. Suppose further that $0 < f(x) < 1$ for all $x \in (a, a')$. (Such functions exist—and in abundance!) Prove that f is not real analytic near $x = a$. What about near $x = a'$?*

Chapter 3
Vector Bundles

3.1 Transition Functions for the Tangent Bundle

Right now we would like to understand how the transition functions $T\psi_{\beta\alpha}$ give us the tangent bundle. And this in turn will motivate the introduction of vector bundles. It is no exaggeration to say that a clear understanding of just the one example of the tangent bundle suffices to give one a clear understanding of vector bundles in general.

Starting from equation (2.12), we immediately have

$$T\psi_{\beta\alpha}(\phi_\alpha(x), v) = (\phi_\beta(x), D\psi_{\beta\alpha}(\phi_\alpha(x))v) \tag{3.1}$$

for all $x \in U_\alpha \cap U_\beta$ and all $v \in \mathbb{R}^m$. Also, recall that

$$\psi_{\beta\alpha} = \phi_\beta \circ \phi_\alpha^{-1} : \phi_\alpha(U_\alpha \cap U_\beta) \to \phi_\beta(U_\beta \cap U_\alpha)$$

is a transition function for the base space manifold M. Since this is an invertible function (whose inverse $\phi_\alpha \circ \phi_\beta^{-1}$ is also smooth), we have that $D\psi_{\beta\alpha}(\phi_\alpha(x))$ is an invertible linear map (by the chain rule) and so lies in $GL(\mathbb{R}^m)$, the *general linear group on* $\mathbb{R}^m$. Explicitly,

$$GL(\mathbb{R}^m) := \{T : \mathbb{R}^m \to \mathbb{R}^m \mid T \text{ is linear and invertible }\}$$

is both a group (algebraic structure) and a smooth manifold (differential structure). The latter assertion follows from the fact that $GL(\mathbb{R}^m)$ can be identified with (that is, mapped homeomorphically onto) an open subset of the Euclidean space $\mathbb{R}^{m^2}$ of

© Springer International Publishing Switzerland 2015
S.B. Sontz, *Principal Bundles*, Universitext, DOI 10.1007/978-3-319-14765-9_3

all $m \times m$ matrices. So $GL(\mathbb{R}^m)$ has an atlas with one chart in it. Moreover, the two structures are compatible; namely, the multiplication μ and inverse i are smooth functions:

$$\mu : GL(\mathbb{R}^m) \times GL(\mathbb{R}^m) \to GL(\mathbb{R}^m) \quad \text{and} \quad i : GL(\mathbb{R}^m) \to GL(\mathbb{R}^m).$$

Here $\mu(A, B) := AB$ and $i(A) := A^{-1}$ for elements $A, B \in GL(\mathbb{R}^m)$. That these two maps are smooth follows from basic formulas from a linear algebra course, namely, the definition of matrix multiplication and Cramer's rule. In each case, the matrix entries of the "output" of the operation are explicitly given smooth functions of the entries of the "inputs" of the operation.

Of course, one has to define a differential structure on the product space $GL(\mathbb{R}^m) \times GL(\mathbb{R}^m)$. There is a general way to do this for the product of two differential manifolds. (Think about it!) However, in this case we need merely note that the product can again be identified with an open subset of a Euclidean space. We have the following important definition.

Definition 3.1 *A* Lie group *is a group G that is also a differential manifold such that the multiplication map $\mu_G : G \times G \to G$ [denoted by $(g, h) \mapsto gh$] and the inverse map $i_G : G \to G$ (denoted by $g \mapsto g^{-1}$) are smooth functions.*

For the time being, we will work with the specific Lie group $G = GL(\mathbb{R}^m)$ but will use the notation G for it in order to prepare ourselves for the future.

Now equation (3.1) is telling us how to paste together the tangent bundle. On the first factor, we use the charts for the given base manifold, while on the second factor (which is the fiber), we use the smooth functions

$$g_{\beta\alpha} := D\psi_{\beta\alpha} \circ \phi_\alpha : U_\alpha \cap U_\beta \to G. \tag{3.2}$$

The properties of these functions are what make the $T\psi_{\beta\alpha}$ into consistent "instructions" for constructing the tangent bundle. We will state these now.

Theorem 3.1 *The functions $g_{\beta\alpha}$ defined in (3.2) satisfy the following properties:*

- $g_{\alpha\alpha}(x) = e$ *for all $x \in U_\alpha$, where $e \in G$ is the identity element of the group.*
- $\left(g_{\beta\alpha}(x)\right)^{-1} = g_{\alpha\beta}(x)$ *for all $x \in U_\alpha \cap U_\beta$, where the exponent -1 on the left hand side is the inverse operation in the group G.*
- $g_{\gamma\alpha}(x) = g_{\gamma\beta}(x)g_{\beta\alpha}(x)$ *for all $x \in U_\alpha \cap U_\beta \cap U_\gamma$, where the concatenation on the right-hand side is the multiplication in the group G.*

Proof: We start with the last property. Now, on the one hand,

$$g_{\gamma\alpha}(x) = D\psi_{\gamma\alpha} \circ \phi_\alpha(x) = D(\phi_\gamma \circ \phi_\alpha^{-1})(\phi_\alpha(x)),$$

while on the other hand,

$$g_{\gamma\beta}(x)g_{\beta\alpha}(x) = \left(D\psi_{\gamma\beta} \circ \phi_\beta(x)\right)\left(D\psi_{\beta\alpha} \circ \phi_\alpha(x)\right)$$

$$= \left(D(\phi_\gamma \circ \phi_\beta^{-1})(\phi_\beta(x))\right)\left(D(\phi_\beta \circ \phi_\alpha^{-1})(\phi_\alpha(x))\right).$$

And the chain rule tells us these two expressions are equal. (The careful reader should understand how the hypothesis $x \in U_\alpha \cap U_\beta \cap U_\gamma$ has been used.)

The first property follows from taking $\alpha = \beta = \gamma$ in the third property, while the second property follows now by taking $\gamma = \alpha$ in the third property and applying the first property. ■

3.2 Cocycles

Now the point of all this is that the construction of the tangent bundle depends exactly on the properties in Theorem 3.1. So now we are ready to jump into generalization mode!

Definition 3.2 *Let M be a differential manifold with a specified smooth atlas $\{U_\alpha, \phi_\alpha\}_{\alpha \in A}$. Let G be a Lie group. Then a* cocycle *on M (with respect to the given atlas) with values in G is a family of smooth functions*

$$g_{\beta\alpha} : U_\alpha \cap U_\beta \to G$$

for all $\alpha, \beta \in A$ satisfying the cocycle condition

$$g_{\gamma\alpha}(x) = g_{\gamma\beta}(x)g_{\beta\alpha}(x) \tag{3.3}$$

for all $x \in U_\alpha \cap U_\beta \cap U_\gamma$.

The other two properties in Theorem 3.1 also hold for any cocycle, as we have already shown in the proof of Theorem 3.1.

Exercise 3.1 *Suppose that M has an atlas consisting of two charts. Describe all the cocycles on M with respect to this atlas. What happens if M has an atlas that has one chart?*

Notice that Theorem 3.1 says in this language that the functions defined in (3.2) are a cocycle. Also, using the particular cocycle in (3.2), we can rewrite (3.1) as

$$T\psi_{\beta\alpha}(\phi_\alpha(x), v) = (\phi_\beta(x), g_{\beta\alpha}(x)v)$$

for all $x \in U_\alpha \cap U_\beta$ and all $v \in \mathbb{R}^m$. To define something similar for a general cocycle, we have to give a meaning to the expression $g_{\beta\alpha}(x)v$ that appears on the right-hand side. In other words, we need some sort of relation between the Lie group G and the vector space where v lives. This sort of relation is known as a *linear action of G on a vector space* or, more commonly, a *representation* of G.

For starters, the model fiber space need no longer be $\mathbb{R}^m$. The simplest (but far from only) generalization is to take $\mathbb{R}^l$ instead, where the integer l is a new parameter. Then we want to consider a map

$$\rho : G \to GL(\mathbb{R}^l),$$

which is both smooth and a group homomorphism. Then we let each element $g \in G$ act as a linear map on $\mathbb{R}^l$ by defining $gv := \rho(g)v$. Sometimes we denote this as $g \cdot v$.

3.3 Smooth Vector Bundles

At last we have all the information for constructing a new smooth manifold: a manifold M with a cocycle $g_{\beta\alpha}$ [with respect to an atlas $\{(U_\alpha, \phi_\alpha)\}_{\alpha \in A}$ on M] with values in a Lie group G and a representation ρ of G acting as linear maps in $\mathbb{R}^l$, namely, $\rho : G \to GL(\mathbb{R}^l)$. Given all this apparatus, we now can define smooth functions $\tilde{g}_{\beta\alpha} : \phi_\alpha(U_\alpha \cap U_\beta) \times \mathbb{R}^l \to \phi_\beta(U_\beta \cap U_\alpha) \times \mathbb{R}^l$ by

$$\tilde{g}_{\beta\alpha}(\phi_\alpha(x), v) := (\phi_\beta(x), g_{\beta\alpha}(x)v) = (\phi_\beta(x), \rho(g_{\beta\alpha}(x))v) \qquad (3.4)$$

for all $x \in U_\alpha \cap U_\beta$ and all $v \in \mathbb{R}^l$. Even though the representation ρ has dropped out of the notation in the first expression for defining the left-hand side, it is there in the second expression. And to be sure, we are using the hypothesis that ρ is a smooth map.

We recognize in the definition (3.4) a structure that we have seen before, for example, in (3.1). The first entry on the right-hand side depends only on the first entry on the left-hand side (i.e., it is independent of the vector v). But the functional relation, which is just the transition function $\phi_\beta \circ \phi_\alpha^{-1}$, can be quite arbitrary (though smooth, of course). The second entry on the left-hand side depends on both of the entries on the right-hand side. However, for each fixed value of x, this second entry depends *linearly* on the vector v.

Now we turn on the same machine as before to produce a new differential manifold E out of the disjoint union of the open sets

$$\phi_\alpha(U_\alpha) \times \mathbb{R}^l \subset \mathbb{R}^m \times \mathbb{R}^l \cong \mathbb{R}^{m+l},$$

modulo the equivalence relation defined by the functions $\tilde{g}_{\beta\alpha}$ of (3.4). This again gives us a special family of charts on E, the *natural charts* (or *local trivializations*), a smooth map $\pi : E \to M$ that is locally the projection onto the first factor $\phi_\alpha(U_\alpha) \times \mathbb{R}^l \to \phi_\alpha(U_\alpha)$, and a naturally defined vector space structure on each *fiber* $E_x := \pi^{-1}(x)$ above every point $x \in M$. Moreover, each fiber is isomorphic (as a vector space) to $\mathbb{R}^l$, though not in general in a canonical way.

Exercise 3.2 *Fill in the details of the previous paragraph. In particular, define a relation on the disjoint union using the functions $\tilde{g}_{\beta\alpha}$, and then show that it is an equivalence relation. Show that*

$$\dim E = \dim M + \dim \mathbb{R}^l = \dim M + l.$$

Also, define the natural charts on E and compute the corresponding transition functions.

Definition 3.3 *Suppose that M is a smooth manifold with a given atlas $\{(U_\alpha, \phi_\alpha)\}$, $g_{\beta\alpha} : U_\alpha \cap U_\beta \to G$ is a cocycle, and $\rho : G \to GL(\mathbb{R}^l)$ is a smooth representation. Then we say that E, as constructed above, is a* smooth vector bundle *over M with model fiber space $\mathbb{R}^l$ and that π is the* vector bundle map *or, simply, the* bundle map. *We call M the* base space *and E the* total space. *A standard notation is this diagram:*

$$\mathbb{R}^l \hookrightarrow E$$
$$\downarrow \pi$$
$$M.$$

The horizontal arrow is not unique. It can be taken to be any of the vector space isomorphisms of $\mathbb{R}^l$ onto any fiber of E.

At this point, as in the case of the definition of the tangent bundle, the gentle reader should be wondering about the role of the choice of the atlas in the definition of vector bundle. Again, Hirzebruch [23] and Husemoller [27] can be consulted.

Again, as in the case of the tangent bundle, we never use general charts on E (that are associated to its structure as a smooth manifold). This is because a general chart does not "see" the fiber structure. For example, the domain of an arbitrary chart need not be diffeomorphic to $U \times \mathbb{R}^l$, where U is open in $\mathbb{R}^m$. Even if it is, it need not preserve fibers. And even if it does, it need not be linear on fibers. In short, the object $\pi : \pi^{-1}(U_\alpha) \to U_\alpha$ calls for a description in terms of its structures. This is an instance of the *categorical imperative*, namely, that the functions between objects with structure should also preserve that structure. The natural charts do exactly that. See Exercise 3.12 for more of this categorical point of view.

Of course, the tangent bundle of M is a vector bundle over M with fibers isomorphic to $\mathbb{R}^m$, where $m = \dim M$. But there had better be plenty of other nontrivial examples of a vector bundle! And there are. In fact, there is a large family of bundles that we can associate with the tangent bundle.

But first we want to note that given a manifold M, there is always at least one smooth vector bundle with base space M and model fiber space $\mathbb{R}^l$. It is the *explicitly trivial* smooth vector bundle

$$\mathbb{R}^l \hookrightarrow M \times \mathbb{R}^l$$
$$\downarrow \pi_1 \, ,$$
$$M$$

where $\pi_1(x, v) := x$ for all $(x, v) \in M \times \mathbb{R}^l$.

Exercise 3.3 *Show that this is a smooth vector bundle by finding a cocycle that defines it.*

3.4   Bundles Associated to the Tangent Bundle

Since the tangent bundle is a vector bundle over the differential manifold M with
a very particular cocycle that defines it, we are going to introduce new notation for
this canonically associated cocycle [cp. equation (3.2)]:

$$t_{\beta\alpha} := D\psi_{\beta\alpha} \circ \phi_\alpha : U_\alpha \cap U_\beta \to GL(\mathbb{R}^m). \tag{3.5}$$

Of course, we used a representation of $GL(\mathbb{R}^m)$ in the construction of the tangent
bundle, though it may have escaped notice. It was $(g, v) \mapsto gv$, where gv is the
usual evaluation of a linear map $g : \mathbb{R}^m \to \mathbb{R}^m$ on a vector $v \in \mathbb{R}^m$. Another way to
say this is that it is the representation $g \mapsto g$ of $GL(\mathbb{R}^m)$ to itself.

But now suppose we consider instead a general smooth representation $\rho :$
$GL(\mathbb{R}^m) \to GL(\mathbb{R}^l)$. Then it is an easy exercise to show that

$$U_\alpha \cap U_\beta \xrightarrow{t_{\beta\alpha}} GL(\mathbb{R}^m) \xrightarrow{\rho} GL(\mathbb{R}^l)$$

is again a cocycle.

Exercise 3.4 *Please do this supposedly easy exercise.*

We denote the vector bundle defined by this cocycle and by the canonical
representation of $GL(\mathbb{R}^l)$ on $\mathbb{R}^l$ as follows:

$$\mathbb{R}^l \hookrightarrow \rho(TM)$$
$$\downarrow \rho(\tau_M)$$
$$M.$$

This is known as the *associated bundle* of the tangent bundle (or sometimes the
bundle associated to the tangent bundle) induced by the representation ρ. The
notation is not common, but we feel it encapsulates what is happening here. Now we
pull out the panoply of standard representations of the general linear group $GL(\mathbb{R}^m)$
from any textbook on the subject (and it turns out they are all smooth), and we then
apply the previous "algorithm" to each one of them.

And this is how tensors come into play in differential geometry. (See the
appendix to this chapter for a crash course on tensor products.) Notice that we
are using linear algebra at the infinitesimal level (that is, on the tangent spaces)
to construct new bundles over M at the global level.

For example, the tensor product of a vector space V with itself, $V \otimes V$, is a
functorial construction. What this boils down to in this context is that for each linear
map $A : V \to V$, there is an associated map $A \otimes A : V \otimes V \to V \otimes V$ with nice
properties (preservation of identity and products). So we get a map

$$\otimes^2 : GL(\mathbb{R}^m) \to GL(\mathbb{R}^m \otimes \mathbb{R}^m) \tag{3.6}$$

given by $A \mapsto A \otimes A$. And the functorial property says that this map is a representation of $GL(\mathbb{R}^m)$. Moreover, it is a smooth map, which again is an exercise for the reader.

Exercise 3.5 *Show that $A \mapsto A \otimes A$ maps $GL(\mathbb{R}^m)$ to $GL(\mathbb{R}^m \otimes \mathbb{R}^m)$, that it is a representation, and that it is smooth.*

This gives us a new cocycle:

$$U_\alpha \cap U_\beta \xrightarrow{t_{\beta\alpha}} GL(\mathbb{R}^m) \xrightarrow{\otimes^2} GL(\mathbb{R}^m \otimes \mathbb{R}^m);$$

namely,

$$x \mapsto t_{\beta\alpha}(x) \otimes t_{\beta\alpha}(x) = D\psi_{\beta\alpha}(\phi_\alpha(x)) \otimes D\psi_{\beta\alpha}(\phi_\alpha(x))$$

for all $x \in U_\alpha \cap U_\beta$. So we get the bundle of *second-order contravariant vectors*:

$$\mathbb{R}^m \otimes \mathbb{R}^m \hookrightarrow TM \otimes TM$$
$$\downarrow$$
$$M.$$

Similarly, by taking the kth tensor product $\otimes^k(V) := V \otimes \cdots \otimes V$ with k factors, we obtain the bundle of *kth-order contravariant vectors*:

$$\otimes^k(\mathbb{R}^m) \hookrightarrow \otimes^k(TM)$$
$$\downarrow$$
$$M$$

whose cocycle is

$$x \mapsto t_{\beta\alpha}(x) \otimes \cdots \otimes t_{\beta\alpha}(x) = D\psi_{\beta\alpha}(\phi_\alpha(x)) \otimes \cdots \otimes D\psi_{\beta\alpha}(\phi_\alpha(x)) \quad (k \text{ factors})$$

for all $x \in U_\alpha \cap U_\beta$.

Exercise 3.6 *Prove that $A \mapsto \otimes^k(A) := A \otimes \cdots \otimes A$ (with k factors) maps $A \in GL(V)$ to $GL(\otimes^k(V))$, that it is a representation, and that it is a smooth map. Exercise 3.5 is the special case of this exercise when $k = 2$.*

Example 3.1

There is another interesting smooth map

$$GL(\mathbb{R}^m) \xrightarrow{*} GL((\mathbb{R}^m)^*)$$

given by sending A to its dual A^* acting on the dual space $(\mathbb{R}^m)^*$. However, this is *not* a representation since $(AB)^* = B^*A^*$, a well-known formula from linear algebra. However, $A \mapsto (A^{-1})^* = (A^*)^{-1}$ is a representation:

$$GL(\mathbb{R}^m) \xrightarrow{(\cdot)^{*-1}} GL((\mathbb{R}^m)^*).$$

Of course, it is smooth too.

Exercise 3.7 *Show that the representation* $A \mapsto (A^{-1})^* = (A^*)^{-1}$ *is smooth.*

The cocycle associated to $t_{\beta\alpha}$ is then

$$(t_{\beta\alpha}^*)^{-1} : U_\alpha \cap U_\beta \to GL((\mathbb{R}^m)^*).$$

Using this cocycle, we construct the *cotangent bundle* of the manifold M:

$$(\mathbb{R}^m)^* \hookrightarrow T^*M$$
$$\downarrow$$
$$M.$$

The cotangent bundle is *not* the tangent bundle. This fact should be carefully examined and thoroughly understood.

Even though the fibers of the cotangent bundle T^*M are vector spaces of dimension m, namely, isomorphic copies of $(\mathbb{R}^m)^*$, their elements are not called vectors, but rather *covectors* or *covariant vectors*. They transform from chart α to chart β according to the linear map $(t_{\beta\alpha}(x)^*)^{-1}$ for $x \in U_\alpha \cap U_\beta$, as noted above. This is *not* how a *vector* (or *tangent vector* or *contravariant vector*) transforms, which is according to the linear map $t_{\beta\alpha}(x)$. [See the remarks just after (2.12).] In physics, one simply says that "a covariant vector is something that transforms like a covariant vector." The translation of this into mathematically acceptable rigorous jargon is

Exercise 3.8 *An exercise for the reader.*

Given all this, we can now tensor up k copies of the cocycle $t_{\beta\alpha}$ and l copies of its dual cocycle $(t_{\beta\alpha}^*)^{-1}$ to get the bundle of kth-order contravariant vectors and lth-order covariant vectors on M:

$$\underbrace{\mathbb{R}^m \otimes \cdots \otimes \mathbb{R}^m}_{k} \otimes \underbrace{(\mathbb{R}^m)^* \otimes \cdots \otimes (\mathbb{R}^m)^*}_{l} \hookrightarrow T^{(k,l)}M$$
$$\downarrow$$
$$M,$$

where there are k copies of $\mathbb{R}^m$ and l copies of $(\mathbb{R}^m)^*$ in the fiber.

Another representation that arises in physics is

$$\rho : GL(\mathbb{R}^m) \to GL(\mathbb{R}^m)$$

given by $\rho(A) := \mathrm{sgn}(\det(A))A$ for $A \in GL(\mathbb{R}^m)$, where $\det(A)$ is the determinant of A and $\mathrm{sgn}(r)$ is the algebraic sign of a nonzero real number; that is, $\mathrm{sgn}(r) = +1$ for $r > 0$ and $\mathrm{sgn}(r) = -1$ for $r < 0$. It turns out that $GL(\mathbb{R}^m)$ for $m \geq 1$ (but not for $m = 0$) has two nonempty connected components: $\det^{-1}(0, \infty)$ and $\det^{-1}(-\infty, 0)$. So $\rho(A) = \pm A$, the sign depending on the component. The corresponding cocycle $\rho(t_{\beta\alpha})$ is *almost* the same as $t_{\beta\alpha}$. It is equal to $t_{\beta\alpha}$ on the component of $GL(\mathbb{R}^m)$ containing the identity, and it is $-t_{\beta\alpha}$ on the other component. The corresponding bundle, even though its fibers are isomorphic copies of $\mathbb{R}^m$, is *not* the tangent bundle. And elements of the fibers are *not* (tangent) vectors, since they transform differently. But the difference is only a minus sign and only when $t_{\beta\alpha}(x)$ has negative determinant. In physics, the elements of the fibers are called *pseudovectors*.

Exercise 3.9 *Understand carefully the previous paragraph. Then consult a physics text to learn a bit about the magnetic field $\vec{B} = (B_1, B_2, B_3)$. Understand why this is a pseudovector and not a vector.*

Exercise 3.10 *It gets worse, or better, depending on your point of view. Consider the maps*

$$\rho_s : GL(\mathbb{R}^m) \to GL(\mathbb{R}^1)$$

$$\rho_{ps} : GL(\mathbb{R}^m) \to GL(\mathbb{R}^1)$$

given by $\rho_s(A) := I$ and $\rho_{ps}(A) := \big(\mathrm{sgn}(\det A)\big)I$ for $A \in GL(\mathbb{R}^m)$. Here $I = (1)$ is the 1×1 identity matrix, which acts on $\mathbb{R}^1 = \mathbb{R}$. The case $m = 0$ is included.

Show that these maps are smooth representations. Show that $\rho_s(\tau_M)$ is trivial. The elements of its fibers are called scalars. *In physics language: A scalar transforms as a scalar, that is, not at all! Peeking ahead for a definition, show that the sections of $\rho_s(\tau_M)$ can be identified with the smooth functions $f : M \to \mathbb{R}$.*

It turns out that $\rho_{ps}(\tau_M)$ need not be trivial. However, its fibers are also isomorphic copies of $\mathbb{R}$. The elements of its fibers are called pseudoscalars. *They do not transform as scalars (which, again, do not transform at all). Rather, a pseudoscalar does change sign (provided that $m \geq 1$) under some changes of charts.*

Often the representation $\rho : GL(\mathbb{R}^m) \to GL(\mathbb{R}^l)$ that we are using is *reducible*; that is, it has a nontrivial invariant vector subspace in $\mathbb{R}^l$. Then ρ restricted to such a subspace gives us another (smooth!) representation of $GL(\mathbb{R}^m)$, and so we can apply our machine to it. As an example, the representation (3.6) has the *symmetric tensors* as an invariant subspace. (The symmetric tensors are those in the subspace generated by all tensors of the form $v \otimes w + w \otimes v$, where $v, w \in \mathbb{R}^m$.) Therefore, there is a corresponding vector bundle of *symmetric second-order tensors* associated to the tangent bundle of M.

A representation that is not reducible is said to be *irreducible*.

3.5 Sections

We should also note at this point that a central role is played in vector bundle theory by the sections of a vector bundle.

Definition 3.4 *Suppose that* $\pi : E \to M$ *is a smooth vector bundle over the differential manifold* M. *Then a smooth function* $s : M \to E$ *is called a* section *if* $s(x) \in E_x$ *holds for all* $x \in M$. *Equivalently,* $\pi \circ s = 1_M$. *Moreover, we define*

$$\Gamma(E) = \Gamma(\pi) := \{s \mid s \text{ is a section of } E\},$$

the space of all sections of $\pi : E \to M$.

Exercise 3.11 *Define a sum and scalar product on* $\Gamma(E)$ *and show that it becomes a vector space over* $\mathbb{R}$. *Show that it typically has infinite dimension.*

The sections of the various tensor bundles defined in the previous section are called *tensor fields*. Sections of the tangent bundle have their own name. They are called *vector fields*. A section of the cotangent bundle is called a *1-form*, though later we will give a different definition (see Chapter 5).

A section g of $T^{(0,2)}M$ gives a bilinear map

$$g_x : T_x M \times T_x M \to \mathbb{R}$$

for every $x \in M$. If this bilinear map is symmetric and nondegenerate, then we say that g is a *metric* on M. If, in addition, each g_x is positive definite, then we say that it is a *Riemannian metric*. Note that the family of maps $\{g_x\}_{x \in M}$ does *not* paste together to give us a globally defined map $TM \times TM \to \mathbb{R}$. The point is that the proposed domain $TM \times TM$ is too big since it contains *all* pairs of tangent vectors. The domain we seek here is called the *fiber product* of TM with itself and is defined as

$$T^{\diamond 2}M := TM \diamond TM := \{(v, w) \in TM \times TM \mid \tau_M(v) = \tau_M(w)\},$$

which is precisely those pairs of tangent vectors that are tangent at the same point in M. (The point of common tangency depends on the pair.) Then the maps $\{g_x\}$ do paste together to give us a global map $g : T^{\diamond 2}M \to \mathbb{R}$. We note that $T^{\diamond 2}M$ becomes a vector bundle over M with projection map $T^{\diamond 2}M \to M$ defined by $(v, w) \mapsto \tau_M(v) = \tau_M(w)$.

3.6 Some Abstract Nonsense

We now indulge in some *abstract nonsense* about smooth vector bundles. If $\pi :$ $E \to M$ and $\pi' : E' \to M'$ are smooth vector bundles over the smooth manifolds M and M', then we say that a *vector bundle map* from E to E' is a pair of smooth functions $f : M \to M'$ and $F : E \to E'$ such that F *covers* f; that is,

$$
\begin{array}{ccc}
E & \xrightarrow{F} & E' \\
\downarrow{\scriptstyle\pi} & & \downarrow{\scriptstyle\pi'} \\
M & \xrightarrow{f} & M'
\end{array}
$$

is commutative [so that, in particular, F maps the fiber above $x \in M$ to the fiber above $f(x) \in M'$]. Moreover, we require that F restricted to any fiber $E_x :=$ $\pi^{-1}(x)$ is a linear map from E_x to $E'_y := (\pi')^{-1}(y)$, where $y = f(x)$.

Exercise 3.12 *In general, the phrase "abstract nonsense" refers to something expressed in the language of category theory. Specifically,*

- *Define the composition of vector bundle maps.*
- *Show that this gives us a category whose objects are the smooth vector bundles and whose morphisms are the vector bundle maps.*
- *Show that the tangent bundle construction $M \mapsto T(M)$ is a functor from the category of smooth manifolds to the category of vector bundles.*
- *For a given manifold M, define a category whose objects are smooth vector bundles over M and whose morphisms cover the identity map $1_M : M \to M$; that is, $M' = M$ and $f = 1_M$ in the above diagram. The exercise is to prove that this indeed is a category.*
  We conclude with a definition: An isomorphism in this category is called an equivalence of smooth vector bundles over M. *We say that two vector bundles over M are* equivalent *if there exists an equivalence between them.*

Definition 3.5 *Suppose that $\pi : E \to M$ is a smooth vector bundle that is equivalent to the explicitly trivial bundle $\pi_1 : M \times \mathbb{R}^l \to M$ for some $l \geq 0$. Then we say that E is a* trivial bundle *and that E is* trivializable.

Exercise 3.13 *Suppose $\pi : E \to M$ is a given trivializable smooth vector bundle whose fiber has dimension l. Then by the definition, there exists an equivalence of vector bundles over M:*

$$
\begin{array}{ccc}
E & \xrightarrow{F} & M \times \mathbb{R}^l \\
\downarrow{\scriptstyle\pi} & & \downarrow{\scriptstyle\pi_1} \\
M & \xrightarrow{1_M} & M.
\end{array}
$$

Show that the set of all such equivalences is in one-to-one and onto correspondence with the set $C^\infty(M, GL(\mathbb{R}^l))$ of all smooth functions from M to $GL(\mathbb{R}^l)$.

3.7   Appendix: Tensor Products

We will now give a quick overview of the tensor product of vector spaces for those who have not seen this before. There are many excellent references for more details. However, an important intuition is easily missed in a rigorous presentation. This intuition goes as follows. We consider a pair of vector spaces V and W over the field $\mathbb{R}$. Then for all vectors $v \in V$ and $w \in W$, we wish to form the most general objects $v \otimes w$ that are bilinear in the vectors v and w. Also, we want these objects $v \otimes w$ to be vectors in some as-yet-unspecified unique vector space, which we denote as $V \otimes W$. Additionally, bilinear means that we have these four identities for $v, v_1, v_2 \in V$, $w, w_1, w_2 \in W$, and $r \in \mathbb{R}$:

- $(v_1 + v_2) \otimes w = v_1 \otimes w + v_2 \otimes w$,
- $v \otimes (w_1 + w_2) = v \otimes w_1 + v \otimes w_2$,
- $(rv) \otimes w = r(v \otimes w)$,
- $v \otimes (rw) = r(v \otimes w)$.

"Most general objects" means that no further identities are imposed on the objects $v \otimes w$. Finally, we want $V \otimes W$ to be the smallest vector space containing these vectors $v \otimes w$. This means that $V \otimes W$ is spanned by all the vectors $v \otimes w$. In particular, if $\{e_\alpha \mid \alpha \in A\}$ is a basis of V and $\{f_\beta \mid \beta \in B\}$ is a basis of W, then we want $\{e_\alpha \otimes f_\beta \mid \alpha \in A, \ \beta \in B\}$ to be a basis of $V \otimes W$. So

$$\dim(V \otimes W) = (\dim V)(\dim W),$$

where the dimensions are cardinal numbers (possibly infinite). When all is said and done, this is most of what one really needs to know in order to work with tensors.

However, the discussion in the previous paragraph is not considered to be rigorous by most mathematicians because the existence of the vectors $v \otimes w$ and that of the spaces $V \otimes W$ have not been proved in the context of a currently accepted theory of sets.

Exercise 3.14 *Show that this discussion can be made rigorous in two easy steps. In the first step, we choose bases $\{e_\alpha \mid \alpha \in A\}$ of V and $\{f_\beta \mid \beta \in B\}$ of W. Then we define $V \otimes W$ to be the vector space whose basis vectors are the expressions $e_\alpha \otimes f_\beta$ for all $\alpha \in A$, $\beta \in B$. The second step consists of showing that the space $V \otimes W$ so defined does not depend on the choice of the two bases, where one has to clarify the meaning of "does not depend" in this setting.*

In any event, we will now give what is generally considered within the mathematics community as the standard, rigorous treatment of the tensor product. While it will fulfill all the intuitions described in the previous discussion, the reader may find it to be counterintuitive or even inelegant. *A chacun son goût*. This rigorous approach is based on a universal property, though this is not how the tensor product was originally defined. The idea begins with bilinear maps.

Definition 3.6 *Let V_1, V_2, and W be vector spaces over the field $\mathbb{R}$. We say that a function $B : V_1 \times V_2 \to W$ is* bilinear *if*

1. *For every $v_1 \in V_1$, the map $V_2 \to W$ defined by $v_2 \mapsto B(v_1, v_2)$ is linear.*
2. *For every $v_2 \in V_2$, the map $V_1 \to W$ defined by $v_1 \mapsto B(v_1, v_2)$ is linear.*

One says that B is linear in each variable separately.

We work here with real vector spaces since $\mathbb{R}$ is the relevant field for our topics. But everything goes through just the same for any field.

Now given vector spaces V_1 and V_2, we look for the most general bilinear map $\iota : V_1 \times V_2 \to W$. Notice that this is a problem whose solution consists of a vector space W and a bilinear map ι. As with any problem where we have to find the "unknowns," we have to consider the questions of existence and uniqueness of the solution. It is important to emphasize that the bilinear map ι is an intrinsic part of the problem. In other words, the tensor product is more than just the equally unknown vector space W. However, in practice, we abuse the terminology and speak of the tensor product of vector spaces without making explicit reference to the associated bilinear map. This is justifiable since that bilinear map always has an obvious form, and so it is implicitly understood to be part of the game.

But we have not yet given a precise meaning to "most general bilinear map." This is what we call the universal property of the tensor product. Here it is:

Definition 3.7 *Let V_1 and V_2 be vector spaces. Then we say that a vector space V_3 together with a bilinear map $\iota : V_1 \times V_2 \to V_3$ has the* universal property *provided that for any bilinear map $B : V_1 \times V_2 \to W$, where W is also a vector space, there exists a unique linear map $L : V_3 \to W$ such that $B = L\iota$. Here is a diagram describing this "factorization" of B through ι:*

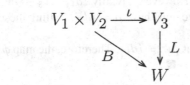

Remarks: Notice that the bilinear map B uniquely determines a linear map L. Also, given any linear map $L : V_3 \to W$, we can define $B := L\iota$, and it turns out that B so defined is bilinear. In short, the theory of bilinear maps is not a brave new world of maps but is part of the familiar world of linear maps. Of course, this is true provided that there exists a space V_3 together with the bilinear map ι that have the universal property.

Now it is a general fact of life that universal properties, such as the one being discussed here, have unique solutions in the sense of the following result.

Proposition 3.1 *Suppose that V_1 and V_2 are vector spaces. Furthermore, suppose that the pairs (V_3, ι) and (V_3', ι') have the universal property. Then there exists a unique (linear!) isomorphism $\phi : V_3 \to V_3'$ such that $\phi\iota = \iota'$.*

Remark: Note that this statement is much stronger than saying that the vector spaces V_3 and V_3' are isomorphic. It says that there is a very specific isomorphism, which also "maps" the bilinear map ι to the bilinear map ι'. (This last sentence is a way to put into words the equation $\phi\iota = \iota'$.) Thus, the map ϕ is an isomorphism of the structure given by the pair (V_3, ι) to the structure given by the pair (V_3', ι'). Also, the structure given by the pair (V_3, ι) is called the tensor product of V_1 and V_2. Similarly, the structure given by the pair (V_3', ι') is also called the tensor product of V_1 and V_2. So this proposition says the tensor product is determined up to a unique isomorphism.

Proof: Since the pair (V_3, ι) satisfies the universal property and ι' is bilinear, there exists a unique linear map $\phi : V_3 \to V_3'$ such that $\phi\iota = \iota'$, as desired. It remains to show that this linear map ϕ is an isomorphism.

Similarly, the pair (V_3', ι') satisfies the universal property and ι' is bilinear. So there exists a unique linear map $\psi : V_3' \to V_3$ such that $\psi\iota' = \iota$. Now we wish to show that this map ψ is the inverse to the map ϕ.

First, let's consider the composition $\psi\phi : V_3 \to V_3$. We have that

$$\psi\phi\iota = \psi\iota' = \iota : V_1 \times V_2 \to V_3,$$

which is a bilinear map. But by the universal property of (V_3, ι) applied to the bilinear map ι itself, we know that there exists a unique linear map $\rho : V_3 \to V_3$ such that $\rho\iota = \iota$. But we have just shown that $\psi\phi$ has the property required of ρ. So it follows that $\rho = \psi\phi$. However, trivially $id_{V_3} : V_3 \to V_3$, the identity map, also satisfies $id_{V_3}\,\iota = \iota$. So we must have $\rho = id_{V_3}$. Putting these identities together yields $\psi\phi = id_{V_3}$.

Similarly, we can show that $\phi\psi = id_{V_3'}$. Therefore, the map ϕ is an isomorphism, its inverse being the map ψ. ∎

This proof is the completely standard way for proving that universal properties determine objects that are unique up to a unique isomorphism. This proposition shows that the solution of the problem of finding a pair (V_3, ι) satisfying the universal property has a unique solution in a very precise sense.

We still have to show that a solution exists. This is typically done by presenting an explicit construction. Unfortunately, this construction tends to scare the wits out of many a beginner. It seems that the explicit details are too cumbersome and a beginner thinks (and fears!) that one must always work with that construction. But that is not so. Once having established the existence via the construction, one never again has to use it. All properties of the tensor product follow from the universal property. This is something like the construction of the real numbers as equivalence classes of Cauchy sequences of rational numbers using a very specific equivalence relation. Once this construction is shown to satisfy the axioms of the real numbers, one never again uses the explicit construction. Actually, in this example the construction is not unique since one can define the real numbers as Dedekind cuts of the rational numbers. But in any case, once the axioms of the real numbers have been established, one uses those axioms as the basic properties of the real numbers. And one never again has to consider the explicit construction of the reals. Given this wordy prologue, here is the result:

Theorem 3.2 *For any vector spaces V_1 and V_2, there exists a pair (V_3, ι) with the universal property.*

Proof: For any set S, we denote by $\mathcal{F}(S)$ the vector space over $\mathbb{R}$ with basis given by the elements $s \in S$. We can think of the vectors in $\mathcal{F}(S)$ as being finite, formal linear combinations $\sum_j a_j s_j$ with $a_j \in \mathbb{R}$ and $s_j \in S$.

We consider $V_1 \times V_2$, the usual cartesian product, to be a *set*. We use the notation $v_1 \times v_2 = (v_1, v_2)$, the ordered pair in $V_1 \times V_2$. So we have the associated vector space $\mathcal{F}(V_1 \times V_2)$. We consider its subspace R generated by all the elements of the forms

- $(v_1 + v_1') \times v_2 - v_1 \times v_2 - v_1' \times v_2$,
- $v_1 \times (v_2 + v_2') - v_1 \times v_2 - v_1 \times v_2'$,
- $(cv_1) \times v_2 - c(v_1 \times v_2)$,
- $v_1 \times (cv_2) - c(v_1 \times v_2)$,

where $c \in \mathbb{R}$, $v_1, v_1' \in V_1$, and $v_2, v_2' \in V_2$. We take the quotient space to define

$$V_3 := \mathcal{F}(V_1 \times V_2)/R.$$

We then denote the equivalence class in the quotient space V_3 of the element $v_1 \times v_2 \in \mathcal{F}(V_1 \times V_2)$ by $v_1 \otimes v_2$. Next, we define $\iota : V_1 \times V_2 \to V_3$ by

$$\iota(v_1 \times v_2) := v_1 \otimes v_2.$$

Because of the definition of R, the map ι is bilinear.

If $B : V_1 \times V_2 \to W$ is any bilinear map, we define $\tilde{L} : \mathcal{F}(V_1 \times V_2) \to W$ by the assignment on the basis vectors $v_1 \times v_2$ given by

$$\tilde{L}(v_1 \times v_2) := B(v_1, v_2).$$

It is important to note that this map $\tilde{L}$ is linear. Since B is bilinear, $\tilde{L}$ maps R to 0. So the map $\tilde{L}$ passes to the quotient space to define a map there denoted by $L : V_3 \to W$. Clearly, $B = L\iota$ and L is the unique linear map $V_3 \to W$ with this property. So this shows that the pair (V_3, ι) satisfies the universal property. ∎

We denote the tensor product of the vector spaces V_1 and V_2 as $V_1 \otimes V_2$. The elements of the vector space $V_1 \otimes V_2$ are called *tensors*. The tensors in $V_1 \otimes V_2$ of the form $v_1 \otimes v_2$ are called *decomposable tensors*. They are characterized in a construction-independent way as being the elements in the range of the bilinear map ι. Clearly, every element in $V_1 \otimes V_2$ is a finite sum of decomposable tensors;

that is, the decomposable tensors generate the vector space $V_1 \otimes V_2$. A fact that is often a mystery for beginners is that in general there are elements in $V_1 \otimes V_2$ that are not decomposable tensors. Perhaps an exercise will help the reader get this straight.

Exercise 3.15 *Suppose that V and W are finite-dimensional vector spaces with* dim $V = n$ *and* dim $W = m$. *Prove that $V \otimes W$ is a finite-dimensional vector space with* $\dim(V \otimes W) = nm$.

The dual space V^* of V also is finite dimensional with dim $V^* = n$. *Also, the vector space* $\mathrm{Hom}(V, W)$ *of all linear maps $V \to W$ is finite dimensional with dimension nm. Show that there is a natural isomorphism*

$$\eta_{V,W} : V^* \otimes W \to \mathrm{Hom}(V, W).$$

Identify the image in $\mathrm{Hom}(V, W)$ (which is essentially a space of matrices) under the isomorphism $\eta_{V,W}$ of the decomposable tensors in $V^* \otimes W$. Show that if $n > 1$ and $m > 1$, then there are elements in $\mathrm{Hom}(V, W)$ that are not in the image under $\eta_{V,W}$ of the decomposable tensors. Consequently, in this case there are tensors in $V^* \otimes W$ that are not decomposable.

The tensor product is an example of something known as a functor. More specifically, it is a functor of two "variables." What this really means is that if we have linear maps $T_1 : V_1 \to W_1$ and $T_2 : V_2 \to W_2$ (where all the objects are vector spaces), then we can define a "nice" map

$$T_1 \otimes T_2 : V_1 \otimes V_2 \to W_1 \otimes W_2.$$

Here "nice" means these two properties:

- $id_{V_1} \otimes id_{V_2} = id_{V_1 \otimes V_2}$.
- For $S_1 : U_1 \to V_1$ and $S_2 : U_2 \to V_2$ as well, we have that

$$(T_1 \otimes T_2)(S_1 \otimes S_2) = (T_1 S_1) \otimes (T_2 S_2).$$

One defines $T_1 \otimes T_2$ on decomposable elements by

$$(T_1 \otimes T_2)(v_1 \otimes v_2) := T_1 v_1 \otimes T_2 v_2. \tag{3.7}$$

The diligent reader may well be suspicious of the validity of this definition. However, it is the usual shorthand way of describing the following rigorous definition. First, one defines the map

$$T_1 \times T_2 : V_1 \times V_2 \to W_1 \otimes W_2$$

by

$$(T_1 \times T_2)(v_1, v_2) := T_1 v_1 \otimes T_2 v_2.$$

Then one checks that $T_1 \times T_2$ is bilinear. Consequently, there exists a unique linear map, denoted $T_1 \otimes T_2$, such that $(T_1 \otimes T_2)\iota = T_1 \times T_2$. Evaluating this equality on (v_1, v_2) then gives (3.7). From now on, we will use such shorthand definitions to define linear maps whose domain is a tensor product.

Exercise 3.16 *Prove the two properties given above of the tensor product of linear maps.*

The next step is to define the tensor product $V_1 \otimes V_2 \otimes \cdots \otimes V_k$ for k vector spaces for any integer $k \geq 2$. The easy, lazy way to do this is by recursion. But this misses an important point, namely, that there exists a map

$$\iota : V_1 \times V_2 \times \cdots \times V_k \to V_1 \otimes V_2 \otimes \cdots \otimes V_k$$

that satisfies a universal property. Here the map ι is *multilinear* and has a universal property with respect to all multilinear maps:

$$M : V_1 \times V_2 \times \cdots \times V_k \to W,$$

where W is a vector space. First, we say that M is multilinear if it is linear in each factor separately. The universal property in this case says that for any such multilinear map M, there exists a unique linear map

$$L : V_1 \otimes V_2 \otimes \cdots \otimes V_k \to W$$

such that $M = L\iota$. Again, this is a problem of solving for two "unknowns." The uniqueness of the solution (up to a unique isomorphism as before) has essentially the same proof. The existence of the solution can be shown as before with a messy, explicit construction. Only now, instead of using an analog of the cumbersome construction given for the case $k = 2$, we can give a more elegant (= easy, lazy) construction using recursion, which is uniquely isomorphic to the other construction since they both satisfy the same universal property.

For example, $V_1 \otimes V_2 \otimes V_3 := (V_1 \otimes V_2) \otimes V_3$ and the trilinear map ι is defined in terms of the universal bilinear maps of the factors. We leave the details of the definition of ι to the reader. We also remark that the alternative definition $V_1 \otimes V_2 \otimes V_3 := V_1 \otimes (V_2 \otimes V_3)$ together with a corresponding alternative definition of ι also gives this triple tensor product since it also satisfies the same universal property. So there is a unique isomorphism between these two constructions. This is again a functor, though now of three "variables." Again, we leave the details to the interested reader.

Also, since we are more accustomed to counting from zero (following the Mayas instead of the Romans), we would like to define the tensor product for the cases of $k = 0$ and $k = 1$ factors. Multilinear maps in the case $k = 1$ are simply linear maps, and the universal pair for a given vector space V_1 is (V_1, id_{V_1}). Then we assign to every (multi-)linear map $M : V_1 \to W$ the linear map $L : V_1 \to W$ given by $L = M$. Of course, while this is a trivial case, it does make the picture of tensor products more complete.

The case of $k = 0$ factors even gives some experts a mild headache. Somehow this case is too trivial! Let's simply state that the answer is the underlying field, $\mathbb{R}$.

Chapter 4
Vectors, Covectors, and All That

In this chapter we will study the material presented so far, but now in the context of a manifold that is an open subset U in the Euclidean space $\mathbb{R}^n$. This amounts to an extended exercise in calculus of several real variables. Of course, this is also the same thing as the local theory of a manifold of dimension n by the very definition of manifold. Calculus of several variables is a theory that is traditionally presented by writing everything in terms of the standard variables $x_1, \ldots, x_n$ of $\mathbb{R}^n$ restricted to U. In our treatment these variables are the local coordinates on U in the very specific choice of the chart $id_U : U \to U$. No chart is more fundamental than any other, but in this context it clearly is convenient to use this extremely simple chart. It is so simple that one typically does not even realize that it is there when first learning calculus of several variables. Since a coordinate-free treatment in terms of an arbitrary chart on U is not going to help us see the relation to calculus of several variables, we will use the standard (local as well as global) coordinates $x_1, \ldots, x_n$. This makes the relation with the calculus of several variables immediate. This chapter is important since its message to the reader is that the previous chapters are nothing more or less than a review of known material cleverly disguised (old wine in new bottles).

4.1 Jacobi Matrix

Let $\phi : U \to V$ be a diffeomorphism (an invertible C^∞-function with a C^∞-inverse) from U onto V, where $U, V \subset \mathbb{R}^n$ are open. Another common name for ϕ is a *change of variables*. We will maintain this notation throughout this chapter. As we remarked earlier, we can represent its derivative $D\phi$ as a matrix in terms of the standard coordinates in U and V as follows:

Let $x = (x_1, x_2, \ldots, x_n) \in U$ and $\phi(x) = (\phi_1(x), \phi_2(x), \ldots, \phi_n(x)) \in V$ be our notation. Then

© Springer International Publishing Switzerland 2015
S.B. Sontz, *Principal Bundles*, Universitext, DOI 10.1007/978-3-319-14765-9_4

$$J_\phi(x) = \begin{pmatrix} \frac{\partial \phi_1}{\partial x_1} & \cdots & \frac{\partial \phi_1}{\partial x_n} \\ \\ \frac{\partial \phi_2}{\partial x_1} & \cdots & \frac{\partial \phi_2}{\partial x_n} \\ \cdot & & \cdot \\ \cdot & \cdots & \cdot \\ \cdot & & \cdot \\ \frac{\partial \phi_n}{\partial x_1} & \cdots & \frac{\partial \phi_n}{\partial x_n} \end{pmatrix}$$

is the Jacobi matrix associated to the linear map $D\phi(x) : \mathbb{R}^n \to \mathbb{R}^n$, using the standard coordinates. We sometimes denote the matrix with the same notation as the linear map, that is, $D\phi(x)$. In particular, its (j, k)th entry (row $= j$, column $= k$) is

$$(J_\phi(x))_{jk} = (D\phi(x))_{jk} = \frac{\partial \phi_j}{\partial x_k}.$$

Note that, as is usual in analysis, we do not explicitly write the dependence of the partial derivative on the point $x \in U$. Of course,

$$\frac{\partial \phi_j}{\partial x_k}(x)$$

is the correct, complete notation.

Also note that the Jacobi matrix is *not* the transpose of the above matrix. Since this is a source of an enormous amount of confusion, we give some comments about what is going on here. In general, if $\psi : U_1 \to U_2$ is a C^∞-function with $U_1 \subset \mathbb{R}^n$ and $U_2 \subset \mathbb{R}^m$, then at each point $x \in U_1$, the derivative $D\psi(x)$ of ψ at x gives the best linear approximation to ψ at the point x. So we have the linear map

$$D\psi(x) : \mathbb{R}^n \to \mathbb{R}^m$$

going from the tangent space at $x \in U_1$ to the tangent space at $\psi(x) \in U_2$. In short, the derivative $D\psi(x)$ goes in the same direction as the function ψ itself.

Now the Jacobi matrix associated to $D\psi(x)$ for $x \in U_1$ is

$$J_\psi(x) = \begin{pmatrix} \frac{\partial \psi_1}{\partial x_1} & \cdots & \frac{\partial \psi_1}{\partial x_n} \\ \cdot & & \cdot \\ \cdot & \cdots & \cdot \\ \cdot & & \cdot \\ \frac{\partial \psi_m}{\partial x_1} & \cdots & \frac{\partial \psi_m}{\partial x_n} \end{pmatrix},$$

where $\psi(x) = (\psi_1(x), \psi_2(x), \ldots, \psi_m(x)) \in U_2 \subset \mathbb{R}^m$. This is an $m \times n$ matrix (namely, a linear map from column vectors in $\mathbb{R}^n$ to column vectors in $\mathbb{R}^m$), acting on the left by matrix multiplication on the column vectors in $\mathbb{R}^n$ and producing column vectors in $\mathbb{R}^m$. This means that $J_\psi(x) : \mathbb{R}^n \to \mathbb{R}^m$, which is the correct direction. However, the transposed matrix, being an $n \times m$ matrix, maps $\mathbb{R}^m$ to $\mathbb{R}^n$ (viewed again as spaces of column vectors), which is opposite to the direction of the derivative of ψ.

Notice that we have named the coordinates $x_1, \ldots, x_n$ on the domain space $U_1 \subset \mathbb{R}^n$ of ψ, but the coordinates on the codomain space $U_2 \subset \mathbb{R}^m$ were unnamed since it was not necessary to name them, as the reader can now see. Traditionally, these coordinates are written as $y_1, \ldots, y_m$ and so the generic point in U_2 is written as $y = (y_1, \ldots, y_m) \in U_2$. Then our rigorously correct notation

$$\psi(x) = (\psi_1(x), \psi_2(x), \ldots, \psi_m(x)) \in U_2$$

is rewritten as $y = \psi(x)$ or as

$$y = (y_1, \ldots, y_m) = (\psi_1(x), \psi_2(x), \ldots, \psi_m(x)).$$

Or again as $y_j = \psi_j(x)$ for each $j = 1, \ldots, m$. This is such a standard abuse of notation that it is rarely recognized as such. The problem is that the coordinates $y_1, \ldots, y_m$ on U_2 are being used as the values of the function ψ. For example, with this abuse of notation, we have this change in notation:

$$\text{Write} \quad \frac{\partial y_j}{\partial x_k} \quad \text{instead of} \quad \frac{\partial \psi_j}{\partial x_k}.$$

Of course, it makes no sense whatsoever to take the partial derivative of a coordinate on U_2 with respect to a coordinate on U_1, so that literally the expression $\partial y_j / \partial x_k$ is meaningless. In practice, however, one is aware that there is a specific function being discussed (say ψ) and that we are using $y_j = \psi_j(x)$. And so we are not using the y_j as a coordinate on U_2. But the reader's favorite text on calculus of several variables may well abuse notation this way. So to relate what is presented here with one's favorite text, one might have to be quite aware of the consequences of these comments.

4.2 Curves

Let J be an open nonempty interval in $\mathbb{R}$. Then a *curve* in $U \subset \mathbb{R}^n$ is a function (of class C^r with $1 \le r \le \infty$):

$$\gamma : J \to U,$$

Note that a curve is a function and is *not* the range of that function. Using standard coordinates, we have for all $t \in J$ that

$$\gamma(t) = (\gamma_1(t), \ldots, \gamma_n(t)).$$

By elementary calculus, the derivative of γ is

$$\gamma'(t) = (\gamma_1'(t), \ldots, \gamma_n'(t)), \tag{4.1}$$

and we say that $\gamma'(t) \in \mathbb{R}^n$ is a vector. But is it a tangent vector? That is, does it follow the transformation rule for a (tangent) vector under a change of coordinates? In short, in the standard, confusing jargon, is the vector $\gamma'(t) \in \mathbb{R}^n$ a vector?

With the change of coordinates $\phi : U \to V \subset \mathbb{R}^n$ (with V open), the curve $\gamma : J \to U$ transforms into the curve $\tilde{\gamma} := \phi \circ \gamma : J \to V$, or in other words, $\tilde{\gamma}(t) := \phi(\gamma(t))$ for every $t \in J$. Using the coordinates in V, we have that

$$\tilde{\gamma}(t) = (\tilde{\gamma}_1(t), \ldots, \tilde{\gamma}_n(t)) \text{ and } \phi(\gamma(t)) = (\phi_1(\gamma(t)), \ldots, \phi_n(\gamma(t))),$$

from which it follows that

$$\tilde{\gamma}_j(t) = \phi_j(\gamma(t)) = \phi_j(\gamma_1(t), \ldots, \gamma_n(t))$$

for $j = 1, \ldots, n$. Then, using the standard coordinates $x_1, \ldots, x_n$ on U, we take the derivative by using the chain rule and get

$$\tilde{\gamma}_j'(t) = \sum_{k=1}^{n} \frac{\partial \phi_j}{\partial x_k}(\gamma(t)) \, \gamma_k'(t) = \sum_{k=1}^{n} (D\phi(\gamma(t)))_{jk} \gamma_k'(t),$$

or equivalently in matrix form,

$$\tilde{\gamma}'(t) = D\phi(\gamma(t))\gamma'(t),$$

where $\gamma'(t)$ and $\tilde{\gamma}'(t)$ are now written as column vectors (namely, $n \times 1$ matrices) and $D\phi$ is the $n \times n$ Jacobi matrix. The last formula says that the derivative of a curve is a tangent vector, namely, that it transforms using the derivative of ϕ. (Compare with the remarks on the equivalence relation $\approx$ following Exercise 2.11.) In short, the derivative of a curve is a vector.

Exercise 4.1 *For each tangent vector $v \in T_x(U)$ where $x \in U$, find a curve $\gamma : J \to U$ with $0 \in J$ such that $\gamma(0) = x$ and $\gamma'(0) = v$. Moral: Tangent vectors are tangents of curves.*

4.3 The Dual of Curves

But what happens for a function $f : U \to \mathbb{R}$ of class C^r? In some sense, this is the case dual to that of a curve. Here $f(x) = f(x_1, \ldots, x_n)$ in standard coordinates. Therefore, we can define the gradient just as in a calculus course:

$$\mathrm{grad}\, f = \left(\frac{\partial f}{\partial x_1}, \ldots, \frac{\partial f}{\partial x_n} \right). \tag{4.2}$$

It seems also to be a vector in $\mathbb{R}^n$. But we have to see how this object transforms under a change of coordinates $\phi : U \to V$.

Note carefully that the change of coordinates is exactly the same as in the preceding case of a curve. When all is said and done, a change of coordinates is a concept in and of itself which we can use in order to study various "objects" (such as the derivative of a curve or the gradient of a function) in terms of their transformation rules. The function $f : U \to \mathbb{R}$ is transformed to $\tilde{f} : V \to \mathbb{R}$ in the new system of coordinates, where $\tilde{f} \circ \phi = f$ or, equivalently, $\tilde{f} = f \circ \phi^{-1}$.

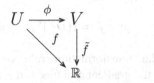

Yet again, but with even more notation, we have for all $x \in U$ that

$$f(x) = \tilde{f}(\phi(x)) = \tilde{f}(\phi_1(x), \ldots, \phi_n(x)).$$

Consequently, by the chain rule,

$$\frac{\partial f}{\partial x_j} = \sum_k \frac{\partial \tilde{f}}{\partial y_k} \frac{\partial \phi_k}{\partial x_j} = \sum_k (D\phi)_{kj} \frac{\partial \tilde{f}}{\partial y_k}. \tag{4.3}$$

(Again, we do not write the point where we have to evaluate the partial derivatives. It is simply understood implicitly. We are also using the notation $y_1, \ldots, y_n$ for the standard coordinates in V.)

We note two important things:

- The entry of the Jacobi matrix that appears here in equation (4.3) is *not* $(D\phi)_{jk}$, as was the case for a curve, but rather $(D\phi)_{kj}$.
- This formula (4.3) still does not give us the entries $\partial \tilde{f}/\partial y_k$ of the gradient of $\tilde{f}$ (which is the object in the new system of coordinates in V) in terms of the entries $\partial f/\partial x_j$ in the original system associated to U.

With regard to the first comment, we are obviously dealing with the transposed matrix $(D\phi)^t$, where $(D\phi)^t_{jk} = (D\phi)_{kj}$. Using this, we can write equation (4.3) in matrix form:

$$\operatorname{grad} f = (D\phi)^t \operatorname{grad} \tilde{f}.$$

The fact that ϕ is invertible implies that its derivative $D\phi$ is an invertible linear map. Then, by linear algebra, it follows that $(D\phi)^t$ is also an invertible linear map. So we can now answer the second point and solve in terms of $\tilde{f}$:

$$\operatorname{grad} \tilde{f} = ((D\phi)^t)^{-1} \operatorname{grad} f.$$

In general, an invertible matrix A does not satisfy $(A^t)^{-1} = A$, and therefore the transformation rule for the gradient is *not* the same as that for a (tangent) vector. But the preceding formula does give us a transformation rule. We now investigate what it is in more detail.

First, a linear map $A : \mathbb{R}^n \to \mathbb{R}^n$ (which we associate to an $n \times n$ matrix) induces a dual map $A^* : (\mathbb{R}^n)^* \to (\mathbb{R}^n)^*$. But, using the standard inner product in $\mathbb{R}^n$, we can identify the dual space $(\mathbb{R}^n)^*$ with $\mathbb{R}^n$, and so we have $A^* : \mathbb{R}^n \to \mathbb{R}^n$. Viewed this way, it turns out that $A^* = A^t$ as matrices. This implies that we can write the transformation rule of the gradient as

$$\operatorname{grad} \tilde{f} = ((D\phi)^*)^{-1} \operatorname{grad} f,$$

which is precisely the transformation rule of a covector. (Compare with Example 3.1.) In short, the gradient of a function is a covector:

$$(\operatorname{grad} f)(x) \in T_x^*(U)$$

for every $x \in U$, where we have now put the point where we evaluate back into the formulas.

4.4   The Duality

We now wish to introduce the infinitesimal notation that is used so much in vector calculus:

$$df(x) := \frac{\partial f}{\partial x_1} dx_1 + \cdots + \frac{\partial f}{\partial x_n} dx_n \in T_x^*(U). \qquad (4.4)$$

While this is related to equation (4.2), it is not the same since for the moment (4.4) is a formal, intuitive notation that comes from calculus. Its rigorous meaning will be given next.

This notation is dual to an infinitesimal notation for tangent vectors. First, we write equation (4.1) of the derivative of the curve $\gamma(t)$ in terms of the standard basis $e_1, \ldots, e_n$ of $T_x(U) \cong \mathbb{R}^n$ as follows:

$$\gamma'(t) = \gamma_1'(t)e_1 + \cdots + \gamma_n'(t)e_n.$$

Here $e_1 = (1, 0, \ldots, 0)$, $e_2 = (0, 1, \ldots, 0)$, and so forth. Now we make an amazing change in notation without giving any motivation. However, we will justify this notation a bit later on. What we do is to write $\partial/\partial x_j$ instead of e_j. So we get

$$\gamma'(t) = \gamma_1'(t)\frac{\partial}{\partial x_1} + \cdots + \gamma_n'(t)\frac{\partial}{\partial x_n} \in T_x(U), \qquad (4.5)$$

which in turn is yet another way of writing equation (4.1). Then the formal "infinitesimals" $dx_1, \ldots, dx_n$ in (4.4) are *defined* to be the dual basis of $T_x^*(U)$ of the basis

$$\left\{ \frac{\partial}{\partial x_1}, \ldots, \frac{\partial}{\partial x_n} \right\}$$

of the tangent space $T_x(U)$, where $x \in U$. This means that

$$\left\langle dx_k, \frac{\partial}{\partial x_j} \right\rangle = \delta_{kj} = \begin{cases} 1 \text{ if } k = j, \\ 0 \text{ if } k \neq j, \end{cases} \qquad (4.6)$$

which is the Kronecker delta. Of course, $\langle \cdot, \cdot \rangle$ denotes the pairing between the dual vector space and the vector space itself, namely, $\langle l, w \rangle = l(w)$, where $l \in W^*$ and $w \in W$ for any vector space W.

The reader should clearly understand that each $\partial/\partial x_j$ is a tangent vector ($\equiv$ vector) and each dx_k is a cotangent vector ($\equiv$ covector). Even though we use the word "infinitesimal" to refer to this notation, the objects $\partial/\partial x_j$ and dx_k are not infinitesimally small. In fact, in the norms associated to the standard inner product on $\mathbb{R}^n$, these are vectors whose norm is 1.

For example, provided that $x = \gamma(t)$, we can derive from equations (4.4), (4.5), and (4.6) that

$$\langle df(x), \gamma'(t) \rangle = \sum_k \frac{\partial f}{\partial x_k}(x) \, \gamma_k'(t) = (f \circ \gamma)'(t).$$

[We make a parenthetical remark here based on material from Chapter 6. Suppose that γ is an integral curve of a vector field X with initial condition $\gamma(t) = x$. Then the right side of the preceding equation becomes

$$Xf(x) = L_X f(x),$$

the Lie derivative of f with respect to the field X evaluated in the point x.]

The "infinitesimal" notation for the bases of the tangent space $T_x(U)$ and cotangent space $T_x^*(U)$, where $x \in U$, deserves some motivation. We will come back to the cotangent space in the next chapter. The tangent vectors at $x \in U$ can be realized as equivalence classes of curves $\gamma : (-a, a) \to U$, where $a > 0$ and $\gamma(0) = x$. (The parameter a can depend on γ.) The equivalence relation is called *first-order contact at x* and is defined as $\gamma_1 \sim \gamma_2$ for curves γ_1 and γ_2 with $\gamma_1(0) = \gamma_2(0) = x$ and $\gamma_1'(0) = \gamma_2'(0)$. The tangent vector defined by such an equivalence class is $\gamma'(0)$, where γ is any representative of the equivalence class.

For example, let $e_1, \ldots, e_n$ be the standard basis of $\mathbb{R}^n$. Then the curve $\gamma_j(t) = x + t e_j$ represents the tangent vector e_j in $T_x(U)$, since $\gamma_j(0) = x$ and $\gamma_j'(0) = e_j$. Of course, this representation is not unique since any other curve that has first-order contact with γ_j at x will also represent the same tangent vector. Nonetheless, let's work with this rather simple representative. How does this tangent vector e_j pair with a smooth function $f : U \to \mathbb{R}$? In other words, we have to calculate

$$\langle df(x), e_j \rangle = \left\langle df(\gamma_j(0)), \gamma_j'(0) \right\rangle$$

$$= (f \circ \gamma_j)'(0)$$

$$= \lim_{h \to 0} \frac{f(\gamma_j(h)) - f(\gamma_j(0))}{h}$$

$$= \lim_{h \to 0} \frac{f(x + he_j) - f(x)}{h} = \frac{\partial f}{\partial x_j}(x),$$

where the last equality is the definition of the partial derivative of f with respect to x_j. Therefore, the tangent vector e_j in $T_x(U)$ acts on f as the partial derivative $\partial/\partial x_j$ at the point $x \in U$. So we have expressed this directly in the notation by writing $\partial/\partial x_j$ instead of e_j. Note that the point $x \in U$ is suppressed from the notation $\partial/\partial x_j$.

Exercise 4.2 *(a) If A is an $n \times n$ invertible matrix, prove that A^t is invertible with $(A^t)^{-1} = (A^{-1})^t$.*
(b) Find an invertible 2×2 matrix such that $(A^t)^{-1} \neq A$.
(c) Recall the name of this set:

$$\{A \mid A \text{ is an } n \times n \text{ invertible matrix}, (A^t)^{-1} = A\}.$$

What sort of structure does it have? Does it have a dimension?

Remark: A change of coordinates (sometimes called a change of variables) in the context of differential structures (including differential functions) always implies an application of the chain rule.

In a course on elementary calculus, we learn that the gradient is a vector. In fact, it is. And this is because there one is using another definition of the gradient. The object we have defined is denoted by df in differential geometry and is called the differential of f. [See equation (4.4).] As we shall see, it can be viewed as a 1-form and can be thought of as a section of the cotangent bundle. It turns out that grad f is a vector field and can be thought of as a section of the tangent bundle.

Exercise 4.3 *Let $w \in T_y^*(U)$ be an arbitrary covector. Prove that there exists an $f : U \to \mathbb{R}$ such that $df(y) = w$. (This is dual to Exercise 4.1.)*

For the next exercise, one has to understand a little of the language of k-forms, at least for $k = 1$ and 2. You may wish to come back to this exercise after studying the next chapter.

Exercise 4.4 *(a) If M is a differential manifold and $f : M \to \mathbb{R}$ is smooth, show that $df : M \to T^*M$ is a smooth section of the cotangent bundle. (It is a special case of a 1-form defined on M, as we shall see in the next chapter.) Part of this problem is to define df in this context.*
(b) Continuing to use the notation from part (a), set $\omega = df$. Show that $d\omega = 0$. (So every ω that is exact in M is closed.)

(c) *If $M = B_r(0) := \{x \in \mathbb{R}^n \mid ||x|| < r\}$ is the open ball of radius $r > 0$ in $\mathbb{R}^n$ and ω is a 1-form in M with $d\omega = 0$, show that $\omega = df$ for some $f : M \to \mathbb{R}$. Is f unique? (So every ω that is closed in $B_r(0)$ is exact.)*

(d) *We say that a subset $S \subset \mathbb{R}^n$ is* star-shaped *if there exists a point $p \in S$ such that for every $x \in S$, the straight-line segment between x and p lies in S. We also say that such a point $p \in S$ is a* star center *for S. Show that every convex subset in $\mathbb{R}^n$ is star-shaped, but that there are star-shaped subsets that are not convex.*

(e) *(Poincaré's lemma) Suppose that $U \subset \mathbb{R}^n$ is an open star-shaped subset and that ω is a 1-form in U with $d\omega = 0$. Prove that $\omega = df$ for some $f : U \to \mathbb{R}$. Is f unique? (So every ω closed in an open star-shaped U is exact.) By the way, part (c) is a particular case of this part.*

(f) *Find an open $U \subset \mathbb{R}^2$, the Euclidean plane, such that there exists a 1-form ω in U with $d\omega = 0$, but there does not exist any $f : U \to \mathbb{R}$ such that $\omega = df$. (So there is an ω that is closed in U that is not exact. Obviously, U cannot be star-shaped. There is a very common example of this that is often given in a course on calculus of several variables.)*

Chapter 5
Exterior Algebra and Differential Forms

We give a brief treatment of these topics. The crucial concept here is that of a *differential k-form* on a manifold. But before discussing that, we have to present a bit of exterior algebra.

5.1 Exterior Algebra

The standard, rigorous treatments of exterior algebra are almost guaranteed to obscure the basic, simple intuition behind this theory. And here is that intuition. We consider any vector space V over the field $\mathbb{R}$. Then for $v, w \in V$, we want to form the objects $v \wedge w$, which are bilinear in v and w and *antisymmetric*, namely, that satisfy the relation

$$v \wedge w = -w \wedge v.$$

Moreover, there should not be any other independent relations. We say that $v \wedge w$ is the *wedge product* of v and w. These objects $v \wedge w$ will be vectors in $V \wedge V$ (a vector space that must also be specified), and this will be the smallest such vector space. Of course, this sounds a lot like the intuition that we introduced earlier for the tensor product. But we wish to ignore this relation to the tensor product. Instead, we start with an arbitrary *ordered* basis $\{e_\alpha \mid \alpha \in A\}$ of V. ("Ordered basis" means that the index set A comes equipped with a linear order. In this book, V is always finite dimensional, say dim $V = n$, in which case we will take $A = \{1, 2, \ldots, n\}$ with its usual order.) Then we consider all the formal objects $e_\alpha \wedge e_\beta$ with $\alpha < \beta$ and let $V \wedge V$ be the vector space whose basis is exactly these objects. (Note that $v \wedge v = -v \wedge v$ holds and so $v \wedge v = 0$ follows.)

The original version of this chapter was revised: Definition 5.4 was replaced. The correction to this chapter is available at https://doi.org/10.1007/978-3-319-14765-9_18

© Springer International Publishing Switzerland 2015

S.B. Sontz, *Principal Bundles*, Universitext, DOI 10.1007/978-3-319-14765-9_5

Take vectors $v, w \in V$. Writing $v = \sum_\alpha r_\alpha e_\alpha$ and $w = \sum_\beta s_\beta e_\beta$, where $r_\alpha, s_\beta \in \mathbb{R}$ with only finitely many of them being distinct from zero, we formally calculate

$$
\begin{aligned}
v \wedge w &= \sum_{\alpha, \beta} r_\alpha s_\beta (e_\alpha \wedge e_\beta) \\
&= \sum_{\alpha < \beta} r_\alpha s_\beta (e_\alpha \wedge e_\beta) + \sum_{\alpha > \beta} r_\alpha s_\beta (e_\alpha \wedge e_\beta) \\
&= \sum_{\alpha < \beta} r_\alpha s_\beta (e_\alpha \wedge e_\beta) - \sum_{\alpha > \beta} r_\alpha s_\beta (e_\beta \wedge e_\alpha) \\
&= \sum_{\alpha < \beta} r_\alpha s_\beta (e_\alpha \wedge e_\beta) - \sum_{\alpha < \beta} r_\beta s_\alpha (e_\alpha \wedge e_\beta) \\
&= \sum_{\alpha < \beta} (r_\alpha s_\beta - r_\beta s_\alpha)(e_\alpha \wedge e_\beta),
\end{aligned}
$$

where we used bilinearity in the first equality, $e_\alpha \wedge e_\alpha = 0$ in the second, and antisymmetry in the third. We now turn this "calculation" around and use it to define $v \wedge w$ as

$$
v \wedge w := \sum_{\alpha < \beta} (r_\alpha s_\beta - r_\beta s_\alpha)(e_\alpha \wedge e_\beta),
$$

where v and w have the above expansions in the basis. Then it is an exercise to prove that the map $(v, w) \mapsto v \wedge w$ is bilinear and antisymmetric. This is the first step in developing this theory. The second step is showing that all the definitions do not depend on the particular choice of the ordered basis $\{e_\alpha \mid \alpha \in A\}$.

Exercise 5.1 *Prove that $v \wedge w \neq 0$ if and only if the vectors v and w are linearly independent.*

We also call $V \wedge V$ the *second exterior power of V* and denote it by $\Lambda^2 V$. We want to extend this construction to the *kth exterior power of V*, denoted by $\Lambda^k V$, for all integers $k \geq 0$. For $k \geq 2$, we let $\Lambda^k V$ be the vector space with basis elements

$$
e_{\alpha_1} \wedge e_{\alpha_2} \wedge \cdots \wedge e_{\alpha_k}, \tag{5.1}
$$

where $\alpha_1 < \alpha_2 < \cdots < \alpha_k$. We define a multilinear and antisymmetric map

$$
V \times \cdots \times V \ (k \text{ factors}) \rightarrow \Lambda^k V \tag{5.2}
$$

by taking $(v_1, \ldots, v_k)$, then expanding each entry in the standard basis, and writing out $v_1 \wedge \cdots \wedge v_k$ formally. Here "antisymmetric" means that the image of the map changes sign with the interchange of any pair of its arguments. Then this formal calculation is taken to be the definition of the desired map in (5.2). We explicitly did the case when $k = 2$ above. For $k \geq 2$, this calculation will require decomposing a sum into $k!$ expressions with one expression for every element of the permutation group of k letters. This makes the general calculation rather messy, but the basic idea is already there is the case when $k = 2$. We leave the detailed calculation to the reader.

Exercise 5.2 *Prove that $v_1 \wedge \cdots \wedge v_k \neq 0$ if and only if the vectors $v_1, \ldots, v_k$ are linearly independent.*

For the case when the index set is $\{1, 2, \ldots, n\}$ with its usual order (that is, $\dim V = n$), we see by counting the number of basis elements defined in (5.1) that for $2 \leq k \leq n$, we have

$$\dim \Lambda^k V = \frac{n!}{k!(n-k)!} = \binom{n}{k}, \tag{5.3}$$

the binomial coefficient. Also, for $k > n$, the basis we defined for $\Lambda^k V$ in (5.1) is the empty set, which is the basis for the zero-dimensional (but nonempty) vector space. Since for $k > n$ the binomial coefficient is conventionally defined to be zero, we have that (5.3) is valid for all integers $k \geq 2$. It now is reasonable to define $\Lambda^k V$ for the cases $k = 0$ and $k = 1$ so that (5.3) is also true in those cases. For $k = 0$, this gives us $\dim \Lambda^0 V = 1$. So we define

$$\Lambda^0 V := \mathbb{R}.$$

For $k = 1$, this gives us $\dim \Lambda^1 V = n$. We thus define

$$\Lambda^1 V := V.$$

Now we form the *graded vector space*

$$\Lambda^* V = \Lambda V := \oplus_{k \geq 0} \Lambda^k V.$$

The elements of $w \in \Lambda^k V$ are said to be *homogeneous of degree k*. We write this as $\deg w = k$. We now make $\Lambda^* V$ into a *graded algebra* by defining for each $j, k \geq 0$ a bilinear map

$$\Lambda^j V \times \Lambda^k V \rightarrow \Lambda^{j+k} V.$$

Note that we have already defined this for the case $j = k = 1$. Since this is to be a bilinear map, it suffices to define it on a pair of basis elements. So let $w_1 = e_{\alpha_1} \wedge \cdots \wedge e_{\alpha_j}$ and $w_2 = e_{\beta_1} \wedge \cdots \wedge e_{\beta_k}$ be basis elements for $\Lambda^j V$ and $\Lambda^k V$, respectively. Then define their *wedge product* to be

$$w_1 \wedge w_2 := e_{\alpha_1} \wedge \cdots \wedge e_{\alpha_j} \wedge e_{\beta_1} \wedge \cdots \wedge e_{\beta_k}, \tag{5.4}$$

where the right side is the multilinear map introduced in (5.2) acting on
$(e_{\alpha_1}, \cdots, e_{\alpha_j}, e_{\beta_1}, \cdots, e_{\beta_k})$. First of all, this does agree with our previous definition
in the case $j = k = 1$. The case when either $j = 0$ or $k = 0$ (or both!) might
be perplexing for the reader. For example, what does $e_{\alpha_1} \wedge \cdots \wedge e_{\alpha_j}$ mean when
$j = 0$? Like it or not, it means the element $1 \in \mathbb{R}$. So, if $\deg w_1 = 0$, then we
write $w_1 = r \in \mathbb{R}$ and we have $w_1 \wedge w_2 = r w_2$. If the reader does not fancy this
argument, then we can just say that this defines $w_1 \wedge w_2$ in this special case. Similar
remarks hold when $\deg w_2 = k = 0$. For the case $\deg w_1 = \deg w_2 = 0$, the wedge
product $w_1 \wedge w_2$ reduces to the usual product of real numbers.

Exercise 5.3 *For the reader who is keenly alert to hidden details, it is clear that we
have assumed that the basis vector* $e_{\alpha_1} \wedge \cdots \wedge e_{\alpha_j}$ *(with* $\alpha_1 < \cdots < \alpha_k$*) of* $\Lambda^k V$
uniquely determines the ordered sequence $e_{\alpha_1}, \ldots, e_{\alpha_j}$ *of basis vectors of* V*. If you
have not already realized this, think again about the definition in (5.4). And then for
everyone, prove that the hidden assumption is true.*

Note also that the right side of (5.4) need not be one of our standard basis vectors
in $\Lambda^{j+k} V$, though there are only three mutually exclusive possibilities:

- $w_1 \wedge w_2$ is a standard basis element.
- $w_1 \wedge w_2$ is -1 times a standard basis element.
- $w_1 \wedge w_2 = 0$.

For example, the last case occurs if and only if the sequence of subindices

$$\alpha_1, \ldots, \alpha_j, \beta_1, \ldots, \beta_k$$

has at least one repeated index. In particular, if $j + k > n$, there must be a repetition,
and so $w_1 \wedge w_2 = 0$. The reader is encouraged to think about the first two cases in
further detail.

We now state a basic result whose proof we leave to the reader.

Proposition 5.1 *The graded vector space* $\Lambda^* V = \oplus_{k \geq 0} \Lambda^k V$ *equipped with the
wedge product is a graded associative algebra with identity* $1 \in \mathbb{R} = \Lambda^0 V$.

But what about commutativity? That holds too, but in a graded sense.

Proposition 5.2 *Suppose that* w_1 *and* w_2 *are homogeneous elements in* $\Lambda^* V$ *with*
$\deg w_1 = j$ *and* $\deg w_2 = k$*. Then we have the identity*

$$w_1 \wedge w_2 = (-1)^{jk} w_2 \wedge w_1. \tag{5.5}$$

Remark: The identity (5.5) is called *graded commutativity*. This could well be
the major result of the theory of the exterior algebra. There are two quite different
ways to think about this result. One way, favored by beginners, is that this is *not*
commutativity but rather some weird, ugly formula that tells us that commutativity
fails in this setting. The other way, that comes with time, is that this is the correct
generalization to the setting of graded algebras of the concept of commutativity in
the setting of ungraded algebras. Actually, since any ungraded algebra $\mathcal{A}$ can be

considered a graded algebra $G(\mathcal{A}) = \oplus_{k\geq0}\mathcal{A}_k$ with $\mathcal{A}_0 = \mathcal{A}$ and $\mathcal{A}_k = 0$ for all $k \geq 1$, we see that the concept of graded commutativity *is* a generalization of commutativity. More specifically, $\mathcal{A}$ is commutative if and only if $G(\mathcal{A})$ is graded commutative. One eventually realizes that (5.5) is a beautiful, natural formula.

Proof: Since both sides of (5.5) are bilinear in the pair (w_1, w_2), it suffices to prove this result for basis elements. So we let $w_1 = e_{\alpha_1}\wedge\cdots\wedge e_{\alpha_j}$ and $w_2 = e_{\beta_1}\wedge\cdots\wedge e_{\beta_k}$ be basis elements. Then one shows (5.5) by a direct calculation using the antisymmetry of the wedge product. ∎

The construction of the kth exterior algebra $\Lambda^k V$ from the vector space V is the first part of a functor. The second part consists of defining for every linear map $T : V \to W$ another linear map

$$\Lambda^k T : \Lambda^k V \to \Lambda^k W.$$

This is defined on the standard basis elements by

$$\Lambda^k T (e_{\alpha_1} \wedge \cdots \wedge e_{\alpha_k}) := Te_{\alpha_1} \wedge \cdots \wedge Te_{\alpha_k}.$$

The result in the next exercise implies that this definition does not depend on the particular choice of basis.

Exercise 5.4 $\Lambda^k T (v_1 \wedge \cdots \wedge v_k) = Tv_1 \wedge \cdots \wedge Tv_k$ *for all* $v_1, \ldots v_k \in V$.

The maps $\Lambda^k T$ can be put together to get an algebra map of the exterior algebras. Explicitly,

$$\Lambda^* T : \Lambda^* V \to \Lambda^* W$$

is defined by $\Lambda^* T := \oplus_{k\geq0}\Lambda^k T$; that is, $\Lambda^* T$ is defined to be $\Lambda^k T$ on the summand $\Lambda^k V$ of $\Lambda^* V$.

Exercise 5.5 *Prove that $\Lambda^* T$ is a morphism of graded algebras with identity.*

Exercise 5.6 *Prove that the constructions Λ^k and Λ^* are functors. Identify carefully the target category in each case.*

Exercise 5.7 *Suppose that V is a finite-dimensional real vector space with* dim $V = n \geq 1$. *Let $T : V \to V$ be linear. Note that $\Lambda^n V$ is a vector space of dimension 1 with basis $w = e_1 \wedge e_2 \wedge \cdots \wedge e_n \neq 0$ provided that $\{e_1, e_2, \ldots, e_n\}$ is a basis of V. Then $\Lambda^n T : \Lambda^n V \to \Lambda^n V$ is determined by its action on this unique basis element. So we can write*

$$\Lambda^n T(w) = r\, w$$

for some real number r. Identify r in terms of T.

Decomposable elements in $\Lambda^k V$ are defined as those elements of the form $v_1 \wedge \cdots \wedge v_k$, where each $v_j \in V$.

Exercise 5.8 *Let V be a finite-dimensional vector space with* $\dim V = n$. *Show that the elements in* $\Lambda^k V$ *are decomposable for* $k \in \{0, 1, n-1, n\}$.

Accordingly, the smallest value of n for having nondecomposable elements satisfies $n \geq 4$. And for $n = 4$, the only possible value for k is $k = 2$.

Show that for $n = 4$ and $k = 2$, there do exist nondecomposable elements.

The approach given in this section is considered by some to be rather inelegant. Whether one subscribes to this opinion or not, this approach is surely incomplete, since it remains to prove that the theory does not depend on the particular choice of the ordered basis of V. Of course, this is true though a bit uncomfortable to prove. We leave those details to the reader. The standard treatment also has its messy moments though the end result is the same. Again, we feel that our approach reveals more clearly the intuition behind exterior algebra.

Exercise 5.9 *The vector space $\Lambda^k(V)$, the kth exterior power of V, satisfies a universal property. Find it. Do the same for the exterior algebra $\Lambda^*(V)$.*

5.2   Relation to Tensors

This section is extremely unimportant and is included merely to indicate compatibility with the usual approach. In most texts the exterior algebra of a vector space is constructed from the tensor algebra of that vector space. But I think that hides the basic structure of the exterior algebra as an object in its own right. Nonetheless, as an afterthought, we can see that there is a relation between these algebras. Here is a sketch of some of that with proofs left to the reader.

First, for any integer $k \geq 1$, we let S_k denote the symmetric group that consists of all permutations of the finite set $\{1, 2, \ldots, k\}$. Also, we introduce the notation

$$\otimes^k V = V \otimes \cdots \otimes V \quad (k \text{ factors}).$$

We define an action of S_k on $\otimes^k V$ by taking $\sigma \in S_k$ and a decomposable vector $v_1 \otimes \cdots \otimes v_k \in \otimes^k V$ and defining

$$\sigma(v_1 \otimes \cdots \otimes v_k) := v_{\sigma(1)} \otimes \cdots \otimes v_{\sigma(k)}.$$

If you are interested, figure out for yourself whether this is a right or left action. But it does not matter!

Also, there is a well-known group morphism sgn : $S_k \to \{-1, +1\}$, where the codomain is the two-element subgroup of the multiplicative group $\mathbb{R}^\times$ of nonzero real numbers. If we write $\sigma \in S_k$ as a product of 2-cycles (interchanges of two numbers) as we always can, then the parity of the number of factors in that product (i.e., $+1$ if even and -1 if odd) does not depend on the particular factorization. Then $\text{sgn}(\sigma)$ is that parity. See any text on finite group theory for everything in detail.

Now it is clear how to define antisymmetric tensors.

Definition 5.1 *Let V be a vector space and $k \geq 1$ an integer. Define the vector space of* alternating k-vectors *(or more simply* antisymmetric tensors*) to be*

$$\Lambda^k(V) := \{\omega \in \otimes^k V \mid \sigma(\omega) = \text{sgn}(\sigma)\,\omega \text{ for all } \sigma \in S_k\}.$$

For $k = 0$, we define $\Lambda^0(V) := \mathbb{R}$ as before.

So $\Lambda^k(V)$ is a subset (even subspace) of $\otimes^k(V)$ and for those who associate "doing mathematics right" with working in the theoretical setting of some (usually unspecified) set theory, this gives a rigorous definition of $\Lambda^k(V)$. But we want $\Lambda^*(V) := \oplus_{k \geq 0} \Lambda^k(V)$ to be an algebra with a wedge product still to be defined. And as we can see from our approach, $\Lambda^*(V)$ is not going to be a subalgebra of $\otimes^*(V) := \oplus_{k \geq 0} \otimes^k (V)$. That is to say, the wedge product is not the restriction of the tensor product to the exterior algebra. For example, take $v_1, v_2 \in V = \Lambda^1(V) = \otimes^1(V)$. Then $v_1 \otimes v_2 \notin \Lambda^2(V)$ in general. This is why we say that the exterior algebra has a structure in its own right. Of course, one can go into the details of relating these two multiplications, but this question only arises because one has put oneself into a straightjacket to begin with by defining the exterior algebra as a subset of the tensor algebra.

One can also realize $\Lambda^k(V)$ as a quotient space of $\otimes^k(V)$ by finding a projection $A_k : \otimes^k(V) \to \otimes^k(V)$ whose image is $\Lambda^k(V)$. This approach has the advantage that the wedge product is the quotient of the tensor product. This property can be used in fact to define the wedge product, as is done in Helgason's famous text [21], where it is called the *exterior product*. In that approach, one has to prove a lemma of Chevalley that says $\oplus_k(\ker A_k)$ is a two-sided ideal in $\oplus_k\left(\otimes^k(V)\right)$. The reader might like trying to prove that lemma. However, we will not continue along this line. See [21] for the details.

Exercise 5.10 *Viewing $\Lambda^k(V)$ as a subspace of $\otimes^k(V)$, identify all of the decomposable vectors $v_1 \otimes \cdots \otimes v_k$ in $\otimes^k(V)$ that are also in $\Lambda^k(V)$.*

The point of this exercise is to show that the concept of decomposable element for one of these algebras has little to do with the same concept for the other algebra when using this realization of $\Lambda^k(V)$.

5.3 Hodge Star

This section is a brief overview of the Hodge star operation. This is a rather basic topic in linear algebra, and so we will only present the results that we consider interesting. Many proofs are left to the reader.

The Hodge star operation is an explicit isomorphism between the spaces $\Lambda^k V$ and $\Lambda^{n-k} V$, where V is a finite-dimensional vector space over $\mathbb{R}$ with $\dim V = n \geq 1$ and $0 \leq k \leq n$. We already know that these two spaces have the same dimension, namely, the equal binomial coefficients:

$$\binom{n}{k} = \binom{n}{n-k} = \frac{n!}{(n-k)!k!}.$$

So we know these two vector spaces are isomorphic. The tricky bit is to select one specific isomorphism in a "functorial" way, that is, in a way that is independent of the choice of V.

Here is the basic idea, which we will see does not work! We take an ordered basis $\{e_i \mid 1 \le i \le n\}$ of V. Then a typical basis element of $\Lambda^k V$ is $\omega = e_{i_1} \wedge e_{i_2} \wedge \ldots \wedge e_{i_k}$, where the subindices are increasing, $i_1 < i_2 < \cdots < i_k$. We map this into the element $\omega' = \pm e_{j_1} \wedge e_{j_2} \wedge \ldots \wedge e_{j_{n-k}}$ of $\Lambda^{n-k} V$, where $j_1 < j_2 < \cdots < j_{n-k}$ is the complementary increasing sequence we get by deleting $i_1 < i_2 < \cdots < i_k$ from $1 < 2 < \cdots < n$. We pick the sign of ω' so that

$$\omega \wedge \omega' = e_1 \wedge e_2 \wedge \cdots \wedge e_n.$$

Another way of specifying the sign of ω' is to note that it is the sign of the permutation that maps the juxtaposed sequence of n indices

$$i_1, i_2, \ldots, i_k, j_1, j_2, \ldots, j_{n-k}$$

to the sequence $1, 2, \ldots, n$. It should be rather clear that this defines a linear isomorphism from $\Lambda^k V$ onto $\Lambda^{n-k} V$. But the resulting isomorphism depends on the choice of the ordered basis of V. Hence, as far as it goes, this definition is okay. But we want more than this!

Let's see what goes wrong in more detail even in the simple case when $\dim V = 2$. In this example, we also take $k = 1$. Let $\{e_1, e_2\}$ be an ordered basis for V. So we want to consider the isomorphism of $\Lambda^1 V = V$ with itself that is associated with this ordered basis. Then with the above definition, we have that $e_1 \mapsto e_2$ and $e_2 \mapsto -e_1$ defines the selected isomorphism $\Lambda^1 V$ to $\Lambda^1 V$. Watch out! Even though we are mapping a vector space to itself, the selected isomorphism here is not the identity. But if we take any vector $e_3 \in V$ that is linearly independent of e_1 and with $e_3 \ne e_2$, then $\{e_1, e_3\}$ is also an ordered basis for V. With this new ordered basis, the selected isomorphism satisfies $e_1 \mapsto e_3 \ne e_2$ and so is not the same as the isomorphism defined by $\{e_1, e_2\}$. So our proposed definition is not "good" since it depends not only on V but also on the choice of ordered basis of V.

This difficulty can *almost* be solved by supposing that V has an inner product (say, positive definite for now) and that $\{e_i \mid 1 \le i \le n\}$ is an ordered, orthonormal basis of V. The point is that the standard basis element $e_{i_1} \wedge e_{i_2} \wedge \ldots \wedge e_{i_k}$ of $\Lambda^k V$ always corresponds to a vector subspace W of dimension k in V, where W is just the subspace spanned by $\{e_{i_1}, e_{i_2}, \ldots, e_{i_k}\}$. This has nothing to do with the inner product. But the image under the proposed Hodge star of this basis element, namely, $\pm e_{j_1} \wedge e_{j_2} \wedge \ldots \wedge e_{j_{n-k}}$ as defined above, now corresponds to $W^\perp$, the orthogonal complement of W in V. It turns out that this definition is not quite independent of the choice of the ordered, orthonormal basis of V, but nearly.

Here is what is happening. If we pick a pair of ordered, orthonormal bases of V, say $\{e_i \mid 1 \leq i \leq n\}$ and $\{f_i \mid 1 \leq i \leq n\}$, then we have a uniquely defined isomorphism $T : V \to V$ defined by $Te_i := f_i$ for $1 \leq i \leq n$. Note that this isomorphism is unique because each basis comes with its own order. For unordered bases, we cannot define T *uniquely* this way. In turns out that T is an orthogonal transformation; that is, it preserves the inner product of V. As such, we know from linear algebra that there are exactly two possibilities (since $n \geq 1$): $\det T = 1$ or $\det T = -1$. Now here's the deal. If we pick some specific ordered, orthonormal basis $\{e_i \mid 1 \leq i \leq n\}$, then the definition of the linear isomorphism $\Lambda^k V \to \Lambda^{n-k} V$ will be the same if we use instead another ordered, orthonormal basis for which $\det T = 1$. So the ordered, orthonormal bases divide into two classes depending on the sign of $\det T$, and the choice of one of these classes is called an *orientation* of V.

Now a mathematical curiosity appears. The imposition of a positive definite inner product on an oriented vector space is sufficient for defining the Hodge star. But it is not necessary. We can do much the same in the more general case when V has any symmetric, nondegenerate bilinear form $g : V \times V \to \mathbb{R}$. Recall that symmetric means

$$g(v, w) = g(w, v)$$

holds for all $v, w \in V$. Also, nondegenerate means that if a vector $w \in V$ satisfies $g(v, w) = 0$ for all $v \in V$, then we must have that $w = 0$.

And this mathematical curiosity is a necessity in physics! The inner product on $\mathbb{R}^4$ in special relativity theory is such a symmetric, nondegenerate bilinear form. And it is not positive definite! It is defined by

$$g(v, w) := v_0 w_0 - v_1 w_1 - v_2 w_2 - v_3 w_3,$$

where $v = (v_0, v_1, v_3, v_4)$ and $w = (w_0, w_1, w_3, w_4)$ are in $\mathbb{R}^4$. When we equip $\mathbb{R}^4$ with this bilinear form, we say that $\mathbb{R}^4$ is a *Minkowski space*. This is also called *spacetime* in the particular case when the component v_0 represents the time as measured in an inertial frame of reference, while v_1, v_2, v_3 are the three spatial coordinates in that frame. However, the same bilinear form is also present on the "Fourier transform space" of spacetime, in which case the first component is energy and the remaining three are the components of the (linear) momentum vector. The essential point here is that the laws of classical physics are invariant under linear transformations of spacetime that preserve this bilinear form g. The set of all such linear transformations is a Lie group, which is called the *Lorentz group*. So we are not engaging in generalization for generalization's sake.

Thus, from now on, $\langle \cdot, \cdot \rangle$ will denote a symmetric, nondegenerate bilinear form on a finite-dimensional vector space V over the real numbers with $\dim V \geq 1$. Such a form will be called an *inner product*. Then for each integer k with $1 \leq k \leq n$, we define an associated inner product on $\Lambda^k V$ on decomposable elements as

$$\langle v_1 \wedge \cdots \wedge v_k, w_1 \wedge \cdots \wedge w_k \rangle_k := \det(\langle v_i, w_j \rangle),$$

where $v_1, \ldots, v_k, w_1, \ldots w_k \in V$. So $(\langle v_i, w_j \rangle)$ is a $k \times k$ matrix with real entries, whose determinant is again a real number. We claim that this is a symmetric, nondegenerate bilinear form on $\Lambda^k V$. That this form is bilinear and symmetric is easy enough. But before proving it is also nondegenerate, we'll have a brief interlude.

Next, we define an *orthonormal basis* $\{e_1, \ldots, e_n\}$ of a vector space V with inner product to be a basis with

$$\langle e_j, e_k \rangle = 0 \quad \text{if } j \neq k \quad \text{(orthogonal)},$$

$$\langle e_j, e_j \rangle = \pm 1 \qquad \qquad \text{(normal)}.$$

There is a theorem of Sylvester that is relevant here. It is known as Sylvester's law of inertia and says that a vector space V with inner product always has as orthonormal basis. It also says that the integer

$$s := \text{card}\{ \, j \mid \langle e_j, e_j \rangle = -1 \, \}$$

does not depend on the choice of the orthonormal basis.

Taking any orthonormal basis $\{e_1, \ldots, e_n\}$ of V, consider the set of vectors $\{e_{i_1} \wedge \cdots \wedge e_{i_k}\}$ in $\Lambda^k V$ with increasing subindices $i_1 < \cdots < i_k$. One computes the inner product of any pair of these and thereby verifies that this is an orthonormal basis of $\Lambda^k V$. But the existence of an orthonormal basis for the inner product implies that the inner product is nondegenerate. And this proves the claim made above.

The definition of an orientation of a vector space V does not depend on an inner product. Here's how. One says that any $0 \neq w \in \Lambda^n V$, where $n = \dim V$ defines an *orientation* of V. Then an ordered basis $e_1, \ldots, e_n$ of V is said to be *positively oriented* with respect to w if $e_1 \wedge \cdots \wedge w_n = rw$ for a real number $r > 0$. Otherwise, one says that the ordered basis is *negatively oriented*.

We let $\{e_1, \ldots, e_n\}$ be a positively oriented, ordered orthonormal basis of V. The idea again is that $e_{i_1} \wedge e_{i_2} \wedge \ldots \wedge e_{i_k}$, where $i_1 < i_2 < \cdots < i_k$, is mapped into the element $\pm e_{j_1} \wedge e_{j_2} \wedge \ldots \wedge e_{j_{n-k}}$, where $j_1 < \cdots < j_{n-k}$ is the complementary increasing sequence. The "tricky bit" is picking the sign in a way that is independent of the choice of the orthonormal basis. We will do exactly that in the next result.

To prove the next theorem, we note that the n-form

$$\text{Vol} := e_1 \wedge e_2 \wedge \cdots \wedge e_n,$$

known as the *volume element*, does not depend on the choice of the ordered, positively oriented orthonormal basis $\{e_i \mid 1 \leq i \leq n\}$ of V.

So here is the result.

Theorem 5.1 *Let V be a real vector space with $1 \leq \dim V = n < \infty$ and with a given orientation and inner product. Then for each integer $0 \leq k \leq n$, there exists a unique linear isomorphism*

$$* : \Lambda^k V \to \Lambda^{n-k} V,$$

called the Hodge star *and denoted as* $\omega \mapsto *\omega$, *such that*

$$\omega \wedge \rho = \langle *\omega, \rho \rangle_{n-k} \text{Vol}$$

for all $\omega \in \Lambda^k V$ *and all* $\rho \in \Lambda^{n-k} V$. *In particular, we have that*

$$\omega \wedge *\omega = \langle *\omega, *\omega \rangle_{n-k} \text{Vol}$$

for all $\omega \in \Lambda^k V$.

Remark: We are working in the context of vector spaces over $\mathbb{R}$, the real numbers. The concept of antilinear map does not apply here as it does for vector spaces over $\mathbb{C}$, the complex numbers. The Hodge star is a linear map. It is not a conjugation.

Proof: Take an element $\omega \in \Lambda^k V$ and define the linear map $\Lambda^{n-k} V \to \Lambda^n V$ for $\rho \in \Lambda^{n-k} V$ by

$$\rho \mapsto \omega \wedge \rho.$$

Since $\dim \Lambda^n V = 1$ and Vol is a nonzero vector in $\Lambda^n V$ (and so forms a basis of it), we can write

$$\omega \wedge \rho = c(\rho) \text{Vol} \tag{5.6}$$

for a unique real number $c(\rho) \in \mathbb{R}$. Clearly, $c : \Lambda^{n-k} V \to \mathbb{R}$ is a linear map. Now the Riesz representation theorem holds in this setting. Consequently, there exists a unique $*\omega \in \Lambda^{n-k} V$ such that

$$c(\rho) = \langle *\omega, \rho \rangle_{n-k}$$

for all $\rho \in \Lambda^{n-k} V$. Substituting this back into (5.6) gives

$$\omega \wedge \rho = \langle *\omega, \rho \rangle_{n-k} \text{Vol},$$

as desired. The uniqueness of the element $*\omega$ satisfying this equality for all $\rho \in \Lambda^{n-k} V$ follows from the nondegeneracy of $\langle \cdot, \cdot \rangle_{n-k}$. $\blacksquare$

Here is the result that we use in practice to calculate the Hodge star.

Corollary 5.1 *Suppose the same hypotheses as in Theorem 5.1 and that* $\{e_1, \ldots, e_n\}$ *is a positively oriented, ordered orthonormal basis of* V. *Take* $\omega = e_{i_1} \wedge \cdots \wedge e_{i_k} \in \Lambda^k V$ *with increasing subindices. Then*

$$*\omega = \sigma \, e_{j_1} \wedge \cdots \wedge e_{j_{n-k}} \in \Lambda^{n-k} V,$$

where $j_1 < \cdots < j_{n-k}$ *is the complementary sequence in* $1 < \cdots < n$ *of* $i_1 < \cdots < i_k$ *and where the sign* $\sigma = \pm 1$ *is chosen so that*

$$e_{i_1} \wedge \cdots \wedge e_{i_k} \wedge e_{j_1} \wedge \cdots \wedge e_{j_{n-k}} = \sigma \, \langle e_{j_1} \wedge \cdots \wedge e_{j_{n-k}}, e_{j_1} \wedge \cdots \wedge e_{j_{n-k}} \rangle_{n-k} \text{Vol}.$$

Remark: Notice that by its definition as a determinant, we have

$$\langle e_{j_1} \wedge \cdots \wedge e_{j_{n-k}}, e_{j_1} \wedge \cdots \wedge e_{j_{n-k}} \rangle_{n-k} = \langle e_{j_1}, e_{j_1} \rangle \langle e_{j_2}, e_{j_2} \rangle \cdots \langle e_{j_{n-k}}, e_{j_{n-k}} \rangle$$

since in this case the matrix is diagonal. This gives us a closed formula for the sign σ:

$$\sigma = \text{sign}(\tau) \langle e_{j_1}, e_{j_1} \rangle \cdots \langle e_{j_{n-k}}, e_{j_{n-k}} \rangle,$$

where τ is the permutation that maps $i_1, \ldots i_k, j_1, \ldots j_{n-k}$ to $1, 2, \ldots, n$.

What happens if we take the Hodge star twice?

Proposition 5.3 *Suppose that V has a positive definite inner product. Then for* $\dim V = n$ *and* $\omega \in \Lambda^k V$ *with* $0 \le k \le n$, *we have*

$$* * \omega = (-1)^{k(n-k)} \omega.$$

An important special case deals with 2-forms on a four-dimensional Euclidean space. In that case, we have $k = 2$ and $n = 4$, so that $* * \omega = \omega$. Now it is well known that a linear operator A on a finite-dimensional space that satisfies $A^2 = I$ has eigenvalues in the set $\{-1, +1\}$. In the case of the Hodge star, both of these numbers occur as eigenvalues.

Exercise 5.11 *Compute the multiplicity of these two eigenvalues.*

Definition 5.2 *Suppose that V is an oriented, positive definite inner product vector space with* $\dim V = 4$. *Then we say a 2-form* $\omega \in \Lambda^2 V$ *is self-dual (SD) if* $* \omega = \omega$. *Similarly, we say that ω is anti-self-dual (ASD) if* $* \omega = -\omega$.

If we take an oriented, inner product vector space V with $\dim V = 4$ and put on it the opposite orientation, then the Hodge star will change, but only by a minus sign. The SD (resp., ASD) 2-forms on V with respect to the new Hodge star will be exactly the ASD (resp., SD) 2-forms with respect to the original Hodge star. In short, the spaces of SD and ASD 2-forms are interchanged when we reverse the orientation of V.

Proposition 5.4 *Let V be an oriented, positive definite inner product vector space with* $\dim V = 4$. *Then for every* $\omega \in \Lambda^2 V$, *there exist unique 2-forms* ω_{SD} *and* ω_{ASD} *that are SD and ASD, respectively, and such that*

$$\omega = \omega_{SD} + \omega_{ASD}.$$

Exercise 5.12 *Recall from vector calculus the definition of the cross product, also known as the vector product, $u \times v \in \mathbb{R}^3$ for $u, v \in \mathbb{R}^3$. Give $\mathbb{R}^3$ its standard inner product, denoted by $u \cdot v$, and standard orientation. Prove that for $u, v, w \in \mathbb{R}^3$, we have*

$$u \times v = *(u \wedge v),$$

$$u \cdot (v \times w) = *(u \wedge v \wedge w).$$

The left side of the second equation is called the triple product *of the three vectors u, v, and w. The right-hand side of that equation makes clearer this choice of name.*

5.4 Differential Forms

We let $k \geq 1$ be an integer. We start with the kth power of the fiber product of the tangent bundle TM with itself:

$$T^{\diamond k}(M) := \{(v_1, \ldots, v_k) \in TM \times \cdots \times TM \mid \tau_M(v_1) = \cdots = \tau_M(v_k)\},$$

that is, the space of all k-tuples of vectors tangent to M at the *same* point of M (with the point depending on the k-tuple). This is a vector bundle over M, of course.

Definition 5.3 *We define a* (differential) k-form *on M to be a smooth map*

$$\eta : T^{\diamond k}(M) \to \mathbb{R}$$

such that η restricted to each fiber of $T^{\diamond k}(M)$ is k-multilinear; that is,

$$w \mapsto \eta(v_1, \ldots, v_{j-1}, w, v_{j+1}, \ldots, v_k)$$

is linear for each $j = 1, \ldots, k$, and antisymmetric, *that is,*

$$\eta(v_1, \ldots, v_i, \ldots, v_j, \ldots, v_k) = -\eta(v_1, \ldots, v_j, \ldots, v_i, \ldots, v_k),$$

for all $1 \leq i < j \leq k$.

Here all the tangent vectors $w, v_1, \cdots, v_k$ are in the same tangent space $T_p(M)$ for some point $p \in M$.

We also denote the vector space of all differential k-forms on M as

$$\Omega^k(M) := \{\eta : T^{\diamond k}(M) \to \mathbb{R} \mid \eta \text{ is a } k\text{-form}\}.$$

One of the most basic properties of differential forms is that they can be *pulled back*. This is what that means:

The construction $T^{\diamond k}(M)$ for every smooth manifold M is the first part of the definition of a covariant functor. The second part is a vector bundle morphism $T^{\diamond k} f : T^{\diamond k}(M) \to T^{\diamond k}(N)$ for every smooth function $f : M \to N$ of smooth manifolds M and N. This is defined by

$$T^{\diamond k} f(v_1, \ldots, v_k) := (T(f)v_1, \ldots, T(f)v_k) \in T^{\diamond k}_{f(x)}(N) \qquad \text{for } k \geq 1,$$

where $(v_1, \ldots, v_k) \in T_x^{\diamond k}(M)$ for some $x \in M$ (which means $v_j \in T_x(M)$ for all $1 \le j \le k$) and $T(f) : T(M) \to T(N)$ is the vector bundle morphism induced by $f : M \to N$ on the tangent bundles. The fact that this is a vector bundle morphism says that this diagram commutes:

$$
\begin{array}{ccc}
T^{\diamond k}(M) & \xrightarrow{\;T^{\diamond k}(f)\;} & T^{\diamond k}(N) \\
\downarrow & & \downarrow \\
M & \xrightarrow{\;\;f\;\;} & N
\end{array}
$$

We often write $f_* = T^{\diamond k} f$ in order to spare the reader from excessive notation, though at the expense of some slight ambiguity. Note this is consistent with the previously introduced notation $f_* = T(f)$, which is the special case here when $k = 1$.

Exercise: *Prove that $T^{\diamond k}$ is a covariant functor.*

Definition 5.4 *Suppose $f : M \longrightarrow N$ is a smooth function of smooth manifolds M and N. Let $\eta : T^{\diamond k}(N) \longrightarrow \mathbb{R}$ be a k-form on N. Then the pullback of η by f is defined as*

$$ f^*(\eta) := \eta \circ f_* : T^{\diamond k}(M) \longrightarrow \mathbb{R}. $$

namely, the composition

$$ T^{\diamond k}(M) \xrightarrow{\;f_*\;} T^{\diamond k}(N) \xrightarrow{\;\eta\;} \mathbb{R}. $$

Exercise: *For $k \ge 1$ prove that $f^*(\eta)$ is a k-form on M and that $f^* : \Omega^k(N) \longrightarrow \Omega^k(M)$ is linear. Then extend all this to the case $k = 0$.*

So $f^*\eta$ is a k-form on N. Note that since k-forms are infinitesimal objects, the pullback by f involves the infinitesimal $f_* := Tf$ associated to f, rather than f itself. An alternative way to define $\Omega^k(M)$ is to consider the smooth representation

$$ \rho_k : GL(\mathbb{R}^{m*}) \to GL(\Lambda^k(\mathbb{R}^{m*})) $$

given by $T \mapsto \Lambda^k(T) = T \wedge \cdots \wedge T$ (k times), where $\Lambda^k(\mathbb{R}^{m*})$ is the wedge product $\mathbb{R}^{m*} \wedge \cdots \wedge \mathbb{R}^{m*}$ (k times). Also, $\mathbb{R}^{m*}$, the dual space of $\mathbb{R}^m$, is the model fiber space for the cotangent bundle of M.

Then we define a *(differential) k-form* as a section of the vector bundle $\rho_k(\tau_M^*)$ associated to the cotangent bundle $\tau_M^* : T^*(M) \to M$ induced by the representation ρ_k, and $\Omega^k(M)$ is defined as the space of all k-forms, that is, all of its sections,

$$ \Omega^k(M) := \Gamma(\rho_k(\tau_M^*)). $$

We mention this alternative definition in part because it is the usual definition in other books.

Exercise 5.13 *Prove that these two definitions of $\Omega^k(M)$ are equivalent.*

It is also convenient to define $\Omega^k(M)$ for $k = 0$. First, we define a 0-form to be a C^∞-function $f : M \to \mathbb{R}$. Then we define $\Omega^0(M)$ to be the vector space of all the 0-forms, namely, $\Omega^0(M) := C^\infty(M)$.

The family of vector spaces $\Omega^k(M)$ for $k \geq 0$ has a multiplication, namely, a bilinear map

$$\Omega^k(M) \times \Omega^l(M) \xrightarrow{\wedge} \Omega^{k+l}(M)$$

defined for each pairs of integers $k, l \geq 0$. This is known as the *wedge product* of differential forms. It is induced by the wedge product defined for the family of vector spaces $\Lambda^k(\mathbb{R}^{m*})$.

Proposition 5.5 *The graded vector space $\Omega^*(M) = \oplus_{k \geq 0}\Omega^k(M)$ equipped with the wedge product is an associative, graded commutative graded algebra whose identity element is the constant function $1 \in \mathbb{R} = \Omega^0(M) = C^\infty(M)$.*

The assignment of objects $M \to \Omega^*(M)$ together with that of smooth functions $f \to f^*$ is a (contravariant) functor from the category of smooth manifolds to the category of associative graded algebras with identity.

5.5 Exterior Differential Calculus

We can think of a k-form as a creature born to be both differentiated and integrated. This is sometimes referred to as *de Rham theory* or as the *exterior differential calculus* of a differentiable manifold. One standard reference for this material is [15]. Of course, there are many other fine books on these topics, such as [1]. A more abstract approach is taken in [21].

While we will not go into the theory of integration here, we make a few remarks on the theory of derivation. It turns out that for any differentiable manifold M and for every integer $k \geq 0$, one can define linear maps

$$d_k : \Omega^k(M) \to \Omega^{k+1}(M).$$

By an extremely common abuse of notation, we use the one symbol d to represent all of these linear maps. Notice that M is also omitted from this notation. And we refer to these maps as the *exterior derivative* of M.

Here is the idea in a nutshell. Since derivatives in calculus are local operators (meaning that the value of a derivative of a function at a point only depends on values of the function very near that point), it suffices to define d in terms of the local coordinates of a chart. So we start with the case when U is an open set in $\mathbb{R}^m$ and use the standard coordinates $(x_1, \ldots, x_m)$ there. Then $\Omega^0(U) = C^\infty(U)$, so that a 0-form is a C^∞-function $f : U \to \mathbb{R}$. We then define $df \in \Omega^1(U)$ by

$$df := \frac{\partial f}{\partial x_1} dx_1 + \frac{\partial f}{\partial x_2} dx_2 + \cdots + \frac{\partial f}{\partial x_m} dx_m = \sum_{j=1}^{m} \frac{\partial f}{\partial x_j} dx_j.$$

This is a formula that we have seen before in (4.4). It has been used for centuries, usually with an interpretation in terms of "infinitesimal" objects. Here it has a rigorous meaning. In particular, for the coordinate *function* $x_k : U \to \mathbb{R}$, where $k = 1, \ldots, m$, we have that

$$d(x_k) = dx_k. \tag{5.7}$$

This is *not* a consequence of the identity $a = a$. The left side is the exterior derivative of the function x_k, while the right side is the independently defined 1-form dx_k. However, it does motivate the notation for the 1-form dx_k.

Exercise 5.14 *Prove (5.7).*

More generally, suppose that $\omega \in \Omega^k(U)$ has the form

$$\omega = f \, dx_{i_1} \wedge dx_{i_2} \wedge \cdots \wedge dx_{i_k}, \tag{5.8}$$

where all the subindices lie in the set $\{1, \ldots, m\}$. Since every k-form can be written uniquely as a finite sum of such k-forms and we want d to be linear, it suffices to define d on these particular k-forms. Notice that we take $k \geq 0$ here so that the previous discussion is a special case of this case. The definition is

$$d\omega := df \wedge dx_{i_1} \wedge dx_{i_2} \wedge \cdots \wedge dx_{i_k}$$

$$= \sum_{j=1}^{m} \frac{\partial f}{\partial x_j} dx_j \wedge dx_{i_1} \wedge dx_{i_2} \wedge \cdots \wedge dx_{i_k}.$$

Maybe the reader has never seen this formula before, but it is not very subtle. It simply says that the exterior derivative acts nontrivially only on the coefficient f on the right side of (5.8). In particular, for the 1-form dx_j, we see that this definition tells us that $d(dx_j) = 0$; that is, the exterior derivative of the 1-form dx_j is zero. This is logically different from $d(d(x_j)) = 0$, where now we are starting with the coordinate function x_j and taking its exterior derivative twice.

Exercise 5.15 *Prove that the exterior derivative applied twice to the coordinate function x_j is zero; that is, $d(d(x_j)) = 0$.*

That finishes the promised nutshell. But details abound to get this to work for a manifold M. What do we have to do? Well, first we define $d\omega$ for a k-form ω at a point in M using a chart containing that point. This is essentially what we presented above. Second, we have to show that the object we have defined does not depend on the choice of the chart. That is, we have to show that the representation changes as a section of the bundle $\Lambda^k(T^*M)$ is supposed to change. It is exactly at this point in the theory that we see why we have to use the cotangent bundle and not the tangent bundle. This has to be so since the difference between the tangent and the cotangent bundles is how their fibers are transformed under a change of local coordinates in the base space. And that corresponds to how their sections change under a change of chart. Unfortunately, this part of the theory relies on a surfeit of coordinate notation. And so we prefer not to go into all the details. However, the moral of the story is that we can always calculate in local coordinates and get the correct results. See Lang's text [32] for more on this state of affairs.

We will need some properties of the exterior differential calculus.

Theorem 5.2 *The exterior derivative* $d : \Omega^k(M) \to \Omega^{k+1}(M)$ *satisfies the following properties:*

- *The exterior derivative* d *is a linear map of vector spaces over* $\mathbb{R}$.
- $d^2 = 0$. *More specifically, this means that for every* $k \geq 0$, *the following composition is zero:*

$$\Omega^k(M) \xrightarrow{d_k} \Omega^{k+1}(M) \xrightarrow{d_{k+1}} \Omega^{k+2}(M),$$

  where we have put subindices on the exterior derivatives for a clearer notation. This is then equivalent to $\operatorname{Im} d_k \subset \operatorname{Ker} d_{k+1}$ *for every* $k \geq 0$. *One says that the sequence of spaces* $\Omega^k(M)$ *together with the maps* d_k *forms a* cochain complex.
- *For* $\omega_1 \in \Omega^j(M)$ *and* $\omega_2 \in \Omega^k(M)$, *we have the* graded Leibniz rule

$$d(\omega_1 \wedge \omega_2) = d\omega_1 \wedge \omega_2 + (-1)^j \omega_1 \wedge d\omega_2.$$

  Notice that all of these terms lie in $\Omega^{j+k+1}(M)$.
- *For any* 0-*form* $f \in \Omega^0(M)$, *that is, any smooth* $f : M \to \mathbb{R}$, *the* 1-*form* df *satisfies* $df(X) = Xf$ *for all vector fields* X *on* M.

We will not prove this. In particular, the last property uses the definition of the action of a vector X field on a smooth function f, which we will see in the next chapter. However, here's what's happening with the second property. We consider any 0-form $f \in \Omega^0(U) = C^\infty(U)$, where U is an open set in $\mathbb{R}^m$. Then we have

$$d^2 f = d(df) = d\left(\sum_{j=1}^m \frac{\partial f}{\partial x_j} dx_j\right) = \sum_{j,k=1}^m \frac{\partial^2 f}{\partial x_k \partial x_j} dx_k \wedge dx_j.$$

Now each term in this last sum with $j = k$ is equal to zero since $dx_j \wedge dx_j = 0$. The terms with $j \neq k$ come in pairs. But each such pair sums to zero since the second mixed partial derivative is symmetric under the interchange $j \leftrightarrow k$ while $dx_k \wedge dx_j$ changes sign under that interchange. So we see that $d^2 f = 0$. Basically, the same argument shows that $d^2\omega = 0$ in local coordinates for any k-form $\omega \in \Omega^k(U)$. But having the result in any chart gives us the result for any manifold M. The other properties are also proved first locally.

The approach in [21] is to take the properties listed in Theorem 5.2 and show there exists a unique set of maps $d : \Omega^k(M) \to \Omega^{k+1}(M)$ satisfying them. The proof consists first of the uniqueness proof in which an explicit formula for d is produced assuming that the properties in Theorem 5.2 hold. Then the proof continues by taking that formula to be the definition of d and proving that the properties in Theorem 5.2 do hold. For the details, see Chapter I, Section 2.4 in [21].

Another basic property of the exterior derivative is that it commutes with pullbacks. Explicitly, this means

Proposition 5.6 *Let* $f : N \to M$ *be a smooth function of manifolds and let* $\omega \in \Omega^k(M)$ *be a k-form. Then*

$$f^*(d\omega) = d(f^*\omega),$$

where we use the same symbol d for the exterior derivative on both manifolds.

This result is also simply written as $f^*d = df^*$. The proof is straightforward and left to the reader.

At this point, one can introduce cohomology though it is not too easy to study it in general without developing more tools.

Definition 5.5 *Let* M *be a differential manifold. For every integer* $k \geq 0$, *define its* kth de Rham cohomology space *as the quotient space*

$$H^k_{dr}(M) := \frac{\operatorname{Ker} d_{k+1}}{\operatorname{Im} d_k}.$$

This is a vector space over $\mathbb{R}$.

It turns out that under mild hypotheses on M, its de Rham cohomology spaces are finite-dimensional vector spaces. The novice reader is allowed to think of this result as an unexpected bolt from the blue. But it appears much more natural as one learns algebraic topology.

Another identity we will use is

$$d\omega(X, Y) = X(\omega(Y)) - Y(\omega(X)) - \omega([X, Y]) \tag{5.9}$$

for every 1-form ω and every pair of vector fields X, Y.

Exercise 5.16 *Prove (5.9).*

5.6   The Classical Example

We have already seen the classical example. It is an open set U in $\mathbb{R}^3$. It is also a special case of our motivating example. We will use the coordinates x_1, x_2, x_3 on U, though the reader can easily replace them with the more common notation x, y, z.

We consider first $f \in \Omega^0(U) = C^\infty(U)$. We know that $df \in \Omega^1(U)$ is a section of the trivial bundle $\Lambda^1(U) = U \times (\mathbb{R}^3)^*$ and so is of the form $df : p \mapsto (p, \omega(p))$.

By a standard abuse of notation, we write $df(p)$ instead of $\omega(p)$. Then, as we have already seen,

$$df = \frac{\partial f}{\partial x_1}dx_1 + \frac{\partial f}{\partial x_2}dx_2 + \frac{\partial f}{\partial x_3}dx_3.$$

As usual, we have omitted the dependence of df and of the partial derivatives on $p \in U$. Also, we recall that dx_1, dx_2, dx_3 is simply imaginative notation for the basis of $(\mathbb{R}^3)^*$ that is dual to the standard basis $\varepsilon_1, \varepsilon_2, \varepsilon_3$ of $\mathbb{R}^3$, where

$$\varepsilon_1 = (1, 0, 0),$$

$$\varepsilon_2 = (0, 1, 0),$$

$$\varepsilon_3 = (0, 0, 1).$$

We have also seen that this basis of $\mathbb{R}^3$ has the equally imaginative notation $\partial/\partial x_k = \varepsilon_k$, especially when we consider these to be vector fields on U. If we also endow $\mathbb{R}^3$ with the standard inner product [which makes $\varepsilon_1, \varepsilon_2, \varepsilon_3$ into an orthonormal basis (ONB)], then we have a unique isomorphism $\mathbb{R}^3 \cong (\mathbb{R}^3)^*$ given by $\varepsilon_k \to dx_k$ and that does not depend on the choice of ONB. In this case, the differential df is identified via this isomorphism with the *gradient* of f as studied in a course on functions of several (or even just three) variables:

$$\frac{\partial f}{\partial x_1}\varepsilon_1 + \frac{\partial f}{\partial x_2}\varepsilon_2 + \frac{\partial f}{\partial x_3}\varepsilon_3 = \left(\frac{\partial f}{\partial x_1}, \frac{\partial f}{\partial x_2}, \frac{\partial f}{\partial x_3}\right) = \text{grad } f = \nabla f,$$

where we use the vector differential operator from vector calculus

$$\nabla = \left(\frac{\partial}{\partial x_1}, \frac{\partial}{\partial x_2}, \frac{\partial}{\partial x_3}\right).$$

Next, we consider a 1-form $\omega = f_1 dx_1 + f_2 dx_2 + f_3 dx_3 \in \Omega^1(U)$, where each $f_k : U \to \mathbb{R}$ is C^∞. The definition then gives

$$dw = df_1\, dx_1 + df_2\, dx_2 + df_3\, dx_3$$

$$= \left(\frac{\partial f_1}{\partial x_2}dx_2 + \frac{\partial f_1}{\partial x_3}dx_3\right) \wedge dx_1 + \left(\frac{\partial f_2}{\partial x_1}dx_1 + \frac{\partial f_2}{\partial x_3}dx_3\right) \wedge dx_2$$

$$+ \left(\frac{\partial f_3}{\partial x_1}dx_1 + \frac{\partial f_3}{\partial x_2}dx_2\right) \wedge dx_3,$$

where we have used $dx_1 \wedge dx_1 = 0$, et cetera. Now, rearranging terms to get an expansion in the standard basis of $\Omega^2(U)$, we have

$$dw = \tag{5.10}$$

$$\left(\frac{\partial f_3}{\partial x_2} - \frac{\partial f_2}{\partial x_3}\right)dx_2 \wedge dx_3 + \left(\frac{\partial f_1}{\partial x_3} - \frac{\partial f_3}{\partial x_1}\right)dx_3 \wedge dx_1 + \left(\frac{\partial f_2}{\partial x_1} - \frac{\partial f_1}{\partial x_2}\right)dx_1 \wedge dx_2.$$

Here the reader will recognize the curl of $\mathbf{f} := (f_1, f_2, f_2)$ in hiding, where we use the usual identification of $\mathbf{f} : U \to \mathbb{R}^3$ as a vector field on U. However, this is not quite the curl since the curl of a vector is a vector, as we learn in the calculus of three variables. In this slightly more detailed presentation, what we want to say is that the curl of a vector field is a vector field. But recall the Hodge star

$$* : \Lambda^2((\mathbb{R}^3)^*) \to \Lambda^1((\mathbb{R}^3)^*),$$

which exists if we have an inner product and orientation on $(\mathbb{R}^3)^*$. The way one does this in physics (and everywhere else, I guess) is to put a positive definite inner product and an orientation on $\mathbb{R}^3$. And these in turn naturally induce a positive definite inner product and an orientation on $(\mathbb{R}^3)^*$. Having done this so that $\varepsilon_1, \varepsilon_2, \varepsilon_3$ is an orthonormal positively oriented basis of $\mathbb{R}^3$, we have that dx_1, dx_2, dx_3 is an orthonormal positively oriented basis of $(\mathbb{R}^3)^*$. Then by using Corollary 5.1, we calculate these Hodge star operations:

$$* dx_1 = dx_2 \wedge dx_3 \quad * dx_2 = dx_3 \wedge dx_1 \quad * dx_3 = dx_1 \wedge dx_2.$$

Now Proposition 5.3 implies that $**$ is the identity map on $\Lambda^1((\mathbb{R}^3)^*)$. Using that or applying Corollary 5.1 again, we see that

$$dx_1 = *(dx_2 \wedge dx_3) \quad dx_2 = *(dx_3 \wedge dx_1) \quad dx_3 = *(dx_1 \wedge dx_2).$$

So applying the Hodge star to (5.10) gives us a 1-form. Next, we now use the canonical isomorphism α of the dual space with the original space (which arises from the inner product) to convert this 1-form into a vector field. And that vector field *is* the curl of $\mathbf{f}$:

$$\alpha(* dw) = \operatorname{curl}\mathbf{f} = \nabla \times \mathbf{f}.$$

Finally, let's consider a 2-form on U:

$$\omega = f_1 dx_2 \wedge dx_3 + f_2 dx_3 \wedge dx_1 + f_3 dx_1 \wedge dx_2 \in \Omega^2(U),$$

again with each $f_k : U \to \mathbb{R}$ being a C^∞-function. Then using the definition of the exterior derivative and eliminating terms that are zero because of a repetition of some dx_k, we see that

$$dw = \frac{\partial f_1}{\partial x_1} dx_1 \wedge dx_2 \wedge dx_3 + \frac{\partial f_2}{\partial x_2} dx_2 \wedge dx_3 \wedge dx_1 + \frac{\partial f_3}{\partial x_3} dx_3 \wedge dx_1 \wedge dx_2$$

$$= \left(\frac{\partial f_1}{\partial x_1} + \frac{\partial f_2}{\partial x_2} + \frac{\partial f_3}{\partial x_3}\right) dx_1 \wedge dx_2 \wedge dx_3.$$

Again thinking of the vector field $\mathbf{f} := (f_1, f_2, f_2)$ on U, we recognize here the divergence of $\mathbf{f}$ in this expression. But in the calculus of three variables, the divergence of a vector is a scalar (that is, the divergence of a vector field is a scalar-valued function). And here $d\omega$ is a 3-form. Again, the Hodge star straightens all this out provided that we have an oriented, inner product space. If dx_1, dx_2, dx_3 is an orthonormal oriented basis of $(\mathbb{R}^3)^*$, then the isomorphism

$$\Lambda^3((\mathbb{R}^3)^*) \to \Lambda^0((\mathbb{R}^3)^*) \cong \mathbb{R}$$

is given on the basis element by $*(dx_1 \wedge dx_2 \wedge dx_3) = 1$. So then we have

$$* d\omega = \frac{\partial f_1}{\partial x_1} + \frac{\partial f_2}{\partial x_2} + \frac{\partial f_3}{\partial x_3} = \operatorname{div} \mathbf{f} = \nabla \cdot \mathbf{f}.$$

We now write out the de Rham complex for U, using these identifications of the exterior derivative:

$$C^\infty(U) = \Omega^0(U) \xrightarrow{\text{grad}} \Omega^1(U) \xrightarrow{\text{curl}} \Omega^2(U) \xrightarrow{\text{div}} \Omega^3(U) \to 0.$$

Using $d^2 = 0$, we immediately have these two identities from vector calculus:

$$\operatorname{curl}(\operatorname{grad} f) = \nabla \times (\nabla f) = 0,$$
$$\operatorname{div}(\operatorname{curl} \mathbf{f}) = \nabla \cdot (\nabla \times \mathbf{f}) = 0,$$

where $f : U \to \mathbb{R}$ and $\mathbf{f} : U \to \mathbb{R}^3$ are C^∞.

These three differential operators (grad, curl, and div) are commonly used in the theory of classical electrodynamics. See the second, and reputably best, edition of the canonical reference [28] on this central topic in physics. As we shall see later in Chapter 12, this section allows us to translate the equations of that theory into the language of differential forms.

Chapter 6
Lie Derivatives

Lie derivatives are discussed in this brief chapter, which I have included in part because it is so traditional. There are two more concrete reasons for this diversion. The first is that Lie derivatives offer some sort of introduction to the idea behind the Frobenius theorem. The second is that they give us an inadequate way of transporting vectors along curves. Why inadequate? This is a technicality, difficult to describe for now. But what we really need to transport vectors in a "parallel" manner is a connection. This is what a connection will do for us. I also discuss why integral curves are not always so important in physics. But be warned that my point of view here is rather heretical.

6.1 Integral Curves

As we have already noted, in some sense vector bundles exist so that we can study their sections. In this sense, the tangent bundle in particular exists so that we can study its sections. As we have already noted, these turn out to have their own name.

Definition 6.1 *A smooth section* $X : M \to TM$ *of the tangent bundle* $\tau_M : TM \to M$ *is called a* vector field on M.

This means that X is a smooth function and that $\tau_M \circ X = 1_M$. An assertion equivalent to $\tau_M \circ X = 1_M$ is $X(p) \in \tau_M^{-1}(p)$ for every $p \in M$, namely, that $X(p)$ is tangent to M at the point $p \in M$. We express this relation in the following diagram:

$$
\begin{array}{c}
TM \\
X \Big\uparrow \Big\downarrow \tau_M \\
M
\end{array}
$$

© Springer International Publishing Switzerland 2015

S.B. Sontz, *Principal Bundles*, Universitext, DOI 10.1007/978-3-319-14765-9_6

In a diagram of this sort, it is always understood (unless stated otherwise) that the composition of first going up and then going down is the identity on the lower space.

In what may seem to the reader to be a never-ending descent, in some sense the concept of vector field exists so that we can study integral curves.

Definition 6.2 *Let X be a vector field on M. Then we say that a smooth function* $\gamma : (a, b) \to M$ *is an integral curve for X provided that*

$$\gamma'(t) = X(\gamma(t)) \tag{6.1}$$

for all $t \in (a, b)$. *Here* (a, b) *is any nonempty open interval in* $\mathbb{R}$. *(We allow* $a = -\infty$ *as well as* $b = +\infty$.)

Of course, as the notation suggests, γ' is the derivative of γ. But this must be explained in more detail since for us the (functorial!) derivative of $\gamma : (a, b) \to M$ is $T\gamma : T(a, b) = (a, b) \times \mathbb{R} \to TM$. So we define γ' by

$$\gamma'(t) := T\gamma(t, 1)$$

for $t \in (a, b)$. Diagrammatically,

where $\pi_1 : (a, b) \times \mathbb{R} \to (a, b)$ is the projection onto the first coordinate and $s_1 : (a, b) \to (a, b) \times \mathbb{R}$ is the smooth function $t \mapsto (t, 1)$. The defining condition (6.1) for an integral curve then says that the lower triangle, starting in the lower left corner, is commutative.

To understand these concepts, we examine what they mean locally. By using a chart, this is equivalent to studying the case where $M = U$ is an open subset of $\mathbb{R}^n$. The previous diagram now becomes the following:

Since $\tau_U(x, v) = x$ by the definition (2.5), the condition $\tau_U \circ X = 1_U$ means that $X(p) = (p, Y(p))$ for every $p \in U$, where $Y : U \to \mathbb{R}^n$ is smooth. Clearly, every such smooth function Y gives us a section of $TU = U \times \mathbb{R}^n$. This gives us a natural vector space isomorphism $\Gamma(TU) \cong C^\infty(U, \mathbb{R}^n)$, which justifies our saying that Y is a vector field.

We have by definition that $T\gamma : (a, b) \times \mathbb{R} \to U \times \mathbb{R}^n$ is given by

$$T\gamma(t, v) = (\gamma(t), D\gamma(t)v)$$

for $t \in (a, b)$ and $v \in \mathbb{R}$. As we already noted in Chapter 4, $\gamma(t) \in U \subset \mathbb{R}^n$ implies that $\gamma'(t) \in \mathbb{R}^n$ [see equation (4.1)]. The linear map

$$D\gamma(t) : \mathbb{R} \to \mathbb{R}^n$$

(namely, the derivative of γ evaluated at t) is completely determined by its action on the basis $\{1\}$ of $\mathbb{R}$, which consists of exactly one element, of course, since this is a one-dimensional space. And this linear map is related to the derivative $\gamma'(t) \in \mathbb{R}^n$ by $D\gamma(t) : 1 \mapsto \gamma'(t)$. This is just a special case of the natural identification $\mathcal{L}(\mathbb{R}, W) \cong W$ for any vector space W over the reals. [As an aside, let's note that $D\gamma(t)v = v\gamma'(t)$ for all $v \in \mathbb{R}$ follows immediately. It is merely an oddity of notation that the scalar v appears first to the right of one derivative and then to the left of another derivative.] Then we have

$$\gamma'(t) = T\gamma(t, 1) = (\gamma(t), D\gamma(t)1) = (\gamma(t), \gamma'(t))$$

and the reader is now aware that we are engaging in a truly dreadful abuse of notation, namely, that γ' has been given two inequivalent definitions. However, this abuse is "justified" on the grounds that the function on the left side here is "really" just the second entry of the right side, as is the case in general when dealing with vector bundles locally.

Now the condition that γ is an integral curve is equivalent to

$$\gamma'(t) = Y(\gamma(t)) \tag{6.2}$$

as elements in $\mathbb{R}^n$. [The derivative here is that of equation (4.1), so that $\gamma'(t) \in \mathbb{R}^n$.] This is an ordinary differential equation of first order in $\mathbb{R}^n$ or, if we wish to think in terms of scalar equations, it is a system of n coupled first-order ordinary differential equations. In general, these are nonlinear equations. Typically, the vector field Y is given and there are n scalar unknowns that we have to solve for, namely, $(y_1, \ldots, y_n) \equiv (y_1(t), \ldots, y_n(t))$. Notice that the notations $y = (y_1, \ldots, y_n)$, though completely traditional, are not needed at all.

Now we have arrived at the point where we see that the study of integral curves is equivalent to the study of certain ordinary differential equations. Since the latter study is a central issue in an abundance of applications of mathematics to all sorts of problems in mathematics itself as well as in other disciplines such as physics, we see that what might have appeared to be a never-ending descent of explanations for our interest in these topics does in fact come to an end rather quickly.

At this point in our discussion, the theory of the local existence of solutions of systems of ordinary differential equations, such as that in equation (6.2), becomes applicable and gives the next result, whose detailed proof can be found in [4, 32, 33] and other standard references.

Theorem 6.1 *Let X be a vector field on the manifold M, say $p \in M$. Then there exist real numbers $a_p = a_p(X) < 0$ and $b_p = b_p(X) > 0$ such that there is a unique integral curve $\gamma_p : (a_p, b_p) \to M$ of X such that $\gamma_p(0) = p$.*

Of course, $\gamma_p(0) = p$ is exactly the initial condition required for the first-order ordinary differential equation.

It turns out that we can pick a_p and b_p such that there is no integral curve defined on any strictly larger open interval. In such a case, we say that (a_p, b_p) is the maximal interval for the integral curve passing through p. From now on, we shall only consider the maximal interval, for which we use the same notation as above: $(a_p, b_p) = (a_p(X), b_p(X))$. It is possible that one or both endpoints are infinite: $a_p = -\infty$ or $b_p = +\infty$. The notation γ_p for the integral curve unfortunately does not include a reference to the vector field X. We say that γ_p is *the integral curve* of X passing through p at (time) $t = 0$.

Next, we define a set that will be the domain of σ (whose definition follows):

$$D(X) := \{ (t, p) \in \mathbb{R} \times M \mid t \in (a_p(X), b_p(X)) \}.$$

Then we define $\sigma = \sigma_X : D(X) \to M$ by

$$\sigma_X(t, p) := \gamma_p(t)$$

for all $(t, p) \in D(X)$.

Another immediate consequence from the theory of ordinary differential equations is the following theorem.

Theorem 6.2 *$D(X)$ is an open subset of $\mathbb{R} \times M$. Moreover, σ_X is of class C^∞ (provided that both M and X are of class C^∞ as we are always assuming).*

Exercise 6.1 *Prove the previous two theorems.*

Exercise 6.2 *Assume that $(a_p, b_p) = (-\infty, \infty) = \mathbb{R}$. [Hence, $D(X) = \mathbb{R} \times M$.] Define a function $\sigma_t : M \to M$ for every $t \in \mathbb{R}$ by*

$$\sigma_t(p) := \sigma_X(t, p) = \gamma_p(t),$$

where $p \in M$. The function σ_t is called the flow *of X at time t. The notation σ_t is a bit unfortunate since the vector field X does not appear in it. Another notation is often used and has the advantage of including the vector field X. It is $\exp(tX) := \sigma_t$. Note that we can think of t as time or as a real parameter.*

Prove that σ_t is of class C^∞ and that we have these properties with respect to the parameter t:

$$\sigma_0 = 1_M,$$

$$\sigma_s \circ \sigma_t = \sigma_{s+t} \quad \text{for all } s, t \in \mathbb{R},$$

$$\sigma_{-t} = (\sigma_t)^{-1} \text{ for every } t \in \mathbb{R}.$$

Notice that the last property implies that each σ_t is a diffeomorphism of M onto itself.

6.2 Examples

The physically most intuitive example of a vector field is given as a field of velocities. Think of a fluid in steady-state motion in an open region U in three-dimensional space. Here "steady state" means that the velocity vector of the fluid at a point $p \in U$ does not depend on time, but only on p. An example is a waterfall under typical circumstances, but not when a wave suddenly arrives from a rupture in a dam upstream. So we have a function $v : U \to \mathbb{R}^3$ that gives this velocity vector. This in turn is equivalent to a vector field $X : U \to TU = U \times \mathbb{R}^3$ given by $X(p) = (p, v(p))$. We say that such a vector field X is a *velocity vector field*. Then an integral curve in this case is a trajectory of a massive particle of (or in) the fluid. By definition, a *trajectory* of a particle is the curve as a function of time that the particle actually follows. And "massive" means that the particle has positive mass. This is the basic solution of the dynamics of the fluid, but it does not proceed from the fundamental equation of motion of physics (Newton's second law), which is based on the sum of all the forces acting on the particle in the fluid and which is proportional to the acceleration of the particle. (Here I am trying to say $F = ma$ in a lot of words.) Rather, the velocity field results from a first integration of Newton's second law. A second integration (i.e., finding all the integral curves) then solves the problem.

In science fiction one hears a lot about force fields. And there are force fields in (nonfiction) science too. For example, suppose that electric charges have been put at various locations and are kept fixed there. Then a "test" particle with a nonzero electric charge q will have an electric force acting on it at any point in the surrounding region. There could be other forces also acting on the test particle. But, anyway, this defines an electric force at each point in a certain open region U. And this force only depends on the charge q and the point $p \in U$, as one ascertains experimentally. Writing this force as $F(p) \in \mathbb{R}^3$, we have a *force field $X : U \to TU = U \times \mathbb{R}^3$* given by $X(p) = (p, F(p))$. As is customary, we consider $F(p)$ instead of $X(p)$ as the force field. It turns out experimentally that this force is proportional to q and so we can write $F(p) = qE(p)$, where $E(p) \in \mathbb{R}^3$ is called the *electric field*. The integral curves of the vector field E are called *electric field lines*. (In physics texts, one typically finds sketches of the *images* of these integral curves.) But these integral curves are not the trajectories of motion of the test particle.

To find the motion of the test particle, one has to integrate the equation of motion $F = ma$, which turns out to be a differential equation, which is second order in time, in the unknown position as a function of time. The integral curves of $E(p)$ arise too as the solution of a differential equation, but it is of first order in time, and so it is not the equation of motion of a test particle. Never. So, physically speaking, the integral curves in this case are solutions of the wrong differential equation. While some people get an intuition from the electric field lines, they are not the solution for which we eventually are looking.

Magnetic fields are another well-known example of vector fields. Everyone seems to know that the Earth has a magnetic field. And so do bar magnetics. This is something that can be defined using test magnets. For example, the Earth's magnetic field can be defined using compasses. The integral curves of these vector fields are known as *magnetic field lines* though what one sees in physics texts are the images of these integral curves. This example is a real mess since the magnetic field is not a vector in $\mathbb{R}^3$ but rather in $\Lambda^2(\mathbb{R}^3)$, which is also a vector space of dimension 3. And this space can be identified with $\mathbb{R}^3$, provided that $\mathbb{R}^3$ has been given an orientation. If you took a physics course that covered magnetism, you probably heard of the right-hand rule, which is just a way of imposing an orientation on $\mathbb{R}^3$. Or maybe you heard of the left-hand rule, which imposes the opposite orientation! Anyway, after a lot of juggling around, one does end up with a *magnetic vector field* $B(p) \in \mathbb{R}^3$ with p in some open region U. But again, its integral curves do not give the trajectories of particles moving in the magnetic field.

The magnetic force on a charged particle depends not only on its position, but also on its velocity. The formula for the magnetic force is $F = q(v \times B)$, where q is the electric charge of the particle, v is its instantaneous velocity vector, $\times$ is the vector cross product in $\mathbb{R}^3$, and B is the magnetic field vector. This formula for the force is the result of lots and lots of scientists trying to find the correct way to describe the magnetic force, but the formula is named for Lorentz. It may well be counterintuitive to physicists as well as to mathematicians. One of its consequences is that the magnetic force is zero for a charged particle at rest, no matter how large the magnetic field is. This is quite unlike the situation with the electric force $F = qE$. Again, the magnetic field lines do give some kind of intuition, but they are solutions of the wrong differential equation if we want to find the trajectory of the particle. And again, the right equation for that is $F = ma$.

Here are some intuitions that one does get from the electric and magnetic field lines. The electric field lines always start and end on the fixed electric charges. The magnetic field lines are always closed curves that have neither a starting point nor an ending point. The intuition is that electric charges are the origin of electric forces, while there are no magnetic charges. At least, no magnetic charges have yet been discovered. The origin of magnetic forces in the absence of magnetic charges will be left to the reader's curiosity for the time being.

The history of Faraday's equation is relevant in this context. (We will discuss this equation in more detail in Chapter 12.) Even though Michael Faraday was a genius with extraordinary insight into physics and chemistry, he had little mathematical ability and understood electricity and magnetism in terms of their field lines, which for him were primary objects of physical reality and not solutions to a differential equation. His point of view was generally not appreciated by his contemporaries. It was Maxwell who later translated the geometric intuitions of Faraday, which were based on his and others' experiments, into the differential equation that is now known as Faraday's law. But the idea is due to Faraday.

When we consider systems with electric fields E and magnetic fields B that vary in time, we analyze them using their own equations of motion, not with $F = ma$. After all, we think of E and B as fields that produce forces, not as massive objects on which forces act. Their equations of motion, which are Maxwell's equations, couple

E and B, meaning that in general we have no way of studying the time evolutions of E and B separately. We speak of their combination as the electromagnetic field. Even though these equations are the product of many scientists besides Maxwell, he is rather famous for having introduced the last term in them, which is known as the *displacement current*. Collectively with the previously understood equations of the electric and magnetic fields, these are known as Maxwell's equations. We will see much more about these equations in Chapter 12.

My main point in this section is that even when a vector field is rather important in terms of the physics of a situation, its corresponding integral curves could be a side issue. Rather, one wants to solve the appropriate equations of motion, whether it be for a system of massive particles or of fields. Or of a combination of both.

Of course, the Hamiltonian approach to physical problems is a way of introducing a special vector field whose integral curves describe the time evolution of the physical system under consideration. This vector field is defined on a special space, known as the *phase space*, which is never an open subset of $\mathbb{R}^3$. Why not? Because a phase space always has even dimension. In this case, the vector fields are relevant to the physics simply because they were cooked up that way! Mathematically, this method replaces a system of ordinary second-order differential equations with another system with twice as many ordinary first-order differential equations. And physically, one is studying a system in terms of its phase space with $2n$ variables instead of its configuration space with n variables.

6.3 Lie Derivatives

We can now use this formalism to define directional derivatives. Recall from basic calculus that the partial derivatives with respect to the standard coordinates in $\mathbb{R}^n$ are special cases of something called the directional derivative. Of course, we can also do this locally for smooth functions $f : M \to \mathbb{R}$ for any smooth manifold M. We would also like to do this for any globally defined vector field X on M. The result will only depend on the local behavior of X, of course. The idea for doing this goes back to the very beginnings of calculus, where the derivative of a function f of a real variable at a point p in its domain is introduced as the limit of the expression

$$\frac{f(p+h) - f(p)}{h}$$

when we let h tend to 0. We can think of this in general terms as the limit of

$$\frac{f(\text{perturbation of } p) - f(p)}{(\text{size of the perturbation})}$$

when we let the size of the perturbation tend to 0. In our context, we can take the perturbation of the point $p \in M$ to be $\gamma_p(t)$ for values of t near 0, and we can take the size of the perturbation to be t. When the size of the perturbation is 0, we have that the perturbation of p is just equal to p (as it should be): $\gamma_p(0) = p$. So we have arrived at the following definition:

Definition 6.3 *Let $f : M \to \mathbb{R}$ be smooth, X be a vector field on M, and $p \in M$. Then the* Lie derivative *of f with respect to X at $p \in M$ is defined by*

$$(\mathcal{L}_X f)(p) := \lim_{t \to 0} \left(\frac{f(\gamma_p(t)) - f(p)}{t} \right),$$

where $\gamma_p(t)$ is the integral curve for X that passes through p at $t = 0$.

Exercise 6.3 *Show that the limit in the preceding definition always exists.*

Exercise 6.4 *Show that $\mathcal{L}_X f$ is a smooth function.*

Clearly, the Lie derivative of f with respect to X at p only depends on the values of f along the image of the smooth curve $t \mapsto \gamma_p(t)$ for values of t in some small neighborhood of 0. We draw attention again to the fact that γ_p is always uniquely defined in some small neighborhood of $0 \in \mathbb{R}$.

Another common notation for this is $Xf := \mathcal{L}_X f$, which we will comment on and in some sense justify now.

Since the Lie derivative depends only on values of f near the point p, we can study everything in local coordinates without losing generality. So we return to the case where the manifold is an open subset U of $\mathbb{R}^n$ with a vector field $Y : U \to \mathbb{R}^n$ and a smooth function $f : U \to \mathbb{R}$. Then we can expand the integral curve γ_p of Y with initial condition $\gamma_p(0) = p$ as

$$\gamma_p(t) = \gamma_p(0) + t\gamma_p'(0) + o(t)$$
$$= p + tY(p) + o(t).$$

Here we are using the Landau little-o notation $o(t)$, which just means some function of t such that $\lim_{t \to 0} o(t)/t = 0$. Referring to the comments for Exercise 6.4, we then have

$$f(\gamma_p(t)) = f(p + tY(p) + o(t))$$
$$= f(p) + tDf(p) \cdot Y(p) + o(t)$$
$$= f(p) + t \sum_{k=1}^{n} \frac{\partial f}{\partial x_k}(p) Y_k(p) + o(t),$$

and so using the definition of the Lie derivative, we have

$$\mathcal{L}_Y f(p) = \lim_{t \to 0} \left(\frac{f(\gamma_p(t)) - f(p)}{t} \right) = \lim_{t \to 0} \left(\sum_{k=1}^{n} \frac{\partial f}{\partial x_k}(p) Y_k(p) + \frac{o(t)}{t} \right)$$
$$= \sum_{k=1}^{n} Y_k(p) \frac{\partial f}{\partial x_k}(p).$$

The last expression agrees with the usual definition given in a calculus course of the directional derivative of f at p in the direction

$$Y(p) = (Y_1(p), \ldots, Y_n(p)).$$

This formula shows that $\mathcal{L}_Y f(p)$ depends on the local behavior of f near p but only on the value of the vector field Y exactly at the point p, and *not* on its local behavior.

Instead of the preceding equality of real numbers, we can write an equality of smooth functions:

$$\mathcal{L}_Y f = \sum_{k=1}^{n} Y_k \frac{\partial f}{\partial x_k}.$$

And now we can continue this one step further and write this as an equality of operators:

$$\mathcal{L}_Y = \sum_{k=1}^{n} Y_k \frac{\partial}{\partial x_k}.$$

Now the right-hand side can be viewed as a linear combination of vector fields on U with coefficients that are smooth functions on U. As such, the right-hand side is a vector field, and we can easily realize that it is the vector field Y. But we just said that it is also equal to an operator on smooth functions on U; that is, it is equal to

$$\mathcal{L}_Y : C^\infty(U) \to C^\infty(U).$$

So we have arrived at another way of viewing a vector field. We write

$$Y = \sum_{k=1}^{n} Y_k \frac{\partial}{\partial x_k},$$

which we interpret as a linear operator mapping $C^\infty(U)$ to itself. Notice that this formula is not new. In the formalism of Chapter 4, this is just another way of writing $Y = (Y_1, \ldots, Y_n)$. What is new here is the interpretation of this formula.

Moreover, we have Leibniz's rule for this operator.

Exercise 6.5 *Suppose that Y is vector field on the nonempty open subset U in $\mathbb{R}^n$. Verify Leibniz's rule $Y(fg) = (Yf)g + f(Yg)$ for all $f, g \in C^\infty(U)$.*

Suppose that $Z : C^\infty(U) \to C^\infty(U)$ is a linear operator satisfying Leibniz's rule. Is it true that $Z = Y$ for some vector field Y on U?

Exercise 6.6 *Now is a good time to reflect on how all this works on a general manifold, that is, why all of this is independent of coordinates.*

Having defined the Lie derivative $\mathcal{L}_Y$ with respect to the vector field Y as an operator on the space of smooth functions, we would like to see if this same "algorithm" can be used to define a Lie derivative of vector fields. We immediately consider the case of a general manifold M in order to illustrate a difficulty in this approach. So we let X be another vector field on M and consider the usual difference quotient:

$$\frac{X(\gamma_p(t)) - X(p)}{t},$$

where γ_p is the integral curve for Y with initial condition $\gamma_p(0) = p$. But now we have a serious conceptual problem with this expression. The point is that $X(p)$ lies in the tangent space $\tau_M^{-1}(p)$ above p while $X(\gamma_p(t))$ lies in the tangent space $\tau_M^{-1}(\gamma_p(t))$ above $\gamma_p(t)$. (Recall that $\tau_M : TM \to M$ is the notation we are using for the tangent bundle of M.) And so the difference of these two vectors is *not* defined since they do not (in general) lie in the same vector space.

One way to confront this difficulty is to use the flow $\sigma_t(p) = \gamma_p(t)$. Since $\sigma_t : p \mapsto \sigma_t(p)$, we have that its derivative satisfies

$$T\sigma_t : \tau_M^{-1}(p) \to \tau_M^{-1}(\sigma_t(p)) = \tau_M^{-1}(\gamma_p(t)).$$

Moreover, this map is an isomorphism with $(T\sigma_t)^{-1} = T(\sigma_{-t})$ and so gives us a way to identify vectors in $\tau_M^{-1}(\sigma_t(p))$ with vectors in $\tau_M^{-1}(p)$. Notice that the latter space does not depend on the parameter t.

We now have the ingredients for the definition.

Definition 6.4 *Let X, Y be vector fields on the smooth manifold M. Then we define the Lie derivative of X with respect to Y at a point $p \in M$ by*

$$\mathcal{L}_Y X(p) := \lim_{t \to 0} \frac{(T\sigma_t)^{-1} \left[X(\gamma_p(t)) \right] - X(p)}{t},$$

where γ_p is the integral curve for the vector field Y with initial condition $\gamma_p(0) = p$ and σ_t is the flow of Y at time t.

Exercise 6.7 *In what space is this limit being taken? What is the topology being used in that space? Does the limit always exist?*

Exercise 6.8 *Show that $\mathcal{L}_Y X$ is a vector field. (Remember that a part of this exercise is to show that $\mathcal{L}_Y X$ is smooth.)*

Exercise 6.9 *If $D(Y) = \mathbb{R} \times M$, then $\sigma_t : M \to M$ is a diffeomorphism, as we have seen, and so Definition 6.4 makes all the sense in the world. Show how to make sense of this definition even when $D(Y) \neq \mathbb{R} \times M$.*

Just to familiarize the reader with the standard jargon of this subject, we introduce the following.

Definition 6.5 *We say that a vector field Y on a differential manifold M is complete if $D(Y) = \mathbb{R} \times M$.*

Exercise 6.10 *Every vector field on a compact manifold is complete.*

Exercise 6.11 *Suppose that $M = U$ is an open subset of $\mathbb{R}^n$ and that the vector fields X and Y are expressed in terms of the standard basis as*

$$X = \sum_{k=1}^{n} X_k \frac{\partial}{\partial x_k} \qquad Y = \sum_{k=1}^{n} Y_k \frac{\partial}{\partial x_k}.$$

Show the following:

$$\mathcal{L}_X Y = \sum_k (\sum_l X_l \frac{\partial Y_k}{\partial x_l} - Y_l \frac{\partial X_k}{\partial x_l}) \frac{\partial}{\partial x_k}$$

$$\mathcal{L}_X Y = -\mathcal{L}_Y X$$

$$\mathcal{L}_X Y = XY - YX.$$

In the last expression, XY and YX are the composition of two first-order differential operators. Note that these composite linear operators are second-order differential operators and do not satisfy Leibniz's rule. But their difference is a first-order differential operator.

The algebraically straightforward result $\mathcal{L}_X Y = -\mathcal{L}_Y X$ should surprise the novice. It says that differentiating Y with respect to X is related to differentiating X with respect to Y, but that the relation is not (in general) equality.

Definition 6.6 *We define the* commutator *of a pair of vector fields X, Y on M by*

$$[X, Y] := XY - YX.$$

It is then an immediate corollary of the preceding that $[X, Y]$ is a vector field.

Now we do at last have an "algorithm" for defining the Lie derivative for every type of tensor field. The key idea is to use the flow to pull the perturbed tensor back to a tensor at a common base point so that the difference quotient makes sense.

Exercise 6.12 *Define the Lie derivative $\mathcal{L}_Y \omega$, where Y is a vector field and ω is a 1-form. Show that*

$$\mathcal{L}_Y (\omega(X)) = (\mathcal{L}_Y \omega)(X) + \omega(\mathcal{L}_Y X),$$

where X is any vector field.

Chapter 7
Lie Groups

We will need a bare minimum of results on Lie groups. More than anything, we will be establishing notation and giving examples.

7.1 Basics

Recall that a Lie group has already been defined in Definition 3.1. We denote the product of a pair of elements $g, h \in G$ as gh and the inverse of an element g by g^{-1}. We also let e denote the identity element in G.

Definition 7.1 *Let G be a Lie group and M be a smooth manifold.*

- *We say that a* right action *of G on M is a smooth map $M \times G \to M$, denoted by $(x, g) \mapsto x \cdot g = xg$ for $x \in M$ and $g \in G$, that satisfies $x \cdot e = x$ and $(x \cdot g) \cdot h = x \cdot (gh)$, for all $x \in M$ and all $g, h \in G$.*
  A left action *of G on M is defined similarly.*
  If there is a right action of G on M, then for each $g \in G$, we define the right action map $R_g : M \to M$ by $R_g(x) := xg$ for all $x \in M$. *This is a diffeomorphism of M whose inverse is $R_{g^{-1}}$.*
  Comment: *If M has more structure than that of a smooth manifold, then we usually are interested in right actions such that R_g preserves the additional structure for every group element g. For example, if the manifold M is also a vector space, then an action that preserves the vector space structure (i.e., every R_g is linear) is called a* linear *action.*
  A manifold M together with a given right action of G is called a right *G-space. If we merely say* action *(resp., G-space, etc.), we mean a right action (resp., a right G-space, etc.).*

The original version of this chapter was revised: equation was modified and new sentences were added. The correction to this chapter is available at https://doi.org/10.1007/978-3-319-14765-9_18

- *The* orbit *of a subset S of M is defined to be*

$$S \cdot G := \{s \cdot g \mid s \in S, g \in G\},$$

which is a subset of M.
- *A subset $S \subset M$ is said to be G-invariant if $S \cdot G \subset S$.*
  Equivalently, $R_g(S) \subset S$ for all $g \in G$.
- *Given right actions of G on the smooth manifolds N and M, we say that a smooth function $f : N \to M$ is G-equivariant or that it is a G-morphism if $f(x \cdot g) = f(x) \cdot g$ holds for all $x \in N$ and all $g \in G$.*
- *We say that an action of G on M is free if $x \cdot g = x$ for some $x \in M$ and some $g \in G$ implies that $g = e$. Equivalently, R_g has no fixed point if $g \neq e$.*

Example: A Lie group G acts on itself from the right by its multiplication map:

$$G \times G \xrightarrow{\mu_G} G.$$

This is a free action. We note that this is an action with respect to the differential structure on G *only*. It is *not* an action with respect to the group structure of G unless $G = \{e\}$. (Compare with the comment in Definition 7.1.) As noted above, we have associated to this right action a right action map $R_g : G \to G$ defined for each $g \in G$. So $R_g(h) := hg$.

Similarly, we use the multiplication map μ_G to define a left action of G on itself. We also have its associated *left action map* $L_g : G \to G$ defined for each $g \in G$ by $L_g(h) := gh$. As we mentioned above, R_g is a diffeomorphism of G onto itself. Similarly, L_g is a diffeomorphism of G onto itself.

Definition 7.2 *Let G be a Lie group. Then the* Lie algebra *of G is the tangent space $T_e(G)$ of G at the identity $e \in G$. Notation: $\mathfrak{g} := T_e(G)$.*

We will show later that this vector space admits a natural product structure that does indeed make it into an algebra, though in general it is not an associative product.

Definition 7.3 *Let G be a Lie group. For each $g \in G$, we define the* adjoint map *$ad_g : G \to G$ by $ad_g(h) := ghg^{-1}$ for all $h \in G$.*

Note that $ad_g = L_g \circ R_{g^{-1}} = R_{g^{-1}} \circ L_g$. Therefore, ad_g is a diffeomorphism of G to itself. Note that $ad_g(e) = e$. Consequently, its derivative at e gives us a linear map $T_e(ad_g) : \mathfrak{g} \to \mathfrak{g}$.

Definition 7.4 *Let G be a Lie group. Define the* adjoint representation *of G by $Ad_g := T_e(ad_g) : \mathfrak{g} \to \mathfrak{g}$ for all $g \in G$.*

Exercise 7.1 *Here are some properties of ad and Ad to be proved:*

- *The maps ad_g for $g \in G$ give a left action of G on itself and that this is an action with respect to the group structure of G (as well as with respect to the differential structure, of course). We say that ad is a left group action. (Compare with the comment in Definition 7.1.)*

- *The function $Ad : G \to GL(\mathfrak{g})$ given by $g \mapsto Ad_g$ is a representation of G as linear operators acting on the left on the vector space $\mathfrak{g}$. In particular, $\mathfrak{g}$ is a (right!) G-space with respect to the action $\mathfrak{g} \times G \to \mathfrak{g}$ defined by $(v, g) \mapsto Ad_{g^{-1}}v$, which is a linear action. (Again, see the comment in Definition 7.1.)*

There are special vector fields on a Lie group G. Recall that for every $g \in G$, we have the left action $L_g : G \to G$. Since L_g is smooth, it can be differentiated, giving $T(L_g) : T(G) \to T(G)$.

Exercise 7.2 *The map $\alpha : G \times T(G) \to T(G)$ defined by*

$$\alpha(g, v) := T(L_g)v$$

is a left action of G on $T(G)$.

Definition 7.5 *Let $X : G \to T(G)$ be a vector field on G. We say X is left invariant if $T(L_g) X = X L_g$ for all $g \in G$. We often write LIVF instead of left invariant vector field. In other words, $X : G \to T(G)$ is a G-equivariant map with respect to the left actions of G on G and $T(G)$.*

Exercise 7.3 *Show that the set $\mathcal{L}(G) := \{X \mid X \text{ is a LIVF on } G\}$ is a vector space over $\mathbb{R}$.*

Exercise 7.4 *Show that $T(L_g) : T_h(G) \to T_{hg}(G)$ is an isomorphism of vector spaces. In particular, identify its inverse isomorphism.*
  Using this show that a left invariant vector field X on G is uniquely determined by its value at the identity element $e \in G$, namely, $X(e) \in T_e(G)$.
  Conversely, show that for every vector $v \in T_e(G)$, there exists (a necessarily unique) left invariant vector field X_v such that $X_v(e) = v$.
  Moreover, the map $v \mapsto X_v$ is an isomorphism of the vector space $T_e(G)$ onto the vector space $\mathcal{L}(G)$.

Since we already have established the notation $\mathfrak{g} = T_e(G)$, we simply say that $\mathfrak{g}$ is the vector space of LIVFs on G. The notation $\mathcal{L}(G)$ for this space will never be used again. A quick way to define the Lie bracket on $\mathfrak{g}$ is by using this identification. So each $X \in \mathfrak{g}$ is a LIVF on G. Now take a pair of LIVFs $X, Y \in \mathfrak{g}$ and consider their commutator $[X, Y] = XY - YX$. As we showed earlier, this is a vector field on G. But it is an exercise to show that $[X, Y]$ is also left invariant. So $[X, Y] \in \mathfrak{g}$.

Exercise 7.5 *Suppose that X and Y are LIVFs on G. Prove that $[X, Y]$ is a LIVF on G.*

The commutator in this setting has a special name.

Definition 7.6 *The Lie bracket of $X, Y \in \mathfrak{g}$ is defined to be*

$$[X, Y] := XY - YX,$$

where X and Y are understood to be LIVFs.

Since $[X, Y] \in \mathfrak{g}$, this defines a multiplication on the vector space $\mathfrak{g}$, thereby making $\mathfrak{g}$ into an algebra. But it is an algebra of a very special type. For example, in general, it is not an associative algebra.

Exercise 7.6 *Show that the Lie bracket is a bilinear map* $\mathfrak{g} \times \mathfrak{g} \to \mathfrak{g}$ *that satisfies the following properties, where* $X, Y, Z \in \mathfrak{g}$:

- *(Antisymmetry)* $[X, Y] = -[Y, X]$.
- *(Jacobi identity)* $[X, [Y, Z]] + [Y, [Z, X]] + [Z, [X, Y]] = 0$.

In summary, $\mathfrak{g}$ is a Lie subalgebra of the Lie algebra of all vector fields on G, where we leave the definition of *Lie subalgebra* to the reader.

The Jacobi identity has many important aspects. For now, let us merely note that it replaces the associative property that one might have expected to hold here.

Definition 7.7 *Let V be a vector space equipped with an antisymmetric, bilinear map $V \times V \to V$ that satisfies the Jacobi identity. Then we say that V with that map is a* Lie algebra.

The product in an abstract Lie algebra is typically denoted by $[\cdot, \cdot]$ and is called the Lie bracket. But this is not always the case, as the next exercise shows.

Exercise 7.7 *Prove that the vector product $\mathbb{R}^3 \times \mathbb{R}^3 \to \mathbb{R}^3$ (as defined in a course of vector calculus; notation: $v \times w$) is a Lie bracket, thereby making $\mathbb{R}^3$ into a Lie algebra.*

Show that the vector product does not satisfy the associative property.

7.2 Examples

The prime example of a Lie group is the general linear group of all $n \times n$ invertible matrices with real entries,

$$GL(n, \mathbb{R}) := \{ A = (a_{ij})_{1 \le i, j \le n} \mid a_{ij} \in \mathbb{R} \text{ and } A^{-1} \text{ exists} \}.$$

Here $n \ge 1$ is an integer. This is a group where the multiplication is matrix multiplication and the identity element is the identity matrix $I = (\delta_{i,j})$. Here $\delta_{i,j}$ is the Kronecker delta.

To put a differential structure on $GL(n, \mathbb{R})$, we first note that $GL(n, \mathbb{R})$ is a subset of $M(n, \mathbb{R})$, the set of all $n \times n$ matrices with real entries, and that we canonically identify $M(n, \mathbb{R})$ with the Euclidean space $\mathbb{R}^{n^2}$. We give $M(n, \mathbb{R})$ the corresponding topology and differential structure. The determinant is actually a continuous function det : $M(n, \mathbb{R}) \to \mathbb{R}$. Then we note that A^{-1} exists if and only if the determinant of A is nonzero, $\det A \ne 0$. Consequently,

$$GL(n, \mathbb{R}) = \det^{-1}\big(\mathbb{R} \setminus \{0\}\big) \subset M(n, \mathbb{R}) \cong \mathbb{R}^{n^2}.$$

shows that $GL(n, \mathbb{R})$ is identified with an open set in $\mathbb{R}^{n^2}$. Here we used that the inverse image under a continuous function of an open set is again open. So we put the differential structure on $GL(n, \mathbb{R})$ that arises from its being an open subset of a Euclidean space.

Exercise 7.8 *Show that with these definitions of the group and differential structures, $GL(n, \mathbb{R})$ is a Lie group.*

Now there are two directions we can take at this time. One way is to define bigger Lie groups such as $GL(n, \mathbb{C})$, the set of all $n \times n$ invertible matrices with complex entries. The other way is to consider Lie subgroups of $GL(n, \mathbb{R})$ and $GL(n, \mathbb{C})$. To do this, we need a definition.

Definition 7.8 *Let G be a Lie group. Then a subset H of G is said to be a Lie subgroup of G if it is a subgroup of G as well as a closed submanifold of G. So H is a Lie group in its own right.*

Here is a very useful, very nontrivial result. See your favorite text on Lie groups or smooth manifolds for a proof. One of my favorites is [33].

Theorem 7.1 *Let H be a closed subgroup of a Lie group G. Then H is a Lie subgroup of G.*

Using these facts, we can generate most of the Lie groups that one meets in real life. However, there are also some Lie groups that do not arise this way. First, we recall that a matrix is *orthogonal* if $A^t A = I$, where A^t is the transpose of the matrix A. Writing $A = (a_{ij})$, we have the definition $(A^t)_{ij} := a_{ji}$. Also, we recall that a matrix is *unitary* if $A^* A = I$, where A^* is the transpose conjugate of the matrix A. Here $(A^*)_{ij} := a_{ji}^*$.

Here is a list of the some of the most commonly used Lie groups. We leave it to the reader to reflect on why each of these is indeed a Lie group. The easy way to see this is to apply Theorem 7.1. However, one can also submit oneself to the discipline of proving directly that in each case we have a submanifold by using Theorem 2.3, which was itself based on the implicit function theorem.

In each of these examples, $n \geq 1$ is an integer.

1. Special linear group with real entries:
   $SL(n, \mathbb{R}) := \{A \in GL(n, \mathbb{R}) \mid \det A = 1\}$.
2. Special linear group with complex entries:
   $SL(n, \mathbb{C}) := \{A \in GL(n, \mathbb{C}) \mid \det A = 1\}$.
3. Orthogonal group:
   $O(n) := \{A \in GL(n, \mathbb{R}) \mid A \text{ is orthogonal}\}$.
4. Special orthogonal group:
   $SO(n) := \{A \in GL(n, \mathbb{R}) \mid A \text{ is orthogonal and } \det A = 1\}$.
5. Unitary group: $U(n) := \{A \in GL(n, \mathbb{C}) \mid A \text{ is unitary}\}$.
6. Special unitary group:
   $SU(n) := \{A \in GL(n, \mathbb{C}) \mid A \text{ is unitary and } \det A = 1\}$.

As is typical of introductory texts, we only consider in this book those Lie groups that arise as closed Lie subgroups of $GL(n, \mathbb{C})$ for some integer $n \geq 1$.

It turns out that a key property of a Lie group concerns whether it is compact as a topological space.

Exercise 7.9 *Identify the compact Lie groups in the above list.*

An important part of the theory of a Lie group concerns the corresponding "infinitesimal" structure, namely, its Lie algebra. We wish to identify the Lie algebra for all of these examples. This will be done after discussing one-parameter subgroups.

We have a very simple, and very useful, fact about left multiplication for Lie groups that are closed Lie subgroups of the matrix group $GL(n, \mathbb{C})$. So we take $A \in G \subset GL(n, \mathbb{C})$. In particular, A is an $n \times n$ invertible matrix with complex entries. Then the left action of A on G, which we denote as usual as

$$L_A(B) = AB$$

for all $B \in G$, is the restriction to G of the left action of A on $GL(n, \mathbb{C})$, for which we use the same notation

$$L_A(B) = AB,$$

but now for $B \in GL(n, \mathbb{C})$, a dense, open subset of $\mathfrak{gl}(n, \mathbb{C}) \cong \mathbb{C}^{n^2}$, which is a vector space. As is well known, although maybe not in this notation, $L_A : \mathfrak{gl}(n, \mathbb{C}) \to \mathfrak{gl}(n, \mathbb{C})$ is a linear map over the field $\mathbb{C}$. Then its derivative is given at each "point" $B \in \mathfrak{gl}(n, \mathbb{C})$ by the same linear map L_A. Putting these words into a formula, we are saying that

$$(DL_A)(B) = L_A$$

for all $B \in \mathfrak{gl}(n, \mathbb{C})$. And everything in this paragraph remains true if we replace the complex field $\mathbb{C}$ with the real field $\mathbb{R}$.

7.3   One-Parameter Subgroups

Let $X \in T_e G = \mathfrak{g}$ determine a LIVF, also denoted by X, on G. Let $\phi(t)$ be the integral curve of the LIVF X passing through $e \in G$ at time $t = 0$, where $t \in (a, b)$ is the maximal domain. So we have that $a < 0 < b$, $\phi(0) = e$ and $\phi'(t) = X(\phi(t))$ for all $t \in (a, b)$. Saying that X is left invariant means that

$$(L_{g*})X = X L_g$$

holds for all $g \in G$. All this follows from the general theory of LIVFs.

Exercise 7.10 *In this context, the maximal domain* (a, b) *of* ϕ *is given by* $a = -\infty$ *and* $b = +\infty$.

We claim that ϕ is a group homomorphism from the group $\mathbb{R}$ under addition to the group G. This means that we have to prove

- $\phi(0) = e$,
- $\phi(s + t) = \phi(s)\phi(t)$ for all $s, t \in \mathbb{R}$.

But $\phi(0) = e$ is already known. For the second identity, let $s \in \mathbb{R}$ be fixed and define these curves for all $t \in \mathbb{R}$:

$$\alpha(t) := \phi(s)\phi(t),$$

$$\beta(t) := \phi(s + t).$$

So we have to prove that $\alpha(t) = \beta(t)$. We do this by using the uniqueness theorem for solutions of the differential equation with initial condition. First, we have that

$$\alpha(0) = \phi(s)\phi(0) = \phi(s)e = \phi(s),$$

$$\beta(0) = \phi(s + 0) = \phi(s).$$

So this shows that these curves have the same initial value at $t = 0$. Second, we compute derivatives. So for $\alpha(t)$, we compute

$$
\begin{aligned}
\alpha'(t) &= \frac{d}{dt}(\phi(s)\phi(t)) \\
&= \frac{d}{dt}(L_{\phi(s)}\phi(t)) \\
&= L_{\phi(s)*}\phi'(t) \qquad \text{by an exercise below} \\
&= L_{\phi(s)*}X(\phi(t)) \\
&= X(L_{\phi(s)}\phi(t)) \qquad \text{by left invariance} \\
&= X(\phi(s)\phi(t)) \\
&= X(\alpha(t)).
\end{aligned}
\tag{7.1}
$$

Next, for $\beta(t)$, we have

$$\beta'(t) = \phi'(s + t) = X(\phi(s + t)) = X(\beta(t)).$$

This not only shows that α and β satisfy the same differential equation with the same initial condition and therefore are equal: $\alpha(t) = \beta(t)$. It also shows that we have found an integral curve passing through the point $\phi(s)$.

What we have discussed so far motivates this definition:

Definition 7.9 *Suppose that G is a Lie group. Then a group homomorphism φ :* $\mathbb{R} \to G$ *is called a* one-parameter subgroup *of G.*

Notice that the range of ϕ is a subgroup of G, but that the definition of a one-parameter subgroup involves a homomorphism into G, not merely a subgroup of G.

We have shown that any LIVF determines a one-parameter subgroup. And conversely, given a one-parameter subgroup ϕ of G, we can define a LIVF on G. We simply note that the element $\phi'(0) \in \mathfrak{g}$ determines a LIVF. Then it is a straightforward exercise to show that these are inverse operations, and so the one-parameter subgroups are in bijective correspondence with the elements in $\mathfrak{g}$.

Exercise 7.11 *To justify one of the steps in (7.1), show that for all* $g \in G$ *and any curve* $\gamma : \mathbb{R} \to G$, *we have*

$$\frac{d}{dt}(L_g\gamma(t)) = L_{g*}\gamma'(t).$$

7.4 The Exponential Map

Now we have enough material developed in order to define the exponential map

$$\exp : \mathfrak{g} \to G$$

from the Lie algebra $\mathfrak{g}$ of G to G itself. We take any $X \in \mathfrak{g}$ and let γ_X be the corresponding one-parameter subgroup of G. Then we define

$$\exp(X) := \gamma_X(1).$$

This definition is deceptively simple. It hides completely from view why in the world this function should be named as it is. And at first sight, it might be a bit mysterious whether this function is smooth, or even continuous.

If $G \subset GL(n, \mathbb{R})$, then we have that $\mathfrak{g} \subset \mathfrak{gl}(n, \mathbb{R})$, the vector space of all $n \times n$ matrices with real entries. For any $X \in \mathfrak{g}$, we define the usual exponential matrix of X by

$$\exp(X) := \sum_{j=0}^{\infty} \frac{1}{j!}X^j = I + X + \frac{1}{2!}X^2 + \frac{1}{3!}X^3 + \cdots,$$

which always converges in the topology induced on G from the usual norm topology defined by the operator norm $||X||_{op} := \sup_{||v||=1} ||Xv||$ for the $n \times n$ matrix $X = (X_{ij}) \in \mathfrak{g}$. Here we are thinking of X as a linear operator $X : \mathbb{R}^n \to \mathbb{R}^n$ and using the usual Euclidean norm $||v||^2 = |v_1|^2 + \cdots + |v_n|^2$, where $v = (v_1, \ldots, v_n) \in \mathbb{R}^n$. Also, I is the $n \times n$ identity matrix.

Exercise 7.12 *Prove that the infinite series defining* $\exp(X)$ *converges.*

Then $\gamma(t) := \exp(tX)$ defines a curve $\gamma : \mathbb{R} \to GL(n, \mathbb{R})$ that satisfies

$$\gamma(0) = \exp(0) = I.$$

Be careful with this notation, which is what one usually sees. The 0 in $\gamma(0)$ is the real number, while the 0 in $\exp(0)$ is the $n \times n$ zero matrix (the matrix whose entries are equal to the real number 0).

Exercise 7.13 *Prove that*

$$\gamma'(t) = \frac{d}{dt} \exp(tX) = X \exp(tX) = X\gamma(t)$$

for all $t \in \mathbb{R}$. In particular, $\gamma'(0) = X$.

Also, show that γ is a one-parameter subgroup of $GL(n, \mathbb{R})$.

This indicates that $\gamma(t)$ is a one-parameter subgroup and that it corresponds to X; that is, it is $\gamma_X(t)$. Then $\gamma_X(1) = \exp(X)$, the usual exponential of the matrix X. This justifies using the same notation and nomenclature in the abstract case of a general Lie group G. However, in this book we will work exclusively with Lie groups that are closed Lie subgroups of a matrix group.

There are many important properties of the exponential map. Here is such a theorem, which we state without giving proof, about the exponential map.

Theorem 7.2 *Let G be a Lie group and* $\exp : \mathfrak{g} \to G$ *its exponential map. Then there exists a neighborhood U of $0 \in \mathfrak{g}$ such that $\exp$ restricted to U is a diffeomorphism of U onto $\exp(U)$, which is a neighborhood of $e \in G$.*

7.5 Examples: Continuation

In this section we discuss the Lie algebras associated to the Lie groups we defined in Section 7.2.

Each of the Lie groups defined in Section 7.2 is a closed Lie subgroup G of $GL(n, \mathbb{C})$ and even in some cases of $GL(n, \mathbb{R})$. So the associated Lie algebra $\mathfrak{g}$ of such a Lie group G is a subalgebra of $\mathfrak{gl}(n, \mathbb{C})$. In particular, this tells us that the Lie bracket in the subalgebra is simply the restriction of the Lie bracket in $\mathfrak{gl}(n, \mathbb{C})$, the latter being nothing other than the commutator of matrices. So this identifies with no further ado the Lie bracket for all of the examples in Section 7.2. Thus, it only remains to identify the vector subspace $\mathfrak{g}$ of $\mathfrak{gl}(n, \mathbb{C})$ for each of those examples.

Exercise: *Prove that the Lie bracket of the Lie algebra $\mathfrak{gl}(n, \mathbb{C})$ is given by the commutator of matrices.*

We start with $SL(n, \mathbb{R}) = \{B \in GL(n, \mathbb{R}) \mid \det B = 1\}$. We have to identify those $A \in \mathfrak{gl}(n, \mathbb{R})$ such that $\exp(tA) \in SL(n, \mathbb{R})$ for all t is some neighborhood of 0. So the condition on A is

$$\det\big(\exp(tA)\big) = 1.$$

But we have this well-known identity for matrices M:

$$\det(\exp M) = \exp(\operatorname{Tr} M),$$

where $\operatorname{Tr} M := \sum_j m_{jj}$ is the trace of the matrix $M = (m_{ij})$. Hence, the condition on A becomes

$$\exp(\operatorname{Tr} tA) = 1,$$

whose only real solution is

$$\operatorname{Tr} tA = 0.$$

Since the trace is linear over the real numbers, we find that $\operatorname{Tr} A = 0$ is a necessary condition for A to be in the Lie algebra of $SL(n, \mathbb{R})$. The reader should now check that this condition is also sufficient. So the Lie algebra of $SL(n, \mathbb{R})$ is

$$\mathfrak{sl}(n, \mathbb{R}) := \{A \in \mathfrak{gl}(n, \mathbb{R}) \mid \operatorname{Tr} A = 0\}.$$

For the rest of the examples in Section 7.2, we identify their associated Lie algebras using the same method. We do this now for

$$SU(n) = \{U \in GL(n, \mathbb{C}) \mid U \text{ is unitary and } \det U = 1\}$$

and leave the rest of the examples as exercises for the reader. Now the one-parameter subgroup $\exp(tA)$ lies in $SU(n)$ for small $|t|$ provided that

$$\exp(tA) \exp(tA)^* = I,$$

$$\det\big(\exp(tA)\big) = 1.$$

These are then shown to be equivalent to the two conditions

$$A + A^* = 0 \qquad \text{and} \qquad \operatorname{Tr} A = 0.$$

The first condition is typically written as $A^* = -A$, and one says that the matrix A is *antihermitian*. For the condition $\operatorname{Tr} A = 0$, one says that A is *traceless*. So the Lie algebra of $SU(n)$ consists of all antihermitian, traceless matrices:

$$\mathfrak{su}(n) := \{A \in \mathfrak{gl}(n, \mathbb{C}) \mid A^* = -A \quad \text{and} \quad \operatorname{Tr} A = 0\}.$$

The Lie algebra for $U(n)$ is similarly given by

$$\mathfrak{u}(n) := \{A \in \mathfrak{gl}(n, \mathbb{C}) \mid A^* = -A\}.$$

There are two main cases that will be used later on in this book. The first case is

$$\mathfrak{su}(2) := \{A \in \mathfrak{gl}(2, \mathbb{C}) \mid A^* = -A \quad \text{and} \quad \text{Tr}\, A = 0\}.$$

These are the matrices of the form

$$\begin{pmatrix} ai & b + ic \\ -b + ic & -ai \end{pmatrix},$$

where $a, b, c \in \mathbb{R}$. Thus, $\dim \mathfrak{su}(2) = 3$.

The second case is

$$\mathfrak{u}(1) = \{A \in \mathfrak{gl}(1, \mathbb{C}) \mid A^* = -A\}.$$

Identifying a 1×1 matrix with its (unique) entry, we see that

$$\mathfrak{u}(1) \cong \{ir \mid r \in \mathbb{R}\} = i\mathbb{R},$$

the pure imaginary numbers.

These results could make the physicists among my readers uncomfortable since they might think that there is an error of a factor of $i = \sqrt{-1}$ here. But what is happening is that there is a difference in convention between mathematicians and physicists in the definition of the Lie algebra. That difference is exactly a factor of $i = \sqrt{-1}$. These two conventions are discussed in detail in Section 14.1. In this section we have used the mathematicians' convention.

7.6 Concluding Comment

The theory of Lie groups is an intellectual industry unto itself. We have not even seen the tip of the iceberg here. A classical but advanced reference is [49]. Another reference more or less at the same level as this book is [20]. A favorite reference of mine with lots of physics is [36]. There are many, many other fine texts on Lie groups and Lie algebras.

Chapter 8
The Frobenius Theorem

It could well be argued that the previous chapters were like an introductory course in a foreign language. Much attention was devoted to learning the new vocabulary and relating it to the vocabulary in a previously known language. Then there were a lot of exercises to learn how the new vocabulary is used. So that's the grammar. But, to continue the analogy, there was no poetry. Or to put it into colloquial mathematical terminology, there were no real theorems. That is about to change. The Frobenius theorem is a real theorem.

8.1 Definitions

The idea behind this theorem is to "integrate" a distribution. First off, a distribution is intuitively a smooth assignment of a vector subspace W_p in the tangent space $T_p M$ at every point $p \in M$. (The rigorous treatment will soon be given.) And to integrate a given distribution at a point $p_0 \in M$ means to find a submanifold N of M such that $p_0 \in N$ and such that the tangent space $T_p N$ for every $p \in N$ is equal to W_p. Saying this again, we require that $T_p N = W_p$ for every $p \in N$. If this sounds something like integrating a vector field (but with the higher-dimensional object W_p replacing the vector field), that is because it *is something like* integrating a vector field, but not exactly! After all, a vector field X defines a one-dimensional subspace of $T_p M$ (namely, its span) at every point $p \in M$ that satisfies $X_p \neq 0$. But it does not define a one-dimensional subspace at those points p where $X_p = 0$. But more importantly, a one-dimensional subspace of $T_p M$ does not single out any particular element in itself to be the value of a vector field at that point.

Here is a simple example that will let you picture this situation. Let $M = \mathbb{R}^3 \setminus \{0\}$ be the three-dimensional Euclidean space minus the origin. At each point $p \in M$, let W_p be the two-dimensional subspace of $T_p M \cong \mathbb{R}^3$ orthogonal to the vector p. Then each sphere of radius $r > 0$ centered at the origin integrates this distribution. More specifically, for any $p_0 \in M$, the sphere S_{r_0} of radius $r_0 = ||p_0||$ and centered

© Springer International Publishing Switzerland 2015

S.B. Sontz, *Principal Bundles*, Universitext, DOI 10.1007/978-3-319-14765-9_8

at the origin integrates the distribution and has $p_0 \in S_{r_0}$. This is simply because the tangent space to such a sphere is orthogonal at each of its points p to the radial vector emanating from the origin to p. Of course, this distribution has been deliberately "cooked up" so that it is clear how to integrate it. But what happens if we jiggle the subspaces W_p a little bit in some funny manner? Will this still be a distribution that we can integrate? All of a sudden geometric intuition goes out the window! But as we shall see, the Frobenius theorem comes in the door to help us.

Before being able to state the Frobenius theorem, we need some rigorous definitions in order to establish exactly what we are talking about. It should be clear that the assignment $p \mapsto W_p \subset T_p M$ for $p \in M$ should be a smooth section of some bundle. This is a typical situation; that is, we have an intuitive idea of what the sections of certain bundle should be before we have a notion as to what the bundle itself is! Sometimes one even says, "B is the bundle whose sections are such-and-such" as if this were a definition of the bundle. Well, it is not. But it is not a meaningless statement either. It is the type of statement that falls under the rubric of "mathematical idea" and as such can be a great guideline for finding the "correct" definition of a bundle. Also, it is a statement than can—and should—become a true statement after the bundle is defined. As such, it will become a theorem though in general its proof will be a triviality simply because the definition has been made so this it *is* a triviality! Actually, the definition itself could be nontrivial. Just one more personal, philosophical aside before returning to mathematics: We have no mathematical definition of what a "mathematical idea" is and I doubt we ever will.

We are going to make the simplifying assumption that M is connected. Then whatever else smoothness of $p \mapsto W_p$ might mean, it should at least imply that the dimension of W_p does not depend on p. So we are led to making the following definition at the infinitesimal level, that is, for vector spaces.

Definition 8.1 Let $0 \leq k \leq n$ be an integer. Then define the Grassmannian of k-planes in $\mathbb{R}^n$ to be

$$G(n,k) := \{W \mid W \subset \mathbb{R}^n \text{ is a subspace with dim } W = k\}. \tag{8.1}$$

This should be the model space for the fibration E over M that we wish to define. That is to say, a typical fiber in E over $p \in M$ should be a copy of $G(n,k)$ except that $\mathbb{R}^n$ will be replaced by $T_p M$, which is, of course, a vector space isomorphic to $\mathbb{R}^n$. In this way, a section of E will be a selection of a k-plane in $T_p M$ for each $p \in M$. Since we will have a smooth structure on E, the sections of E that we will consider will be the smooth sections $s : M \to E$. And such a smooth section will be called a *distribution of k-planes* over M. That is the program. Notice that this accords with the maxim that "smooth is smooth is smooth" in that we are using the one and only (and standard) definition of smooth map for the particular case of the map s. Some authors introduce an ad hoc definition of smooth just for this one situation.

Exercise 8.1 *If you are already familiar with some other definition for the smoothness for a distribution of k-planes, then keep reading and convince yourself that it is equivalent to the definition given below.*

Definition (8.1) of $G(n, k)$ just defines a *set* without any further structure. In order to implement this program of defining the bundle E, we first need to provide $G(n, k)$ with a topology and a smooth structure. Then, using an atlas for M, we will construct a cocycle to glue together appropriate smooth manifolds to define E. The idea that $G(n, k)$ is naturally a smooth manifold should not be too surprising since $G(n, 1)$, being the set of all lines in $\mathbb{R}^n$ passing through the origin, is the real projective space $\mathbb{R}P^{n-1}$ of dimension $n - 1$, which is a quotient of the unit sphere S^{n-1} in $\mathbb{R}^n$ by the $\mathbb{Z}_2$ action that maps $p \in S^{n-1}$ to $-p \in S^{n-1}$. Then the standard topology and smooth structure on S^{n-1} induce a topology and smooth structure on the quotient space $\mathbb{R}P^{n-1}$. We now want do the same sort of thing with $G(n, k)$ for any $0 \le k \le n$.

First, we recall the Lie group $GL(n)$ of all invertible transformations $T : \mathbb{R}^n \to \mathbb{R}^n$. Also, recall that $GL(n)$ can be realized as an open subset of $\mathbb{R}^{n^2}$ and as such is a smooth manifold of dimension n^2. Each such T acts on $G(n, k)$ by mapping the k-plane $W \in G(n, k)$ to $T(W)$, which is again a k-plane since T is invertible. That is, $T(W) \in G(n, k)$. Moreover, this action of $GL(n)$ is *transitive*, meaning that for any fixed $W_0 \in G(n, k)$, the orbit $GL(n) W_0$ of W_0 is all of $G(n, k)$. [That is, for any $V \in G(n, k)$, there exists a linear map $T \in GL(n)$ that maps W_0 to V.] So the set $G(n, k)$ is isomorphic (as a set) to the quotient of $GL(n)$ by the *stabilizer* Σ of W_0,

$$\Sigma := \{T \in GL(n) \mid T(W_0) = W_0\}.$$

Exercise 8.2 *Show that Σ is a closed subgroup of $GL(n)$. Then find the dimension of Σ.*

It is known that one can define a smooth structure on the quotient space $GL(n)/\Sigma$ so that the quotient map $GL(n) \to GL(n)/\Sigma$ is smooth. This is a nontrivial result. In particular, one must show that the topology on $GL(n)/\Sigma$ is Hausdorff.

Exercise 8.3 *Use the bijection between $GL(n)/\Sigma$ and $G(n, k)$ to give $G(n, k)$ a smooth structure so that this bijection is a diffeomorphism.*

What is the dimension of $G(n, k)$?

Finally, it is true that the action map $\sigma : GL(n) \times G(n, k) \to G(n, k)$ [given by $\sigma : (T, V) \mapsto T(V)$] is smooth. Think about this statement for some small values of n and k and see if you are convinced that it is true in those cases. With some further thought, you should be able find a proof.

We now can construct the bundle E. We are given a smooth manifold M to start off with and so we also have the cocycle for its tangent bundle $t_{\alpha\beta} : U_\alpha \cap U_\beta \to GL(n)$, where $n = \dim M$ and $\{(U_\alpha, \phi_\alpha) \mid \alpha \in A\}$ is an atlas of M. In particular,

$\{U_\alpha \mid \alpha \in A\}$ is an open cover of M. Also, recall that $t_{\beta\alpha} = D(\phi_\beta \circ \phi_\alpha^{-1}) \circ \phi_\alpha$. Now for each pair $\alpha, \beta \in A$ with $U_\alpha \cap U_\beta$ nonempty, we define the transition function between smooth manifolds

$$\psi_{\beta\alpha} : \phi_\alpha(U_\alpha \cap U_\beta) \times G(n,k) \to \phi_\beta(U_\alpha \cap U_\beta) \times G(n,k)$$

by

$$\psi_{\beta\alpha}(\phi_\alpha(x), V) := (\phi_\beta(x), t_{\alpha\beta}(x)(V))$$

for all $x \in U_\alpha \cap U_\beta \subset M$ and all $V \in G(n,k)$. Finally, one uses these transition functions to define in the usual way a total space E and a projection map $\pi : E \to M$, which gives us a smooth manifold E that sits "over" M with each fiber $\pi^{-1}(p)$ diffeomorphic to $G(n,k)$ for every $p \in M$. This manifold is called the *Grassmannian bundle of k-planes* over M [notation: $E = G(TM,k)$]. The construction given in this paragraph generalizes. See Exercise 9.3.

Next, a smooth section of $\pi : E \to M$ is called a *(smooth) distribution of k-planes over M*. That is the rigorous definition of a distribution! And it encodes in the language of bundle theory the intuitive idea of smoothly varying k-dimensional subspaces of the tangent spaces T_pM. (Notice that the previous sentence is not, and cannot be, a theorem.) This construction of $G(TM,k)$ is analogous to the construction of vector bundles associated to the tangent bundle. In the latter case, one has a *linear* action of $GL(n)$ on a *vector space*. Instead of that, we now have an action of $GL(n)$ on the compact manifold $G(n,k)$ (cf. Exercise 8.5). So we should also think of $G(TM,k)$ as a bundle (but not a *vector* bundle) associated with the tangent bundle of M. In the regard, we again call your attention to Exercise 9.3.

Exercise 8.4 *Make sure you understand the previous paragraph in detail. In particular, if $N \subset M$ is a submanifold with $\dim N = k$, explain how we can think of the map $p \mapsto T_pN \subset T_pM$ for $p \in N$ as a smooth distribution.*

Exercise 8.5 *Recall the notation $O(n)$ for the Lie group of $n \times n$ orthogonal matrices. Using an argument similar to that given above, but now using $O(n)$ instead of $GL(n)$, show that there is a bijection of sets*

$$G(n,k) \cong O(n)/(O(k) \times O(n-k))$$

for every integer k satisfying $0 \le k \le n$. (Actually, the two cases $k = 0$ and $k = n$ are quite trivial and should not be considered further.) Convince yourself that the smooth structure on $G(n,k)$ that makes this bijection a diffeomorphism is the same smooth structure as described in Exercise 8.3.

Calculate again the dimension of $G(n,k)$ and see if this agrees with the result you obtained in Exercise 8.3.

Finally, conclude that $G(n,k)$ is a compact manifold.

Next, our intuitive idea of what it means to integrate a given distribution of k-planes over M translates directly into a rigorous definition.

Definition 8.2 *Suppose that M is a smooth manifold and W is a smooth section of the bundle $G(TM, k) \to M$. In other words, W is a distribution of k-planes over M. Let $p_0 \in M$ be given. Then we say that a k-dimensional submanifold N of M integrates W with initial condition p_0 provided that $p_0 \in N$ and $T_p N = W_p$ for every $p \in N$. And we say that W is* integrable *if there is such a submanifold N for every $p_0 \in M$.*

8.2 The Integrability Condition

The Frobenius theorem consists of a necessary and sufficient condition for a distribution W to be integrable. So we now introduce that condition.

Definition 8.3 *Let W be a distribution of k-planes over a smooth manifold M. Then we say W that satisfies the* Frobenius integrability condition *in an open set $U \subset M$ provided that for every pair of vector fields X, Y defined in U satisfying $X_p \in W_p$ and $Y_p \in W_p$ for all $p \in U$ we have that their Lie bracket satisfies $[X, Y]_p \in W_p$ for all $p \in U$.*

We also express this by saying that any pair of vector fields lying in W over U has a Lie bracket also lying in W over U.

In general, when we dub a condition to be an integrability condition, one may very well wonder whether this is a definition or a theorem. In some sense, it is both! As we can see in the case of the Frobenius integrability condition, we clearly have a definition. There it is in all of its mathematical rigor! But definitions should arise from mathematical considerations. And important definitions arise from important mathematical considerations. So the justification for using the expression "integrability condition" in the above definition is that it indeed gives a sufficient condition for the integrability of something (which in this case is a distribution). Therefore, it is the crucial hypothesis in a theorem. Without the theorem, the definition would be a hollow formality. In this particular case, the integrability condition is also necessary, but, in general, this need not be so to justify using the phrase "integrability condition" in the wording of a definition.

Here is another way to think about the Frobenius integrability condition. Suppose that the pair of vector fields X and Y commute whenever $X_p \in W_p$ and $Y_p \in W_p$ for all $p \in U$. This means that $[X, Y] = 0$. Then in this case, the Frobenius integrability condition is trivially satisfied. So this gives a sufficient, but no longer necessary, condition for the integrability of W. And this situation does arise in examples. But the Frobenius integrability condition is a weaker condition. It allows some lack of commutativity of X and Y, but not an arbitrary "amount" of noncommutativity. Colloquially speaking, one could say that X and Y commute modulo W. So an alternative manner of thinking about Frobenius' integrability condition is that it is a sort of commutativity condition that is weaker than commutativity as such.

Yet another way to think of Frobenius' integrability condition on W is to note that it is equivalent to saying that the subbundle of $T(M)$ determined by the distribution W has the property that any two of its sections have a Lie bracket that is again a section of that subbundle. However, we have not defined *subbundle*. We will use the concept of subbundle in passing later on, so the reader might wish to look up the definition now.

8.3   Theorem and Comments

We now can state the theorem of this chapter. It is sometimes called *the Frobenius theorem* but is also known as *the Frobenius integrability theorem*.

Theorem 8.1 *Let W be a distribution of k-planes over a smooth manifold M. Then W is integrable if and only if W satisfies the Frobenius integrability condition.*

So for a distribution W to be integrable, a condition that is both necessary and sufficient is the Frobenius integrability condition. The necessity of this condition is rather trivial to prove and so is left to the reader.

On the other hand, the sufficiency of this condition justifies its name according to the above comments. Also, the proof of the sufficiency is more involved. But more important than the proof is an understanding of what is being asserted. Somehow the proof is an anticlimax to the whole story! The ideas that led to the statement of this theorem are important, however.

So we will not present here a proof of the Frobenius integrability theorem, which is, of course, proved in many texts, such as [1] and [32]. Particularly recommended is the crystal-clear proof in [33], where Lee shows that the Frobenius integrability condition implies that the distribution locally has a basis given by *commuting* vector fields. This reduces the argument to a rather easy special case. At a conceptual level, this proof shows that commutativity is what is behind the Frobenius integrability condition.

Chapter 9
Principal Bundles

9.1 Definitions

We now introduce the structures that will be used in our applications to physics. We begin with most of the ingredients for the construction of a vector bundle, but now we will use them to construct something else.

We start with a manifold M with a given atlas $\{(U_\alpha, \phi_\alpha)\}$ of charts and a cocycle $g_{\beta\alpha} : U_\alpha \cap U_\beta \to G$, where G is a given Lie group. However, now we do not necessarily have a representation of G acting on a vector space though we could be using a cocycle that comes from a given vector bundle. Rather, we will use the fact that any Lie group G always acts smoothly on itself by multiplication from the left. (But this is *not* a group action unless $G = \{e\}$.) So we now define smooth functions

$$\tilde{g}_{\beta\alpha} : \phi_\alpha(U_\alpha \cap U_\beta) \times G \to \phi_\beta(U_\beta \cap U_\alpha) \times G$$

by a formula very similar to equation (3.4):

$$\tilde{g}_{\beta\alpha}(\phi_\alpha(x), g) := (\phi_\beta(x), g_{\beta\alpha}(x)g) \qquad (9.1)$$

for all $x \in U_\alpha \cap U_\beta$ and all $g \in G$. We wish to emphasize that the expression $g_{\beta\alpha}(x)g$ is the left action of an image of a cocyle, namely, $g_{\beta\alpha}(x) \in G$, on an element $g \in G$, the model fiber space. In this setting, the group of the cocycle and the model fiber space are the same space. This condition can be replaced with something more general, as we will see shortly in Exercise 9.3.

Theorem 9.1. *Suppose M is a smooth manifold with a given atlas $\{(U_\alpha, \phi_\alpha)\}$ of charts and a cocycle $g_{\beta\alpha} : U_\alpha \cap U_\beta \to G$, where G is a given Lie group. Then the functions in (9.1) define an equivalence relation on the disjoint union*

$$\bigsqcup_{\alpha \in A} (\phi_\alpha(U_\alpha) \times G)$$

© Springer International Publishing Switzerland 2015

S.B. Sontz, *Principal Bundles*, Universitext, DOI 10.1007/978-3-319-14765-9_9

by defining

$$(\phi_\alpha(x), g) \in \phi_\alpha(U_\alpha \cap U_\beta) \times G \subset \phi_\alpha(U_\alpha) \times G$$

to be equivalent to

$$\tilde{g}_{\beta\alpha}(\phi_\alpha(x), g) = (\phi_\beta(x), g_{\beta\alpha}(x)g) \in \phi_\beta(U_\alpha \cap U_\beta) \times G \subset \phi_\beta(U_\beta) \times G.$$

The topological space of its equivalence classes can be made into a smooth manifold, denoted P, which can be provided with a smooth map $\pi : P \to M$ such that there are diffeomorphisms, $\tilde{\phi}_\alpha : \pi^{-1}(U_\alpha) \to \phi_\alpha(U_\alpha) \times G$, called local trivializations, *which make the diagram*

$$
\begin{array}{ccc}
P \supset \pi^{-1}(U_\alpha) & \xrightarrow{\tilde{\phi}_\alpha} & \phi_\alpha(U_\alpha) \times G \\
\downarrow{\scriptstyle \pi} \quad\quad \downarrow{\scriptstyle \pi} & & \downarrow{\scriptstyle \pi_1} \\
M \supset \quad U_\alpha & \xrightarrow{\phi_\alpha} & \phi_\alpha(U_\alpha)
\end{array}
$$

commute for every $\alpha \in A$, where π_1 is the projection onto the first factor. One says that the map $\tilde{\phi}_\alpha$ covers the map ϕ_α or that $\tilde{\phi}_\alpha$ is a covering *of ϕ_α.*

Then the so-called transition functions

$$\tilde{\phi}_\beta \circ \tilde{\phi}_\alpha^{-1} : \phi_\alpha(U_\alpha \cap U_\beta) \times G \to \phi_\beta(U_\beta \cap U_\alpha) \times G$$

associated to the local trivializations are equal to the functions $\tilde{g}_{\beta\alpha}$ in (9.1):

$$\tilde{\phi}_\beta \circ \tilde{\phi}_\alpha^{-1} = \tilde{g}_{\beta\alpha}.$$

For each $x \in M$, we define the *fiber above x* to be $P_x := \pi^{-1}(x)$.

Moreover, there is a smooth action of G acting on the right of P such that each subspace $\pi^{-1}(U_\alpha)$ is G-invariant and such that each local trivialization $\tilde{\phi}_\alpha$ is a G-equivariant map, where the action of G on the space $\phi_\alpha(U_\alpha) \times G$ is by multiplication on the right in the second factor. Explicitly, this right action is

$$(\phi_\alpha(U_\alpha) \times G) \times G \cong \phi_\alpha(U_\alpha) \times (G \times G) \xrightarrow{id \times m_G} \phi_\alpha(U_\alpha) \times G,$$

where m_G is the multiplication of the group G.

Our notation for the right action $\alpha : P \times G \to P$ of G on P is

$$\alpha(x, g) = xg = x \cdot g$$

for $x \in P$ and $g \in G$. This action is free and sends each fiber into itself; that is,

$$
\begin{array}{ccc}
P \times G & \xrightarrow{\alpha} & P \\
\downarrow & & \downarrow \pi \\
M & \xrightarrow{1_M} & M
\end{array}
$$

commutes, where the vertical arrow on the left is projection onto the first factor followed by π.

Definition 9.1 *We say that P is a* principal bundle *over M with (structure) group G. We call M the* base space, *P the* total space, *and π the* bundle map. *The usual notation for this is*

$$
\begin{array}{ccc}
G & \hookrightarrow & P \\
& & \downarrow \pi \\
& & M
\end{array}
$$

Here the horizontal inclusion map is not unique. It is any G-equivariant diffeomorphism of G onto some fiber in P and so is determined by the image of $e \in G$ in P.

Warning: These local trivializations $\tilde{\phi}_\alpha$ are not charts on the total space P in the sense of Definition 2.1, since the range of $\tilde{\phi}_\alpha$ is not necessarily an open subset of a Euclidean space. However, the family $\{\pi^{-1}(U_\alpha), \tilde{\phi}_\alpha\}$ serves much the same purpose as an atlas on P would. What this means, as the reader by now will surely appreciate, is that when we want to study P locally in a way consistent with its structure as a principal bundle, then the local trivializations are what we have to use.

Exercise 9.1 *Prove Theorem 9.1.*
Also, show that $\dim P = \dim M + \dim G$.

Exercise 9.2 *Consider the local trivializations of a vector bundle (which, by the way, we also call* natural charts). *Are they charts? Does it really matter?*

We have defined principal bundles, which will be the objects of a category. So we still have to define what the morphisms of that category will be.

Definition 9.2 *Suppose that $\pi : P_1 \to M_1$ and $\rho : P_2 \to M_2$ are principal bundles with the structure group G. Then a* morphism *of these principal bundles is a pair of smooth maps $F : P_1 \to P_2$ and $f : M_1 \to M_2$ such that F is a G-equivariant map and this diagram commutes:*

$$
\begin{array}{ccc}
P_1 & \xrightarrow{F} & P_2 \\
\pi \downarrow & & \downarrow \rho \\
M_1 & \xrightarrow{f} & M_2
\end{array}
$$

In this situation, we say that F is a covering *of f or that F covers f.*

Sometimes, we abbreviate this in a rather colloquial manner and simply say that F is a morphism of principal bundles covering *f.*

At this point, we omit the traditional exercise requesting the reader to prove that for every Lie group G we indeed do have a category of principal bundles with structure group G. But we will come back to this.

Using this definition, we now can see that the local trivializations $\tilde{\phi}_\alpha$ in Theorem 9.1 are isomorphisms of principal bundles covering ϕ_α.

Exercise 9.3 *There is now an immediate generalization whose details we leave to the reader in this exercise. Let M be a manifold with a given atlas $\{(U_\alpha, \phi_\alpha)\}$ of charts and with a cocycle $g_{\beta\alpha} : U_\alpha \cap U_\beta \to G$ for some given Lie group G. Let F be a left G-space.*

- *Define the* fiber bundle *over M with typical fiber F.*
- *Suppose that some other structure of the fiber space F is preserved by the given left action of G on F. Prove that the fibers in the total space of the fiber bundle can be given this structure using the local trivializations.*
- *Define a morphism of fiber bundles and show that one gets a category.*

Exercise 9.4 *Let $\pi : P \to M$ be a principal bundle with group G.*

(a) *Say $p \in P$. Put $x := \pi(p)$. Prove that the orbit of the action that passes through p is the fiber in which p lies; that is, $p \cdot G = \pi^{-1}(x)$. Show that any fiber of P is diffeomorphic to G but not in general in a canonical way.*

(b) *We know now that each fiber is a diffeomorphic copy of G. But can we give each fiber a group structure in a natural way (and varying smoothly as we vary the fiber) so that it is also isomorphic to G as a (Lie) group? Try defining such a group structure using local trivializations and see whether this makes sense, that is, does not depend on the choice of local trivialization.*

(c) *Say $p_1, p_2 \in P$ are in the same fiber; that is, $\pi(p_1) = \pi(p_2) \equiv x$. Show that there exists a unique element $g \in G$ such that $p_2 = p_1 \cdot g$.*

In the next exercise, we will use the following construction.

Definition 9.3 *Let $G \hookrightarrow P_1 \xrightarrow{\pi^1} M$ and $G \hookrightarrow P_2 \xrightarrow{\pi^2} M$ be principal bundles over M with Lie group G. Then their* fiber product *is defined by*

$$P_1 \diamond P_2 := \{(p_1, p_2) \in P_1 \times P_2 \mid \pi^1(p_1) = \pi^2(p_2)\}.$$

We also write $P^{\diamond 2} := P \diamond P$.

Exercise 9.5 *Let $\pi : P \to M$ be a principal bundle with group G. Consider $x \in M$. One should note that even though the fiber $\pi^{-1}(x)$ does not have a product or an inverse operation, the* affine operation $\pi^{-1}(x) \times \pi^{-1}(x) \to G$, *denoted by $(p_1, p_2) \mapsto p_1^{-1} p_2$, does make sense for $p_1, p_2 \in \pi^{-1}(x)$, where $p_1^{-1} p_2 := g \in G$ is the unique element in G such that $p_2 = p_1 \cdot g$.*

Show that this fiberwise-defined operation induces a smooth map on the fiber product of P with itself: $P^{\diamond 2} \to G$.

Prove the following properties of the affine operation, where p_1, p_2, p_3 all lie in the same fiber, $p \in P$ and $g, g_1, g_2 \in G$:

(a) $p_2 = p_1(p_1^{-1}p_2)$. [Compare this with $(pg^{-1})g = p$.]
(b) $p^{-1}p = e$.
(c) $(p_1^{-1}p_2)^{-1} = p_2^{-1}p_1$.
(d) $(p_1^{-1}p_2)(p_2^{-1}p_3) = p_1^{-1}p_3$.
(e) $(p_1g_1)^{-1}(p_2g_2) = g_1^{-1}(p_1^{-1}p_2)g_2$.
  In particular, $(p_1g)^{-1}(p_2g) = g^{-1}(p_1^{-1}p_2)g$.

(f) Let $G \hookrightarrow P_1 \xrightarrow{\pi^1} M$ and $G \hookrightarrow P_2 \xrightarrow{\pi^2} M$ be principal bundles over M with Lie group G. Suppose that $\Psi : P_1 \to P_2$ is a morphism of principal bundles covering 1_M. Show that Ψ preserves the affine operations.

Exercise 9.6 *As a special case of the preceding exercise, we consider the principal bundle $G \hookrightarrow G \to \{x\}$ over a one-point space $\{x\}$.*

- *Show that the affine operation for this principal bundle is the map that sends $(g_1, g_2) \in G \times G$ to the element $g_1^{-1}g_2 \in G$.*
- *Show that the left multiplication map $L_h : G \to G$ for any $h \in G$ preserves the affine operation on G. [Recall: $L_h(g) = hg$.] Express this as a commutative diagram. What happens with the right multiplication?*
- *Let $P \to M$ be a principal bundle with Lie group G. Show that each fiber of P is isomorphic to G with respect to their affine operations though not, in general, in a canonical, unique way.*

What we call an affine operation is well known. For example, see [27] Ch. 4 §2, where it is defined for any free G-space and is called the *translation function*. If there is any possible novelty here, it could only be in the name of the operation or its notation. But even then that would only be in this context.

Here are a bunch of important definitions, which can be a bit tricky to assimilate all at once. But they are really essential for further understanding of this material, especially gauge theory.

Definition 9.4 *Let P_1 and P_2 be principal bundles over M.*

- *An equivalence from P_1 to P_2 is an isomorphism $P_1 \to P_2$ of principal bundles covering the identity map of M. (As usual, an isomorphism is an invertible morphism. The nontrivial definition of a morphism of principal bundles will be dealt with momentarily in Exercise 9.8.)*
- *P_1 and P_2 are equivalent if an equivalence exists from P_1 to P_2.*
- *The explicitly trivial principal bundle over M with structure group G is $\pi_1 : M \times G \to M$, where π_1 is the projection onto the first factor.*
- *A principal bundle over M with structure Lie group G is said to be trivial if it is equivalent to the explicitly trivial bundle $M \times G$.*

- *A* trivialization *of a trivial bundle* P *with structure group* G *is a specific equivalence from* P *to* $M \times G$.
- *We say that a principal bundle* $\pi : P \to M$ *is* locally trivial *if, for every point* $x \in M$, *there exists an open set* $U \subset M$ *containing* x *such that* $\pi : \pi^{-1}(U) \to U$ *is a trivial principal bundle.*

We wish to emphasize the following important point here: The objects A, B in a category are *isomorphic* if there exists an *isomorphism* $f : A \to B$. However, the isomorphism f need neither be unique nor be specified. In other words, saying that A and B are isomorphic in no way selects an isomorphism between them. This is also true in the particular case when $A = B$. In the previous definition, we adapt the standard convention of saying "equivalence" instead of "isomorphism" when speaking of principal bundles.

Exercise 9.7 *Identify the principal bundle defined by the cocycle* $g_{\beta\alpha}(x) = e$.

Exercise 9.8 *Let* G *be a given Lie group. We propose that the reader do the abstract nonsense that is appropriate at this point in the discussion. In case a bit of exposure to the language of category theory is needed, the basics are given in about six pages in Chap. I, §7 of [31].*

- *Suppose that* A *is an object in a category. Let* Aut(A) *denote the set of all isomorphisms* $f : A \to A$ *(called* automorphisms *in this case). Show that* Aut(A) *has a product that has an identity element and an inverse operation. Show that with these structures,* Aut(A) *is a group, which is called the* automorphism group *of* A.
- *Define the category of all principal bundles with structure group* G, *a given Lie group.*
- *Show that every principal bundle is locally trivial.*

Example: Let M be a smooth manifold of dimension m with its canonically associated cocycle $t_{\beta\alpha}$ for its tangent bundle, as given in (3.5). Then this defines a principal bundle over M with structure group $GL(\mathbb{R}^m)$. It is called the *(principal) bundle of frames* over M. One possible notation is

$$GL(\mathbb{R}^m) \hookrightarrow P(M)$$
$$\downarrow \pi \ .$$
$$M$$

This particular principal bundle and structures associated with it are much studied in classical differential geometry. For example, see [46].

In general, for any given vector bundle $E \to M$, we can produce an associated principal bundle over M with structure group $GL(\mathbb{R}^l)$, where l is the dimension of the typical fiber in E.

9.2 Sections

As in the study of vector bundles, the concept of a section is of central importance. Actually, in some intuitive sense, bundles are defined so that we can define sections of them.

Definition 9.5 *Let* $\pi : P \to M$ *be a principal bundle. Then we say that a smooth function* $s : M \to P$ *is a* (global) section *of* P *if* $\pi \circ s = 1_M$. *For any open subset* $U \subset M$, *we say that any section of* $\pi : \pi^{-1}(U) \to U$ *is a* local section *over* U. *We use the notation* $\Gamma(P)$ *for the set of all sections of* $\pi : P \to M$.

Maybe it is worth commenting that $\pi : \pi^{-1}(U) \to U$ is a principal bundle over U with structure group G. But maybe my kind reader has already realized that this is true.

While nontrivial vector bundles have lots of sections, the situation is quite different for principal bundles, as we shall now see.

Theorem 9.2 *A principal bundle has a section if and only if it is trivial. Moreover, the set of all sections of the bundle is in bijective correspondence with the set of its trivializations.*

Proof: Let $\pi : P \to M$ denote the principal bundle with structure group G. We set up the bijection by defining two functions between the set $\Gamma(P)$ of sections and Triv(P), the set of all trivializations of P. But before doing that, let us quickly note that every trivial bundle (principal or not) always has a section. The converse for principal bundles (but not for other sorts of bundles) follows from the existence of Ψ as given in the next paragraph.

First, we define $\Psi : \Gamma(P) \to \text{Triv}(P)$ as follows. Let $s \in \Gamma(P)$ be a section. Then define $\Psi(s) : P \to M \times G$ by $\Psi(s) : p \mapsto (x, s(x)^{-1}p)$, where $x = \pi(p)$, using the affine operation defined in Exercise 9.5. [Note that $s(x)$ and p lie in the same fiber, namely, the fiber above x.] We also define $s_e : M \to M \times G$ by $s_e(x) := (x, e)$ for $x \in M$, where $e \in G$ is the identity element. Consider

$$
\begin{array}{ccc}
P & \xrightarrow{\Psi(s)} & M \times G \\
\pi \downarrow \uparrow s & & \pi_1 \downarrow \uparrow s_e \\
M & \xrightarrow{1_M} & M
\end{array}
\tag{9.2}
$$

This gives us two commutative diagrams, one where we take the two down arrows and one where we take the two up arrows. Here π_1 denotes the projection onto the first factor.

Exercise 9.9 *Prove that* $\Psi(s) : P \to M \times G$ *is a smooth function and is a trivialization of* P. *Also, prove that the two diagrams in (9.2) commute.*

Another way to understand the definition of $\Psi(s)$ is to note that on $\text{Im}\,s$, the image of the section s, it is given by $s(x) \to (x, e)$. Having defined $\Psi(s)$ on exactly

one element of each fiber above every $x \in M$, we can use G-equivariance to extend
the definition to the entire fiber above x.

Second, we define $\Phi : \mathrm{Triv}(P) \to \Gamma(P)$ as follows. Suppose that we are given a
trivialization F. So this diagram with the down arrows commutes:

$$
\begin{array}{ccc}
P & \xrightarrow{\ F\ } & M \times G \\
\pi \Big\downarrow\Big\uparrow \Phi(F) & & \pi_1 \Big\downarrow\Big\uparrow s_e \\
M & \xrightarrow{\ 1_M\ } & M
\end{array}
$$

Then we define the up arrow $\Phi(F)$ exactly so that the diagram with the up arrows
commutes; that is, $\Phi(F) := F^{-1} \circ s_e$.

Exercise 9.10 *Show that $\Phi(F)$ is a section of P. Prove that Ψ and Φ are inverses
to each other.*

This exercise concludes the proof. ∎

Of course, every principal bundle is locally trivial. So it always has local sections.

9.3 Vertical Vectors

We now proceed to the topic of vertical (tangent) vectors on a principal bundle. We
start with a principal bundle $G \hookrightarrow P \xrightarrow{\pi} M$. As we noted before, the map of
G onto a fiber is not unique. In fact, there is exactly one G-equivariant map for
each $p \in P$.

Definition 9.6 *For each $p \in P$, we define $\iota_p : G \to P$ by $\iota_p(g) := p \cdot g$ for
all $g \in G$.*

It is common to think of the fiber as sitting vertically "above" the point $x = \pi(p)$
in M. So the tangent vectors at $p \in P$ in the direction of the fiber $\iota_p(G) =
\pi^{-1}(x) \subset P$ are thought of as pointing vertically. This image has a bit to do with the
one vertical direction with respect to the surface of the earth. However, here we have
$\dim(G)$ linearly independent vertical directions at each $p \in P$. Having indulged in
this motivational discussion, we now proceed to the mathematical definition.

We first note that $i_p(e) = p$, so that we have an induced map of tangent spaces:

$$
\mathfrak{g} = T_e(G) \xrightarrow{T_e \iota_p} T_p P.
$$

Definition 9.7 *Let $G \hookrightarrow P \xrightarrow{\pi} M$ be a principal bundle. We say that a tangent
vector in $T_p P$ is vertical if it lies in the image of $T_e \iota_p$. Notation for the set of vertical
vectors at p is*

$$
\mathrm{Vert}(T_p P) := \mathrm{Im}\, T_e \iota_p = T_e \iota_p(\mathfrak{g}).
$$

Exercise 9.11 *Prove that the following is a short exact sequence of vector spaces:*

$$0 \to \mathfrak{g} \xrightarrow{T_e \iota_p} T_p P \xrightarrow{T_p \pi} T_x M \to 0. \tag{9.3}$$

See the comments for this exercise in the appendix for the definition of short exact sequence.

It follows that $\dim(\text{Vert}(T_p P)) = \dim G$ *and that*

$$\iota_{p,*} = T_e \iota_p : \mathfrak{g} \to \text{Vert}(T_p P) \tag{9.4}$$

is an isomorphism of vector spaces.

Remark: We have introduced new notation here. If $f : M \to N$ is a smooth map of manifolds, then we put $f_* := Tf : TM \to TN$.

Another immediate consequence of the exactness of the sequence (9.3) is that we can equivalently define the subspace of vertical vectors at $p \in P$ as $\text{Ker}(T_p \pi)$.

Definition 9.8 *For each* $A \in \mathfrak{g}$, *we denote the corresponding vertical vector at* $p \in P$ *by*

$$A^\sharp(p) := T_e \iota_p(A) = \iota_{p,*}(A) \in \text{Vert}(T_p P).$$

Then we define

$$A^\sharp : P \to TP$$

by $p \mapsto A^\sharp(p)$ *for all* $p \in P$. *We say that* $A^\sharp$ *is the* fundamental vector field *on* P *associated to* A.

Exercise 9.12 *Prove the following:*

- $A^\sharp$ *is indeed a vector field on* P. *In particular, it is a smooth map.*
- *The map* : $\mathfrak{g} \xrightarrow{(\cdot)^\sharp} \Gamma(TP)$ *given by* $A \mapsto A^\sharp$ *is linear and preserves the Lie algebra structures:*

$$[A, B]^\sharp = [A^\sharp, B^\sharp]$$

for all $A, B \in \mathfrak{g}$, *where the Lie bracket on the left side is that in the Lie algebra* $\mathfrak{g}$, *while that on the right side is that of vector fields on* P. *In short, this is a morphism of Lie algebras.*

Chapter 10
Connections on Principal Bundles

The topic of this chapter has become standard in modern treatments of differential geometry. The very words of the title have even been incorporated into part of a common cliché: Gauge theory is a connection on a principal bundle. We will come back to this relation between physics and geometry in Chapter 14. But just on the geometry side there has been an impressive amount of results, only a fraction of which we will be able to deal with here. Sometimes we speak of the need to translate geometric terminology into physics terminology. And vice versa. Curiously, there is also a need to translate geometrical terminology developed in one context into geometrical terminology from another context. And that is especially true for this topic. In this regard, the books [4] by Choquet-Bruhat and co-authors and [46] by Spivak are quite helpful references. Also quite readable is Darling's text [6]. In an effort to keep this chapter as efficient as practically possible, we have not presented all the equivalent or closely related ways of approaching this central topic.

10.1 Connection as Horizontal Vectors

We wish to emphasize that every principal bundle has vertical vectors. The exact sequence (9.3) is somehow telling us that the directions in the base space M correspond to horizontal directions. However, unlike the case with TG, the elements in TM are not canonically identified with elements in TP. We find this to be an intuitive, geometric motivation for the following important definition.

Definition 10.1 *Let* $G \hookrightarrow P \xrightarrow{\pi} M$ *be a principal bundle. A connection in* P *is the selection for every* $p \in P$ *of a vector subspace* $\mathrm{Hor}(T_p P) \subset T_p P$ *such that*

1. $\mathrm{Vert}(T_p P) + \mathrm{Hor}(T_p P) = T_p P$.
2. $\mathrm{Vert}(T_p P) \cap \mathrm{Hor}(T_p P) = 0$, *the zero subspace*.
3. *The subspace* $\mathrm{Hor}(T_p P)$ *depends smoothly on* $p \in P$.
4. *The* G-*invariance condition:*

© Springer International Publishing Switzerland 2015
S.B. Sontz, *Principal Bundles*, Universitext, DOI 10.1007/978-3-319-14765-9_10

$R_{g,*}\big(\mathrm{Hor}(T_pP)\big) = \mathrm{Hor}(T_{pg}P)$ *for all* $g \in G$ *and* $p \in P$.

We say that the elements of $\mathrm{Hor}(T_pP)$ *are* horizontal *vectors of* P *at* p *(with respect to the connection).*

In the usual terminology of linear algebra, the first two conditions in this definition say that $\mathrm{Hor}(T_pP)$ is a *complementary subspace* to the vertical subspace $\mathrm{Vert}(T_pP)$ in T_pP. We will write this as

$$\mathrm{Vert}(T_pP) \oplus \mathrm{Hor}(T_pP) = T_pP. \qquad (10.1)$$

However, this is not a direct sum representation of the vector space T_pP in terms of orthogonal subspaces. The space T_pP has not been provided with an inner product and so the concept of orthogonality does not make sense in this context. Rather, we have a direct sum representation in the category of vector spaces. It does follow immediately that $\dim(\mathrm{Hor}(T_pP)) = \dim M$ for all $p \in P$.

The third condition uses the definition of a smooth distribution in a tangent bundle, as explained in the chapter on the Frobenius theorem. We note that $\mathrm{Vert}(T_pP)$ depends smoothly on the point $p \in P$, and so the third condition seems reasonable since it requires the same of its complementary subspace.

The fourth condition will seem more reasonable in a moment. But first we will explain its name. We can define a subspace $\mathrm{Hor}(TP) := \cup_{p\in M}\mathrm{Hor}(T_pP)$ of $TP = \cup_{p\in M}(T_pP)$. This is called the *subbundle of all horizontal vectors*. Now the maps $R_g : P \to P$ induce maps $R_{g,*} : TP \to TP$, and this makes TP into a G-space. Then the fourth condition is equivalent to saying that $\mathrm{Hor}(TP)$ is a G-invariant subbundle of TP, which is both a G-space and a bundle.

Exercise 10.1 *Show that* $T\pi$ *restricted to* $\mathrm{Hor}(T_pP)$ *gives an isomorphism of vector spaces*

$$\mathrm{Hor}(T_pP) \to T_xM$$

for all $p \in P$, *where* $x = \pi(p)$.

The next result shows how two canonical diffeomorphisms of G onto the same fiber are related by the group action.

Proposition 10.1 *Let* $G \hookrightarrow P \xrightarrow{\pi} M$ *be a principal bundle. Then*

$$R_g \circ \iota_p = \iota_{pg} \circ ad_{g^{-1}}$$

for all $g \in G$ *and* $p \in P$. *Equivalently, the following diagram commutes:*

$$
\begin{array}{ccc}
G & \xrightarrow{\;ad_{g^{-1}}\;} & G \\
{\scriptstyle \iota_p}\big\downarrow & & \big\downarrow{\scriptstyle \iota_{pg}} \\
P & \xrightarrow[\;R_g\;]{} & P
\end{array}
\qquad (10.2)
$$

Proof: We pick an arbitrary element $h \in G$ in the object in the upper left corner of the diagram and chase it. First, we go down and then across:

$$h \xrightarrow{\iota_p} ph \xrightarrow{R_g} (ph)g = phg.$$

Next, we go across and then down:

$$h \xrightarrow{ad_{g^{-1}}} g^{-1}hg \xrightarrow{\iota_{pg}} (pg)(g^{-1}hg) = p(hg) = phg.$$

So the chase shows that the diagram (10.2) does commute. ∎

The last proposition shows that $g \mapsto ad_{g^{-1}}$ is the right action of G on G corresponding to the right action $g \mapsto R_g$ of G on P. Now we show that the vertical subspaces always satisfy the G-invariance property.

Proposition 10.2 *The family of vertical vector subspaces of the principal bundle* $G \hookrightarrow P \xrightarrow{\pi} M$ *satisfies the G-invariance condition; namely,*

$$R_{g,*}\big(\mathrm{Vert}(T_p P)\big) = \mathrm{Vert}(T_{pg} P)$$

for all $g \in G$ and $p \in P$. Consequently, $\mathrm{Vert}(TP) := \cup_{p \in P} \mathrm{Vert}(T_p P)$ *is a G-invariant subbundle of TP.*

Proof: Differentiating the functions in diagram (10.2), starting at the point $e \in G$ in the object in the upper left corner, we get this commutative diagram:

$$
\begin{array}{ccc}
\mathfrak{g} & \xrightarrow{\;Ad_{g^{-1}}\;} & \mathfrak{g} \\
{\scriptstyle \iota_{p,*}}\downarrow & & \downarrow{\scriptstyle \iota_{pg,*}} \\
T_p P & \xrightarrow{\;R_{g,*}\;} & T_{pg} P
\end{array}
\qquad (10.3)
$$

By definition, the image of $\iota_{p,*}$ (resp., $\iota_{pg,*}$) is $\mathrm{Vert}(T_p P)$ (resp., $\mathrm{Vert}(T_{pg} P)$). So the commutativity of the diagram (10.3) combined with the fact that the horizontal arrows are isomorphisms implies that

$$R_{g,*}\big(\mathrm{Vert}(T_p P)\big) = \mathrm{Vert}(T_{pg} P). \qquad ∎$$

The last proposition shows that $g \mapsto Ad_{g^{-1}}$ is the right action of G on $\mathfrak{g}$ corresponding to the right action $g \mapsto R_{g*}$ of G on $T(P)$. As always, a commutative diagram can be rewritten in terms of algebraic formulas. We now do this for the preceding proposition.

Proposition 10.3 *Let $G \hookrightarrow P \xrightarrow{\pi} M$ be a principal bundle. Then*

$$R_{g,*}(A^{\sharp}(p)) = (Ad_{g^{-1}}A)^{\sharp}(pg) \qquad (10.4)$$

holds for all $g \in G$, $A \in \mathfrak{g}$, and $p \in P$.

Proof: Just chase $A \in \mathfrak{g}$ in the upper left through the diagram (10.3). ∎

Since any pair of points on the same fiber of P are related by some map R_g, the condition of G-invariance means that the distinguished subspaces (of vertical and horizontal vectors) for all points within one fiber are completely determined by what those spaces are at any single point in the fiber. While G-invariance is an interesting property that vertical vectors happen to have (and remember that the vertical vectors are an intrinsic part of the structure of a principal bundle), it is something that we *require* by definition of the horizontal vectors. Otherwise, a complementary subspace to the vertical subspace could flop about (smoothly, of course) as we move around in a fiber, while the vertical subspace simply cannot do that, per Proposition 10.2.

We would like to remark that a principal bundle is basically a topological structure although in our discussion it is seen more precisely as a differential topological structure. And as we shall see later, a connection is a geometric structure on a principal bundle. This may not be terribly obvious to the reader, especially at first reading, but we point it out since it is at this juncture that geometric ideas enter into the theory.

10.2   Ehresmann Connections

While the next formulation of a connection (as a vector-valued 1-form with values in a Lie algebra) follows directly from the considerations given in the previous section, we do not feel that it is as intuitive an approach. The same geometric ideas are encoded here, but they are not so readily visible! But this approach is extremely important and so had best be learned well.

We note that a connection on a principal bundle gives us the direct sum representation (10.1) and consequently a linear map

$$\mathrm{pr}_p : T_p P \to \mathrm{Vert}(T_p P),$$

namely, the projection onto the first summand. But the codomain of this map is isomorphic to the Lie algebra $\mathfrak{g}$ and via a canonically given isomorphism, as we already have noted in (9.4). Hence, we have a linear map for each $p \in P$:

$$T_p P \xrightarrow{\mathrm{pr}_p} \mathrm{Vert}(T_p P) \xrightarrow{(T_e \iota_p)^{-1}} \mathfrak{g}.$$

Thus, we first define a linear map $\omega_p := (T_e \iota_p)^{-1} \circ \mathrm{pr}_p : T_p P \to \mathfrak{g}$ for $p \in P$. This has somewhat of the flavor of a covector. Whereas a covector at $p \in P$ is a linear map $T_p P \to \mathbb{R}$ (that is, its codomain is the scalars), here we have a linear map on the same domain but whose codomain is a vector space, namely, $\mathfrak{g}$. Having done this for each point $p \in P$, we can next paste these together to get

$$\omega : TP \to \mathfrak{g}$$

defined by $\omega | T_p P := \omega_p$.

Exercise 10.2 *Properties of ω:*

- *Show that $\omega_p(A^\sharp(p)) = A$ for all $A \in \mathfrak{g}$ and $p \in P$.*
  This is also written as $\omega(A^\sharp) = A$ and should be understood as saying that the composition

$$P \xrightarrow{A^\sharp} TP \xrightarrow{\omega} \mathfrak{g}$$

  is the constant map whose value is always equal to $A \in \mathfrak{g}$.
- *Prove that $\ker \omega_p = \mathrm{Hor}(T_p P)$.*
- *Show that ω is a smooth function. Also, ω is linear when restricted to each fiber of TP.*

Hence, except for its codomain, ω is something like a 1-form.

Definition 10.2 *Suppose that N is a smooth manifold, W is a vector space, and $\zeta : TN \to W$ is a smooth map that is linear when restricted to each fiber of TN. Then we say that ζ is a W-valued 1-form on N. We denote the set of all W-valued 1-forms on N by $\Omega^1(N; W)$.*

Exercise 10.3 *Continuing the notation of the previous definition, suppose that $W = \mathbb{R}$. Show that in this case ζ can be identified canonically with a 1-form on N, that is, a section of the cotangent bundle of N.*

Moreover, show that every 1-form on N gives us such a ζ.

In summary, we have $\Omega^1(N; \mathbb{R}) = \Omega^1(N)$.

So a connection on a principal bundle determines (via the construction given above) a $\mathfrak{g}$-valued 1-form ω on the total space P of the bundle. So far, we have encoded the first three defining properties of a connection into the construction of the $\mathfrak{g}$-valued 1-form ω. In order to present the G-invariance property in the usual way, we introduce a definition. Note that this definition is the obvious generalization of the case $k = 1$ of Definition 5.4.

Definition 10.3 *Suppose that N_1 and N_2 are smooth manifolds, ζ is a W-valued 1-form on N_2, and $f : N_1 \to N_2$ is a smooth map. Then the pullback of ζ by f is defined as*

$$f^*(\zeta) := \zeta \circ Tf = \zeta \circ f_* : TN_1 \to W,$$

namely, the composition

$$TN_1 \xrightarrow{Tf = f_*} TN_2 \xrightarrow{\zeta} W.$$

Exercise 10.4 *Show that* $\Omega^1(N; W)$ *is a vector space over* $\mathbb{R}$. *In particular, the reader is asked to define the sum and the scalar multiplication. Show that* $\Omega^1(N; W)$ *is infinite dimensional in general. Explain exactly which cases are exceptional.*

Exercise 10.5 *Show that the pullback is functorial (in the contravariant sense), namely, that these three properties hold:*

- *For every smooth* $f : N_1 \rightarrow N_2$ *and every* $\zeta \in \Omega^1(N_2; W)$, *show that the pullback* $f^*(\zeta)$ *is a* W-*valued 1-form on* N_1. *Moreover, we have that* $f^* : \Omega^1(N_2; W) \rightarrow \Omega^1(N_1; W)$ *is linear. So the excruciatingly correct categorical notation for* f^* *is* $\Omega^1(f; W)$. *But this is just too cumbersome!*
- $1_N^*(\zeta) = \zeta$ *for* $\zeta \in \Omega^1(N; W)$.
- $(f \circ g)^* \zeta = g^*(f^*(\zeta))$ *for* $N_0 \xrightarrow{g} N_1 \xrightarrow{f} N_2$ *and* $\zeta \in \Omega^1(N_2; W)$.

The most widely used notation for pullbacks (of all sorts) is the upper index star. This started out as a generic notation for *contravariant functors* while the lower index star indicated a *covariant functor*. This is why we are now using f_* to denote the covariant functor Tf.

Theorem 10.1 *Let* ω *be the* $\mathfrak{g}$-*valued 1-form on* P *constructed as above from a connection on a principal bundle* $G \hookrightarrow P \xrightarrow{\pi} M$. *Then*

$$R_g^* \omega = Ad_{g^{-1}} \circ \omega \tag{10.5}$$

holds for every $g \in G$.

Proof: We have written the result as formula (10.5) since this is how it is usually presented. However, we prefer diagrams to formulas when the latter are more transparent and easier to remember. So we would like to express equation (10.5) as a diagram. First, we note that by using the definition of pullback, we can rewrite the left-hand side: $R_g^* \omega = \omega \circ R_{g,*}$. [The reader should notice that pullback is an important concept, but we have introduced it here only to write the formula (10.5). And now we are eliminating pullback from the discussion.] Thus, equation (10.5) is equivalent to the commutativity of the following diagram for every $g \in G$:

$$
\begin{array}{ccc}
TP & \xrightarrow{R_{g,*}} & TP \\
\downarrow{\omega} & & \downarrow{\omega} \\
\mathfrak{g} & \xrightarrow{Ad_{g^{-1}}} & \mathfrak{g}
\end{array}
\tag{10.6}
$$

Since the top arrow $R_{g,*}$ in this diagram respects the decomposition

$$TP = \text{Hor}(TP) \oplus \text{Vert}(TP)$$

by Definition 10.1 and Proposition 10.2, it is sufficient (and necessary, too) to show the commutativity of the following two diagrams for every $g \in G$:

$$\begin{array}{ccc}
\mathrm{Hor}(TP) & \xrightarrow{\ R_{g,*}\ } & \mathrm{Hor}(TP) \\
\Big\downarrow \omega & & \Big\downarrow \omega \\
\mathfrak{g} & \xrightarrow{\ Ad_{g^{-1}}\ } & \mathfrak{g}
\end{array} \qquad (10.7)$$

and

$$\begin{array}{ccc}
\mathrm{Vert}(TP) & \xrightarrow{\ R_{g,*}\ } & \mathrm{Vert}(TP) \\
\Big\downarrow \omega & & \Big\downarrow \omega \\
\mathfrak{g} & \xrightarrow{\ Ad_{g^{-1}}\ } & \mathfrak{g}
\end{array}. \qquad (10.8)$$

Now the arrows labeled ω in the diagram (10.7) are zero since the kernel of ω is exactly the horizontal subspace. Hence, the diagram (10.7) commutes.

Next, the diagram (10.8) commutes if and only if

$$\begin{array}{ccc}
\mathrm{Vert}(T_p P) & \xrightarrow{\ R_{g,*}\ } & \mathrm{Vert}(T_{pg} P) \\
\Big\downarrow \omega & & \Big\downarrow \omega \\
\mathfrak{g} & \xrightarrow{\ Ad_{g^{-1}}\ } & \mathfrak{g}
\end{array} \qquad (10.9)$$

commutes for every $p \in P$. Now we propose taking an arbitrary element $A^{\sharp}(p)$ in $\mathrm{Vert}(T_p P)$ (where $A \in \mathfrak{g}$) and chasing it through the diagram. Going down and then across, we have

$$A^{\sharp}(p) \xrightarrow{\ \omega\ } A \xrightarrow{\ Ad_{g^{-1}}\ } Ad_{g^{-1}}(A).$$

Now we go across [using equation (10.4)] and then down:

$$A^{\sharp}(p) \xrightarrow{\ R_{g,*}\ } R_{g,*} A^{\sharp}(p) = (Ad_{g^{-1}}(A))^{\sharp}(pg) \xrightarrow{\ \omega\ } Ad_{g^{-1}}(A).$$

And this finishes the diagram chase argument for the diagram (10.9).

For those who prefer to avoid the algebraic details of a diagram chase, one can prove that (10.9) commutes by noting that the vertical arrows are isomorphisms and that, when they are replaced by their inverses, (10.9) becomes (10.3). The reader should realize that this is really the same argument as the diagram chase.   ∎

We have arrived at the following important definition.

Definition 10.4 *Let* $G \hookrightarrow P \xrightarrow{\ \pi\ } M$ *be a principal bundle. Then we say that an Ehresmann connection* on this bundle *is a* $\mathfrak{g}$-*valued* 1-*form* ω *on* P *that satisfies the* G-*equivariance condition*

$$R_g^* \omega = Ad_{g^{-1}} \circ \omega \qquad (10.10)$$

for all $g \in G$, which is equivalent to the commutativity of the diagram

$$
\begin{array}{ccc}
TP & \stackrel{R_{g,*}}{\longrightarrow} & TP \\
\downarrow{\scriptstyle\omega} & & \downarrow{\scriptstyle\omega} \\
\mathfrak{g} & \stackrel{Ad_{g^{-1}}}{\longrightarrow} & \mathfrak{g}
\end{array}
\tag{10.11}
$$

as well as the condition

$$\omega(A^\sharp) = A$$

for all $A \in \mathfrak{g}$ or, equivalently, that the composition

$$P \stackrel{A^\sharp}{\longrightarrow} TP \stackrel{\omega}{\longrightarrow} \mathfrak{g}$$

is the constant function $P \ni p \to A \in \mathfrak{g}$.

Other names *for an Ehresmann connection are* connection form *and* gauge connection.

Diagram (10.11) is the reason that equation (10.10) comes to be called the *G-equivariance condition.* Since TP and $\mathfrak{g}$ are G-spaces whose right actions are given by the horizontal arrows in the diagram (10.11), the commutativity of this diagram is equivalent to saying that $\omega : TP \to \mathfrak{g}$ is a G-equivariant map. Notice that the presence of g^{-1} in $Ad_{g^{-1}}$ gives us a right action on $\mathfrak{g}$. The assignment $g \mapsto Ad_g$ is a left action.

One important property of Ehresmann connections is preservation under pull-backs.

Theorem 10.2 *Suppose that $\pi_1 : P_1 \to M_1$ and $\pi_2 : P_2 \to M_2$ are principal bundles, each with the same structure Lie group G. Let $F : P_1 \to P_2$ be a morphism of principal G-bundles covering a smooth map $f : M_1 \to M_2$. Say ω is an Ehresmann connection on P_2. Then $F^*\omega$ is an Ehresmann connection on P_1.*

Proof: We already know that $F^*\omega = \omega \circ F_*$ is a $\mathfrak{g}$-valued 1-form on P_2. So it remains to check the two defining properties of an Ehresmann connection.

We take $A \in \mathfrak{g}$ and let $A^{\sharp 1}$ (resp., $A^{\sharp 2}$) denote the fundamental vertical vector field on P_1 (resp., on P_2). Then we take $p \in P_1$ and calculate

$$
\begin{aligned}
F^*\omega(A^{\sharp 1}(p)) &= F^*\omega(\iota_{p*}(A)) \\
&= \omega(F_*\iota_{p*}(A)) \\
&= \omega\big((F\iota_p)_*(A)\big) \\
&= \omega\big((\iota_{F(p)})_*(A)\big) \\
&= \omega\big(A^{\sharp 2}(F(p))\big) \\
&= A,
\end{aligned}
$$

where in the last equation we used the hypothesis that ω is an Ehresmann connection on P_2. We also used $F \iota_p = \iota_{F(p)} : G \to P_2$. So $F^*\omega$ satisfies one of the defining properties of an Ehresmann connection on P_1.

For the other property, let R_g^1 (resp., R_g^2) denote the right action of $g \in G$ on P_1 (resp., P_2). Then we calculate

$$
\begin{aligned}
R_g^{1*}(F^*\omega) &= (F^*\omega)R_{g*}^1 \\
&= \omega F_* R_{g*}^1 \\
&= \omega(FR_g^1)_* \\
&= \omega(R_g^2 F)_* \\
&= \omega R_{g*}^2 F_* \\
&= (R_g^{2*}\omega)F_* \\
&= (Ad_{g^{-1}}\omega)F_* \\
&= Ad_{g^{-1}}(\omega F_*) \\
&= Ad_{g^{-1}}(F^*\omega).
\end{aligned}
$$

Here we used $FR_g^1 = R_g^2 F$, which is true since F is a G-morphism. Also, we used the hypothesis that ω is an Ehresmann connection on P_2 in the seventh equality. This calculation shows that $F^*\omega$ satisfies the other defining property of an Ehresmann connection.

Putting these last two parts of the proof together says that $F^*\omega$ is an Ehresmann connection on P_1. ∎

10.3 Examples

One motivation for the theory of connections on principal bundles comes from classical differential geometry. The example there is the principal bundle associated to the tangent bundle of a smooth manifold. Then objects such as the *Levi–Civita connection* on a semi-Riemannian manifold and any *Koszul connection* are examples *related to* an Ehresmann connection as we have presented it. See [46] for details.

There are also two trivial examples. The first is the trivial Lie group $G = \{e\}$ whose Lie algebra is $\mathfrak{g} = \{0\}$, the trivial vector space. In this case, each fiber contains exactly one point, so that the total space can be identified with the base space. So the principal bundle is $\{e\} \hookrightarrow M \xrightarrow{1_M} M$, where M is an arbitrary smooth manifold. It follows that $\text{Vert}(T_p M) = 0$ for every $p \in M$, the total space. But there is only one complementary subspace to this in $T_p M$, namely, all of $T_p M$.

Moreover, the definition $\text{Hor}(T_pM) := T_pM$ does in fact define a connection; that is, it is smooth and G-invariant. And its associated Ehresmann connection is clearly $\omega : TM \to \{0\}$, which is the unique $\{0\}$-valued 1-form on M. So we end up studying an arbitrary smooth manifold M with no other structure.

The second trivial example is $M = \{x\}$, the smooth manifold consisting of exactly one point, say x. Then there is exactly one fiber in the total space, namely, the fiber above x, and it is a copy of the Lie group. So the total space can be identified with G. Then the principal bundle looks like $G \hookrightarrow G \to \{x\}$. As usual, the map $G \hookrightarrow G$ is *not* unique. Now, of course, we are simply studying an arbitrary Lie group G. The vertical vector space at $g \in G$, the total space, satisfies $\text{Vert}(T_gG) = T_gG$. But there is only one complementary subspace to this in T_gG, namely, the zero subspace. Moreover, the definition $\text{Hor}(T_gG) := \{0\}$ does, in fact, define a connection; that is, it is smooth and G-invariant. Its associated Ehresmann connection is denoted $\omega : TG \to \mathfrak{g}$.

Now we want to evaluate $\omega(v)$ for $v \in T(G)$. Since $\text{Vert}(TG) = TG$, we simply have $\omega(v) = v$, the projection onto the vertical subspace. But if $v \in T_g(G)$, we have $\omega(v) \in T_g(G)$, which is not the correct space. The point is that we have to identify the isomorphic spaces $T_g(G)$ and $\mathfrak{g}$. We do this in the canonical way using the Maurer–Cartan form Ω. Restricted to $T_g(G)$, the Maurer–Cartan form satisfies $\Omega = L_{g^{-1}*}$, the derivative of the map of multiplication on the left by g^{-1}. So we identify the Ehresmann connection as

$$\omega(v) = L_{g^{-1}*}(v) = \Omega(v) \in \mathfrak{g}.$$

This quite brief argument shows that $\omega = \Omega$. In particular, the Maurer–Cartan form Ω is an Ehresmann connection. We remind the reader of the definition from Lie group theory.

Definition 10.5 *Let G be a Lie group with Lie algebra $\mathfrak{g}$. Then there is a $\mathfrak{g}$-valued 1-form on G, denoted by*

$$\Omega : TG \to \mathfrak{g},$$

and defined by $\Omega_g := (L_{g^{-1}})_* : T_gG \to \mathfrak{g}$ for every $g \in G$.
It is called the Maurer–Cartan form *on G.*

We have just proved the next proposition. But since it is a central result in geometry, we will give a direct proof as well.

Proposition 10.4 *Suppose G is a Lie group with Lie algebra $\mathfrak{g}$. Then its Maurer–Cartan form Ω is an Ehresmann connection in the principal bundle $G \hookrightarrow G \to \{x\}$.*

Proof: Clearly, Ω is a $\mathfrak{g}$-valued 1-form. So we have to show the two defining properties of an Ehresmann connection.

First, we consider $\Omega(A^\sharp(g))$ for $g \in G$ and $A \in \mathfrak{g}$. To do this, we consider $g \in G$ as a point in the total space. Then $\iota_g : G \to G$ maps the structure Lie group into the total space. Recalling the definition of ι_g, we see for all $h \in G$ that

$$\iota_g(h) = gh = L_g(h).$$

So $\iota_g = L_g$, the left multiplication map by $g \in G$, which in turn gives us

$$A^\sharp(g) = \iota_{g*}(A) = L_{g*}(A) \in T_g(G).$$

Then

$$\Omega(A^\sharp(g)) = L_{g^{-1}*}\big(L_{g*}(A)\big) = A,$$

which is one of the defining properties of an Ehresmann connection.

For the other defining property of an Ehresmann connection, we take $h \in G$ and consider $(R_h^* \Omega)(v) = \Omega(R_{h*}v)$ for $v \in T_g(G)$. So $R_{h*}v \in T_{gh}(G)$. So we see that

$$\begin{aligned}
(R_h^* \Omega)(v) &= \Omega(R_{h*}v) \\
&= L_{(gh)^{-1}*}(R_{h*}v) \\
&= L_{(h^{-1}g^{-1})*} R_{h*} v \\
&= L_{h^{-1}*} L_{g^{-1}*} R_{h*} v \\
&= L_{h^{-1}*} R_{h*} L_{g^{-1}*} v \\
&= Ad_{h^{-1}} \Omega(v).
\end{aligned}$$

Since this holds for all $v \in T_g(G)$, we have shown the other property of an Ehresmann connection. ∎

The Maurer–Cartan form Ω is closely associated with the fact that the tangent bundle of a Lie group is trivial and, in fact, has a distinguished trivialization. Consider the commutative diagram

$$\begin{array}{ccc}
TG & \xrightarrow{\tilde{\Omega}} & G \times \mathfrak{g} \\
\downarrow{\scriptstyle \tau_G} & & \downarrow{\scriptstyle \pi_1} \\
G & \xrightarrow{1_G} & G,
\end{array}$$

where $\tilde{\Omega}(p) := (\tau_G(p), \Omega(p))$ for all $p \in TG$. To make this diagram commute, we were forced to put the first entry in $\tilde{\Omega}(p)$ equal to $\tau_G(p)$. Thus, the map $\tilde{\Omega}$ is really determined by its second entry, namely, the Maurer–Cartan form Ω.

Exercise 10.6 *Prove that the preceding diagram commutes and that it provides a trivialization of the tangent bundle TG.*

Exercise 10.7 *The maps Ω and $\tilde{\Omega}$ are intrinsically associated with the Lie group G. Actually, we sometimes incorporate this relation into the notation and write Ω^G and $\tilde{\Omega}^G$, respectively, for these maps. Suppose that H is also a Lie group. Let $\mathfrak{h}$ denote its Lie algebra. Suppose that $\phi : G \to H$ is a smooth homomorphism.*

- *Show the commutativity of*

$$TG \xrightarrow{\Omega^G} \mathfrak{g}$$
$$\downarrow T\phi \qquad \downarrow \phi_*$$
$$TH \xrightarrow{\Omega^H} \mathfrak{h}$$

- *Show the commutativity of*

$$TG \xrightarrow{\tilde{\Omega}^G} G \times \mathfrak{g}$$
$$\downarrow T\phi \qquad \downarrow \phi \times \phi_*$$
$$TH \xrightarrow{\tilde{\Omega}^H} H \times \mathfrak{h}$$

- *The two preceding results show the naturality of Ω and $\tilde{\Omega}$. Learn how to say this using the concept of* natural transformation. *(Take a quick look into a book on category theory, if needed.)*

Here is an important result, often overlooked in introductory texts, about Ehresmann connections on a Lie group G. (But this is presented in Mayer's part of [10].) It provides a complete classification of all of the Ehresmann connections on a Lie group.

Theorem 10.3 *Let G be a Lie group. Then the Maurer–Cartan form on G is the only Ehresmann connection on the principal bundle $G \hookrightarrow G \to \{x\}$.*

Proof: We already have the existence of one Ehresmann connection on G, namely, its Maurer–Cartan form. So we only need to show uniqueness. Hence, let $\omega : T(G) \to \mathfrak{g}$ be an Ehresmann connection on G. Then we have this commutative diagram for all $g \in G$:

$$T_e(G) \xrightarrow{R_{g,*}} T_g(G)$$
$$\downarrow \omega \qquad \qquad \downarrow \omega .$$
$$\mathfrak{g} \xrightarrow{Ad_{g^{-1}}} \mathfrak{g}$$

Since the horizontal arrows are isomorphisms, this shows that ω restricted to $T_g(G)$ for any $g \in G$ (the object on the upper right) is completely determined by ω restricted to $T_e(G) \equiv \mathfrak{g}$.

So next take $A \in \mathfrak{g}$. Thinking of G as the total space of the principal bundle $G \to \{*\}$, we have the fundamental vector field $A^\sharp$ on G. Hence, $A^\sharp : G \to T(G)$ is a section of the tangent bundle of the total space. By definition,

$$A^\sharp(e) = \iota_{e*}(A).$$

But $\iota_e : G \to G$ is $\iota_e(g) = ge = g = id(g)$, and so

$$\iota_{e*}(A) = id_*(A) = A.$$

In short, $A = A^\sharp(e)$. Now we have this sequence of equalities:

$$\omega(A) = \omega(A^\sharp(e)) = A.$$

The first equality is okay by what we have just shown. And the second equality holds since ω is assumed to be an Ehresmann connection on G. So the arbitrary Ehresmann connection ω on G is completely determined on $T_e(G)$. And so it is completely determined on all of $T(G)$. Therefore, it is unique, as we claimed. ∎

Exercise 10.8 *The reader is requested to check that the Maurer–Cartan form* $\Omega :$ *$T(G) \to \mathfrak{g}$ when restricted to $T_e(G) \equiv \mathfrak{g}$ is the identity.*

Exercise 10.9 *Evaluate in terms of their matrix entries the Maurer–Cartan forms of the classical matrix Lie groups such as $GL(n), O(n), U(n)$, and so on.*

There are many more examples of principal bundles. One rather large and useful class of them arises by considering a closed Lie subgroup H of a Lie group G. Then the quotient space G/H (say, of left cosets) can be made into a differential manifold, which already is a rather nice theorem. But moreover, the quotient map $\pi : G \to G/H$ is a principal bundle whose structure Lie group is H. This is an extremely pretty result. However, the proofs of these assertions would lead us very far afield, and so we leave it to the interested readers to consider this further on their own.

The Hopf bundles also provide important examples of principal bundles. The Hopf bundle most commonly considered is $\pi : S^3 \to S^2$ with structure Lie group S^1. Here S^k denotes the standard k-dimensional sphere, defined as the unit vectors in $\mathbb{R}^{k+1}$, the $(k+1)$-dimensional Euclidean space equipped with the standard Euclidean norm. In turns out that S^3 can be viewed as the unit quaternions in $\mathbb{H} = \mathbb{R}^4$ and as such is a Lie group. We can also embed the complex plane $\mathbb{C}$ into the quaternions (say, $a + bi \mapsto a + bi$) and use that to embed the Lie subgroup S^1 of $\mathbb{C}$ into $\mathbb{H}$. This will realize S^1 as a closed Lie subgroup of S^3. So the general theory described in the previous paragraph can be applied here.

The "tricky bit" is to identify the quotient space as S^2, which, by the way, is not a Lie group. However, S^3 is the set of unit vectors in $\mathbb{R}^4 \cong \mathbb{C}^2$. And the action of S^1 on S^3 gives orbits that are exactly the equivalence classes that arise in the construction of the complex projective space $\mathbb{C}P^1$ from $\mathbb{C}^2$. Finally, $\mathbb{C}P^1$ can be identified with the one-point compactification of $\mathbb{C}$, which is the sphere S^2. This last step is done by putting the usual coordinates on $\mathbb{C}P^1 \setminus \{*\}$, where $\{*\}$ is any point in $\mathbb{C}P^1$ though the typical choice is to take the point to be the equivalence class of $(1, 0)$ [resp. $(0, 1)$] to get an atlas for $\mathbb{C}P^1$ with 2 charts. Then one sees that $\mathbb{C}P^1 \setminus \{*\}$ is diffeomorphic to the complex plane $\mathbb{C}$ in such a way that the excluded point $\{*\}$ corresponds to the "point at infinity" of the one-point compactification of $\mathbb{C}$, which is S^2.

Recall that, in general, $\mathbb{C}P^n$ for any integer $n \geq 1$ is defined as the space of complex lines in $\mathbb{C}^{n+1}$ passing through the origin. We implicitly used this fact in the previous paragraph in the case $n = 1$. Each such complex line (being actually a copy of the complex plane) intersects the unit sphere S^{2n+1} in $\mathbb{C}^{n+1}$ in a circle, which is a copy of S^1. And the family of these circles in S^{2n+1} is in one-to-one correspondence with the points of $\mathbb{C}P^n$. Actually, these circles are the orbits of the natural action of S^1 on S^{2n+1} given by $x \mapsto \lambda x$ for $x \in S^{2n+1}$ and $\lambda \in S^1$. In sum, we get a principal bundle $S^{2n+1} \to \mathbb{C}P^n$ with structure Lie group S^1. This is called a *Hopf bundle*.

The Hopf bundles are not trivial. This can be seen in the case $n = 1$ by using topological invariants to prove that S^3 and $S^2 \times S^1$ do not even have the same homotopy type and so cannot be diffeomorphic spaces. For example, the first homotopy group of S^3 is trivial, $\pi_1(S^3) = 0$, while the first homotopy group of $S^2 \times S^1$ is isomorphic to the group of integers, $\pi_1(S^2 \times S^1) \cong \mathbb{Z}$. Similar topological arguments can be made for $n > 1$. Again, we leave it to the interested reader to delve into this in greater detail.

10.4 Horizontal Lifts

We continue the discussion of a principal bundle with a given connection by introducing horizontal lifts. The geometric idea behind this is that the various horizontal subspaces in $T(P)$ are considered to be parallel copies of each other. So as we move from point to point in the base M (say, along a curve in M), we can think of the horizontal subspaces above those points as moving with us too. The next theorem and its corollary say just that in a rigorous way but require more machinery even to write things down.

Specifically, it says we can find motions "upstairs" in P that project down to a given motion in M and such that the corresponding velocity in P is a horizontal vector at every moment during the motion. (For motion, read: curve.) These motions in P will be called horizontal lifts. Moreover, if we also require that the motion in P starts at a given point in P, then the horizontal lift is unique. If this all sounds something like existence and uniqueness of solutions to ordinary differential equations (ODEs), that's because that's what it is!

Theorem 10.4 *Let $G \hookrightarrow P \xrightarrow{\pi} M$ be a principal bundle provided with an Ehresmann connection ω. Suppose that $\gamma : [0,1] \to M$ is a smooth curve and $p_0 \in \pi^{-1}(\gamma(0))$ is any point in the fiber above $\gamma(0) \in M$. Then there exists a unique curve $\tilde{\gamma}_{p_0} = \tilde{\gamma} : [0,1] \to P$ such that*

(a) $\tilde{\gamma}(0) = p_0$.
(b) $\pi \circ \tilde{\gamma} = \gamma$ or diagrammatically:

$$\begin{array}{ccc} & & P \\ & \tilde{\gamma} \nearrow & \downarrow \pi \\ [0,1] & \xrightarrow{\gamma} & M \end{array}$$

(One says that $\tilde{\gamma}$ lifts γ.)

(c) $\tilde{\gamma}'(t) \in \mathrm{Hor}(TP)$ *for all* $t \in [0, 1]$. *At the endpoints of* $[0, 1]$, *the derivative is understood to be one-sided. (One says that $\tilde{\gamma}$ is* horizontal.*)*

Proof: The idea is to reduce this to an ODE whose solution is $\tilde{\gamma}$. Then one applies the theory of ODEs to get the result. But what is the ODE? Well, $\tilde{\gamma}(t)$ is horizontal if and only if $\omega(\tilde{\gamma}'(t)) = 0$. We start in the special case when the image of γ is contained in a chart U over which P is trivial. Without loss of generality, we take $\pi^{-1}(U) = U \times G$ and $p_0 = (\gamma(0), e)$. Also, we do not lose generality by assuming that the tangent bundle of M is also trivial over U. At this point, we make yet another assumption (and do lose some generality), namely, that G is a matrix Lie group.

Then any $\tilde{\gamma}$ lifting γ to $U \times G$ can be written as

$$\tilde{\gamma}(t) = (\gamma(t), g(t))$$

for some unique $g(t) \in G$. Putting $t = 0$ into this, we see that $g(0) = I$, the identity matrix. Since the curve γ is given, the unknown we have to solve for is $g(t)$.

We see by differentiating the last equation that

$$\tilde{\gamma}'(t) = (\gamma(t), g(t), \gamma'(t), g'(t)) \in TU \times TG,$$

where we write first the element $(\gamma(t), g(t))$ in the base space $U \times G$ and then the element $(\gamma'(t), g'(t))$ in the vector space fiber above that point.

So applying $\omega : TP = TU \times TG \to \mathfrak{g}$ to the previous equation and using $\omega(\tilde{\gamma}'(t)) = 0$, we get

$$\omega(\gamma(t), g(t), 0, g'(t)) = -\omega(\gamma(t), g(t), \gamma'(t), 0),$$

an equation with both sides in the finite-dimensional vector space $\mathfrak{g}$.

Since $g(t)$ is a vertical curve in $U \times G$, the Ehresmann connection ω acts as the Maurer–Cartan form applied to its tangent vector $(g(t), g'(t)) \in TG$. This result does not depend on $\gamma(t) \in U$. So we get

$$g(t)^{-1}g'(t) = -\omega(\gamma(t), g(t), \gamma'(t), 0).$$

But for the right side we have

$$\omega(\gamma(t), g(t), \gamma'(t), 0) = \omega\big(R_{g(t)*}(\gamma(t), e, \gamma'(t), 0)\big)$$

$$= Ad_{g(t)^{-1}}\big(\omega(\gamma(t), e, \gamma'(t), 0)\big)$$

$$= g(t)^{-1}\big(\omega(\gamma(t), e, \gamma'(t), 0)\big)g(t).$$

Putting these together, we arrive at the ODE for $g(t)$:

$$g'(t) = -\omega(\gamma(t), e, \gamma'(t), 0)g(t). \tag{10.12}$$

This is an equation of $n \times n$ matrices for some integer n. The initial condition again is $g(0) = e = I$, the $n \times n$ identity matrix. This is a *linear* ODE in the unknown $n \times n$ matrix $g(t)$, and by the theory of such equations, we know that there exists a unique solution that is *global* and whose values lie in the vector space of $n \times n$ matrices. (The solution can be written using the formalism of *path-ordered integrals* for those who know what those words mean.)

An issue remaining still is whether the solution of (10.12) lies in the closed Lie subgroup G. Of course, the initial condition $g(0) = I$ guarantees that the solution starts in G. But why should it stay in G? After all, many ODEs with this same initial condition will not have a solution in G. So what is special about (10.12) that makes its solution lie in G? Well, it is because the matrix multiplying $g(t)$ on the right side of (10.12) lies in $\mathfrak{g}$, the Lie algebra of G. The details are left to the reader. ∎

This proof gives the basic idea, but only under additional assumptions. The reader can find the details in the proof of Proposition 3.1 in Volume I of [30].

We say that the curve $\tilde{\gamma}$ is the *horizontal lift of γ* or, more precisely, that $\tilde{\gamma}_{p_0}$ is the *horizontal lift of γ starting at $p_0 \in \pi^{-1}(\gamma(0))$*.

Corollary 10.1 *Using the same hypotheses and notation as in the preceding theorem, we define the map $Par_t^\gamma : \pi^{-1}(\gamma(0)) \to \pi^{-1}(\gamma(t))$ by $p_0 \mapsto \tilde{\gamma}_{p_0}(t)$ for all $t \in [0, 1]$.*

Then Par_t^γ is G-equivariant and maps the fiber $\pi^{-1}(\gamma(0))$ diffeomorphically onto the fiber $\pi^{-1}(\gamma(t))$ in P.

Proof: Since ODEs have solutions that are smooth in the initial conditions, the map Par_t^γ is smooth. The corresponding map that uses the part of the curve from $\gamma(0)$ to $\gamma(t)$, but with the time parameter reversed, gives the smooth inverse of Par_t^γ.

To show the G-equivariance, take $g \in G$ and $p_0 \in \pi^{-1}(\gamma(0))$. Hence, for all $t \in [0, 1]$,

$$Par_t^\gamma(p_0) = \tilde{\gamma}_{p_0}(t) \quad \text{and} \quad Par_t^\gamma(p_0 g) = \tilde{\gamma}_{p_0 g}(t)$$

by the definition of Par_t^γ. But then $t \mapsto R_g Par_t^\gamma(p_0) = R_g \tilde{\gamma}_{p_0}(t)$ is a curve in P covering

$$\pi(R_g \tilde{\gamma}_{p_0}) = \pi(\tilde{\gamma}_{p_0}) = \gamma.$$

Putting $t = 0$ shows that $R_g \tilde{\gamma}_{p_0}(t)$ starts at

$$R_g \tilde{\gamma}_{p_0}(0) = R_g p_0 = p_0 g.$$

Finally,

$$\frac{d}{dt}\left(R_g \tilde{\gamma}_{p_0}(t)\right) = R_{g,*}\left(\tilde{\gamma}_{p_0}'(t)\right)$$

is horizontal for all $t \in [0, 1]$ since $\tilde{\gamma}_{p_0}'(t)$ is horizontal and the horizontal subspaces are invariant under the right action. In short, $R_g \tilde{\gamma}_{p_0}(t)$ satisfies the three properties that uniquely characterize $\tilde{\gamma}_{p_0 g}(t)$. So we then see that $R_g \tilde{\gamma}_{p_0}(t) = \tilde{\gamma}_{p_0 g}(t)$. Equivalently, in the notation of the parallel transport map, we have that

$$R_g Par_t^\gamma (p_0) = Par_t^\gamma (p_0 g) = Par_t^\gamma (R_g p_0).$$

Since this is true for all p_0, we conclude that $R_g Par_t^\gamma = Par_t^\gamma R_g$, which exactly says that Par_t^γ is a G-equivariant map. ∎

The ODE (10.12) has some more interesting structure, which we now present. We do all of this for a globally trivial principal bundle, but, as usual, everything applies equally well in any local trivialization of a general principal bundle. So we consider the trivial principal bundle $\pi_1 : P = M \times G \to M$, where π_1 is the projection onto the first factor. Its canonical section $s : M \to M \times G$ is given by $s(x) = (x, e)$ for $x \in M$. We want to pull back $\omega : T(M \times G) \to \mathfrak{g}$, an arbitrary Ehresmann connection, by s. Thus, first we note that the derivative $s_* : TM \to T(M \times G) = TM \times TG$ is

$$s_*(x, v) = (x, e, v, 0),$$

where we are taking $(x, v) \in TM$ in some chart. Thus, $x \in M$ and $v \in \mathbb{R}^m$ is a tangent vector to M at x for $m = \dim M$. Let $\gamma : [0, 1] \to M$ be a curve. Its derivative $T\gamma : [0, 1] \times \mathbb{R} \to TM$ is then determined by $T\gamma(t, 1) = (\gamma(t), \gamma'(t))$. So we pull back, getting

$$s^*\omega(T\gamma(t, 1)) = \omega(s_*(\gamma(t), \gamma'(t))) = \omega(\gamma(t), e, \gamma'(t), 0).$$

Also, $A_b := s^*\omega : TM \to \mathfrak{g}$ is the gauge field on the base space. But something sneaky happens when this gets translated into the physics formalism. After all, A_b is a 1-form and so is linear in the expression $\gamma'(t)$. So we write

$$A_b(T\gamma(t, 1)) = s^*\omega(\gamma(t), \gamma'(t)) = A(\gamma(t)) \gamma'(t),$$

where $A : M \to \mathrm{Lin}(\mathbb{R}^m, \mathfrak{g})$, the linear maps from $\mathbb{R}^m$ to $\mathfrak{g}$. Summarizing, we can write the ODE (10.12) as

$$g'(t) = -\big(A(\gamma(t)) \gamma'(t)\big) g(t).$$

In other words, the function g lies in the kernel of the first-order differential operator

$$\frac{d}{dt} + A(\gamma(t)) \gamma'(t)$$

But be careful! The derivative here is actually an $n \times n$ matrix that is d/dt times the identity matrix. And the second term is the zero-order differential operator that is given by multiplication by the $n \times n$ matrix $A(\gamma(t)) \gamma'(t)$.

This has all been done in the mathematics convention for the Lie algebra of a Lie group. In the physics convention, the Lie algebra differs by a factor of $i = \sqrt{-1}$, and so the relevant differential operator becomes

$$\frac{d}{dt} \pm iA(\gamma(t))\,\gamma'(t).$$

The sign here depends on whether one multiplies or divides by i when going from the mathematics to the physics convention. There seems to be no uniformly accepted way to do this conversion. This discussion has led to a sort of minimal coupling expression for the appropriate differential operator that replaces d/dt. But note that the curve γ remains, as it must, an essential part of the formalism.

10.5   Koszul Connections

We now use a connection to define a new type of derivative. The point is that an Ehresmann connection gives us a way of identifying tangent vectors based at two distinct points *provided* that a curve is given between those two points. This will allow us to differentiate a vector field (modulo details which follow). But for now we note that this is the same idea behind the definition of the Lie derivative of a vector field (with respect to another vector field). However, beware! The implementation of the same idea here gives us something that has properties decidedly different from those of a Lie derivative.

Here is the idea locally. We continue with the notation of Theorem 10.4, except that we also take $P = M \times G$ to be explicitly trivial. We put $x_0 := \gamma(0)$ and $x_1 := \gamma(1)$. We write

$$\tilde{\gamma}(t) = (\gamma(t), g(t)) \in M \times G = P$$

for the horizontal lift of the curve γ with initial condition $\tilde{\gamma}(0) = (x_0, h)$ for any $h \in G$. The map $g : [0, 1] \to G$ depends on the curve γ and the initial point $g(0) = h$, but these dependencies are omitted from the notation.

Now suppose that the Lie group G acts on the vector space W; that is, there is a morphism of Lie groups $\rho : G \to GL(W)$. The associated vector bundle (in this case) is $\pi_1 : M \times W \to M$, where π_1 is the projection onto the first factor.

We then define the *parallel transport* of $(x_0, w_0) \in M \times W$ in the fiber above $x_0 \in M$ along a curve γ with $x_0 = \gamma(0)$ and $x_1 := \gamma(1)$ to the point $\gamma(t) \in M$ to be

$$(\gamma(t), \rho\big(g(t)h^{-1}\big)w_0) \in \pi_1^{-1}(\gamma(t)).$$

In particular, for $t = 1$, we get $(x_1, \rho(g(1)h^{-1})w_0) \in \pi_1^{-1}(x_1)$. The result is a mapping from the vector space $\pi_1^{-1}(x_0)$ to the vector space $\pi_1^{-1}(x_1)$. This mapping depends only on the choice of the curve γ in M from x_0 to x_1. It does *not* depend on the choice of the initial condition $h \in G$. This mapping is identified with $\rho(g(1)h^{-1})$, which is clearly linear. The notation is

$$\tau^\gamma = \tau^\gamma_{x_0,x_1} : \pi_1^{-1}(x_0) \to \pi_1^{-1}(x_1).$$

By reversing the parameterization of the curve γ, we get the inverse to the linear map τ^γ. Thus, τ^γ is an isomorphism of vector spaces. And this is the geometric essence of parallel transport in a vector bundle; it is a way of identifying distant fibers provided that one picks a curve joining their base points. While very distant fibers may be in distinct path components (and so no parallel transport is possible), one can always path connect points in a (connected!) chart. This leads to the use of parallel transport to define infinitesimal structures.

The rest of the definition consists of showing that this does not depend on the local coordinates used and that this can be extended along a curve whose image does not lie in one coordinate neighborhood. The general result holds for the associated vector bundle $P \times_G W$ over the space M for any principal bundle P, trivial or not. These are the typical lemmas. They will be left to the reader.

One more delicate point is that every vector bundle does indeed arise as the associated vector bundle of a principal bundle. This is not magic. One takes the cocycle that defines the vector bundle, say $g_{\alpha,\beta} : U_\alpha \cap U_\beta \to GL(W)$, where W is the model vector space for the fiber, and uses this as "pasting" information to glue together local trivializations $U_\alpha \times GL(W)$ in place of the local trivializations $U_\alpha \times W$. This works because $GL(W)$ acts on itself via left multiplication.

Now we let X be a vector field on M and $\sigma : M \to E$ be a section of a vector bundle $\hat{\pi} : E \to M$ with model fiber space W. We suppose that the principal bundle that gives rise to E has an Ehresmann connection defined on it. So we have the corresponding parallel transport maps τ^γ on the fibers of E for every curve γ in M. We now define the *covariant derivative* of σ with respect to X. This will be another section of E, denoted $\nabla_X \sigma : M \to E$. It is defined for $x_0 \in M$ as

$$\nabla_X \sigma(x_0) := \lim_{t \to 0} \frac{(\tau^\gamma_t)^{-1}[\sigma(\gamma(t))] - \sigma(x_0)}{t}, \tag{10.13}$$

where γ is any curve in M with $\gamma(0) = x_0$ and $\gamma'(0) = X(x_0)$. Here $(\tau^\gamma_t)^{-1}$ is the parallel transport vector space isomorphism on the fibers of E:

$$(\tau^\gamma_t)^{-1} : \hat{\pi}^{-1}(\gamma(t)) \to \hat{\pi}^{-1}(x_0).$$

Note that both vectors in the numerator of (10.13) lie in the same vector space, namely, in $\hat{\pi}^{-1}(x_0)$, and so their difference is defined and is again an element of that vector space. Additionally, it is worthwhile commenting that this vector space $\hat{\pi}^{-1}(x_0)$ does not depend on t. Also, notice that the curve γ need not be an integral curve of X. The only information from the vector field X that enters this definition is its value $v = X(x_0)$ at the one single point x_0. The values of σ in the direction of v and infinitesimally near x_0 enter the definition. This is a generalization of the idea of directional derivative, here in the direction v, of an object that is smoothly varying in that direction. Because of this, we also say that $\nabla_X \sigma$ is the *covariant derivative of σ in the direction of X*.

An important special case of this is when we take the vector bundle E to be the tangent bundle $TM \to M$. In this case, a section is known as a vector field and a standard notation for it is Y. In this case, $\nabla_X Y$ is again a section of TM, that is, a vector field on M. But we can also immediately take ∇_X of all the sections of the usual tensor bundles, exterior algebra bundles, symmetric tensor bundles, and so on, that we use in differential geometry. All that we have to check is that they are indeed vector bundles—which they are—and that there is an Ehresmann connection hanging around in the background—which there usually is. For example, every Riemannian manifold has a canonically defined connection known as the Levi–Civita connection. The same works for semi-Riemannian manifolds.

Returning to the case $\nabla_X Y$ for vector fields X, Y on M, our comments above indicate an important asymmetry in the roles of X and Y here. Again, the vector field $\nabla_X Y$ at a point $x_0 \in M$ depends only on the valueof $X(x_0)$ but on the values of Y infinitesimally close to x_0. This is reflected in the last two properties in the following result.

Proposition 10.5 *Suppose that X is a vector field, σ is a section of a vector bundle $p : E \to M$, and $f : M \to \mathbb{R}$ is a C^∞-function. Then, given that the above hypotheses hold, we have*

1. *$\nabla_X \sigma : M \to E$ is a smooth section of $p : E \to M$.*
2. *The mapping $(X, \sigma) \to \nabla_X \sigma$ is bilinear over the real numbers.*
3. *$\nabla_{fX}\, \sigma = f \, \nabla_X \sigma$.*
4. *$\nabla_X (f\sigma) = f \, \nabla_X \sigma + (Xf)\sigma$.*

In particular, when we take σ to be a vector field Y on M, the last two properties show the dramatic difference between the roles of X and those of Y in the covariant derivative $\nabla_X Y$. This contrasts sharply with the Lie derivative $\mathcal{L}_X Y$, which depends on the local values of both X and Y.

For the next definition, we use the notation $\Gamma(E)$ for the vector space of all sections of a vector bundle $E \to M$. If we take the special case when $E = TM$ is the tangent bundle, then $\Gamma(E) = \Gamma(TM)$ is the space of vector fields on M.

Definition 10.6 *If $E \to M$ is a vector bundle over a manifold M, then we say that an assignment $(X, \sigma) \to \nabla_X \sigma$ from $\Gamma(TM) \times \Gamma(E)$ to $\Gamma(E)$ is a Koszul connection if the properties of the previous proposition hold.*

Therefore, an Ehresmann connection on a principal bundle induces Koszul connections on all of its associated vector bundles.

It turns out, just as in the discussion before this definition, that any Koszul connection at a point $x_0 \in M$ depends only on the value of X at that point, and not on values near (but not equal to) x_0. This follows from the third property in the previous proposition by taking a sequence of smooth approximate characteristic functions f_n with compact support K_n such that $x_0 \in \text{int}(K_n)$ for all n, $f_n \equiv 1$ on some neighborhood of x_0, and $\cap_n K_n = \{x_0\}$. So we get a bilinear mapping

$$(v, \sigma) \to \nabla_v \sigma$$

from $TM \times \Gamma(E)$ to $\Gamma(E)$. If we keep σ fixed, this gives a map TM to $\Gamma(E)$, which is linear on the fibers of TM. The notation for this is $\nabla\sigma$, and so

$$\nabla\sigma : v \mapsto \nabla_v \sigma.$$

So we have a $\Gamma(E)$-valued 1-form on M. In general, a *section-valued* 1-*form* θ on M is such a 1-form, $\theta : TM \to \Gamma(E)$. One also has a corresponding definition of a *section-valued k-form*. This paragraph has been merely a short interlude in order to introduce some common jargon and show that such objects are to be found in (mathematical) nature.

In my opinion, the word "connection" is overused in this corner of differential geometry. I believe an Ehresmann connection is well named since it gives a way to connect tangent spaces (and other fibers of vector bundles) at distinct points. On the other hand, a Koszul connection ∇_X is a differential operator since it satisfies Leibniz's rule. That makes it an object in analysis. But these names are set in stone even though no one knows where that stone is to be found.

10.6   Covariant Derivatives

There is also a covariant derivative defined on the forms in $\Omega^k(P)$ for any principal bundle $G \hookrightarrow P \xrightarrow{\pi} M$ with an Ehresmann connection ω.

But first for $v \in TP$, we define $v^H \in \text{Hor}(TP)$ to be the horizontal part of v in $\text{Hor}(TP)$ in the decomposition $TP = \text{Vert}(TP) \oplus \text{Hor}(TP)$ associated with the Ehresmann connection ω.

Definition 10.7 *Suppose $G \hookrightarrow P \xrightarrow{\pi} M$ is a principal bundle with an Ehresmann connection ω. Let $\theta \in \Omega^k(P)$ be a k-form. Define $D\theta \in \Omega^{k+1}(P)$ by*

$$D\theta(v_1, \ldots, v_{k+1}) := d\theta(v_1^H, \ldots, v_{k+1}^H),$$

where $v_1, \ldots, v_{k+1}$ are $(k+1)$ vectors in the same tangent space $T_p P$ for some $p \in P$ and d is the exterior derivative.

We say that $D\theta$ is the (exterior) covariant derivative of θ and that D is the (exterior) covariant derivative.

For a W-valued k-form $\theta \in \Omega^k(P; W)$, where W is a finite-dimensional vector space, we can define $D\theta$ component by component with respect to any basis of W.

As expected, one then proves that this does not depend on that particular choice of basis. This will be used in the next chapter for Lie algebra-valued forms.

Since the map $v \mapsto v^H$ is a projection [that is, $(v^H)^H = v^H$], we have $D\theta(v_1, \ldots, v_{k+1}) = D\theta(v_1^H, \ldots, v_{k+1}^H)$. So to calculate a covariant derivative, it suffices to see how it acts on horizontal vectors.

For any $\theta \in \Omega^k(P)$, the covariant derivative $D\theta$ is a horizontal form in the sense of this definition:

Definition 10.8 *Suppose $G \hookrightarrow P \xrightarrow{\pi} M$ is a principal bundle with an Ehresmann connection ω. Let $\eta \in \Omega^k(P)$ be a k-form and $v_1, \ldots, v_k$ be a set of k vectors in $T_p P$ for some point $p \in P$.*

Then we say that η is horizontal *if*

$$\eta(v_1, \ldots, v_k) = 0$$

whenever at least one of the v_js is vertical.
Similarly, we say that η is vertical *if*

$$\eta(v_1, \ldots, v_k) = 0$$

whenever at least one of the v_js is horizontal.

Notice that the definition of horizontal form does not require the existence of an Ehresmann connection, while the definitions of covariant derivative and vertical form do require an Ehresmann connection.

This definition is meant to apply to all k-forms, including 0-forms. But the logic of the case $k = 0$ might be confusing. So let $f : P \to \mathbb{R}$ be a smooth function; that is, f is a 0-form. But the sequence $v_1, \ldots, v_k$ when $k = 0$ is understood to have zero vectors in it; namely, it is the empty sequence. And in that case, we have not defined what it means to evaluate f on the empty sequence. But whatever it might be defined to mean, it is never the case that at least one of the vectors v_j is vertical. So one says that the condition for f to be horizontal is *vacuously* true. This comes from propositional logic. There the implication $p \implies q$ (read: p implies q) has a truth table that gives this implication the value True whenever p has the value False. And that is what is happening here! So we conclude that any 0-form f is horizontal.

But the same sort of argument shows that any 0-form f is vertical! And that is what one has: Every 0-form is both horizontal and vertical. This is not a contradiction. This might sound at first like sophistry, but it is standard logic.

The covariant derivative D is related to the Koszul connections associated to the same Ehresmann connection. As a simple case, we have for a 0-form f that the 1-form Df is given by

$$Df(X) = \nabla_X f \ (= Xf),$$

where X is horizontal. This is an easy argument left to the reader. Also, for a 1-form θ, we have the 2-form $D\theta$ given by

$$D\theta(X,Y) = \nabla_X(\theta(Y)) - \nabla_Y(\theta(X)) - \theta([X,Y]),$$

where X and Y are horizontal. This is an immediate consequence of (5.9). Similar, but more complicated, formulas hold for k-forms.

10.7 Dual Connections as Vertical Covectors

In this section we take a dual point of view. Typically—and pedagogically—we think of vectors as primary objects and covectors as secondary objects defined in terms of vectors. But thinking strictly in terms of mathematical structures, this is misleading at best and dead wrong at worst. The duality between vectors and covectors is symmetric; therefore, neither plays a more fundamental role than the other. So we can speak quite correctly of the duality between covectors and vectors (in that order). It is an accident of language—and history—that we put one of these structures before the other. It has no logical, mathematical justification. However, we have followed the tradition of defining a connection as a structure (namely, horizontal vectors) in a *tangent* space. Now we will discuss the equally valid dual definition in the dual *cotangent* space.

Given a principal bundle $G \hookrightarrow P \xrightarrow{\pi} M$, we have the naturally associated vertical vector spaces of TP, denoted by $\mathrm{Vert}(T_pP)$ for each $p \in P$. By duality, these define subspaces of T^*P, denoted $\mathrm{Hor}(T_p^*P)$ and called the *horizontal subspace* of T_p^*P. The definition is

$$\mathrm{Hor}(T_p^*P) := \{l \in T_p^*P \mid \langle l, v \rangle = 0 \quad \text{for all } v \in \mathrm{Vert}(T_pP)\}.$$

So the horizontal covectors are those that annihilate all the vertical vectors. They are naturally defined for any principal bundle; there is no connection needed.

The next definition is the natural dual of the definition that we have already seen for the tangent bundle TP.

Definition 10.9 *Let $G \hookrightarrow P \xrightarrow{\pi} M$ be a principal bundle. We define a* dual connection *in P to be the selection for every $p \in P$ of a vector subspace $\mathrm{Vert}(T_p^*P) \subset T_p^*P$ such that*

- $\mathrm{Vert}(T_p^*P) + \mathrm{Hor}(T_p^*P) = T_pP$.
- $\mathrm{Vert}(T_p^*P) \cap \mathrm{Hor}(T_p^*P) = 0$, *the zero subspace.*
- *The subspace $\mathrm{Vert}(T_p^*P)$ depends smoothly on $p \in P$.*
- *The G-equivariance condition:*
  $R_g^* \mathrm{Vert}(T_{pg}P) = \mathrm{Vert}(T_pP)$ *for all $g \in G$ and $p \in P$.*

We say that the elements of Vert($T_p^* P$) *are vertical covectors of P at p (with respect to the connection).*

This dual point of view continues step by step in analogy with the theory of connections as given in the previous sections, so we do not feel the need to elaborate on it further. However, one reason we mention this dual viewpoint is that this is what one uses to generalize connections to the noncommutative setting, as described in the companion volume [44].

Exercise 10.10 *Prove that the dimension of the horizontal (resp., vertical) sub-space equals the dimension of the base (resp., fiber) space. Explicitly, show that* $\dim \text{Hor}(T_p^* P) = \dim M$ *and* $\dim \text{Vert}(T_p^* P) = \dim G$.

Chapter 11
Curvature of a Connection

11.1 Definition

In this chapter, the discussion will highlight the reasons behind our saying that a connection is a geometric structure. We begin with the crucial concept of the curvature of a connection. First, we introduce some notation.

Suppose that $G \hookrightarrow P \xrightarrow{\pi} M$ is a principal bundle with Ehresmann connection ω. Suppose that $w \in TP$. Then we write

$$w = w^V + w^H,$$

where $w^V \in \text{Vert}(TP)$ is the vertical part of w and $w^H \in \text{Hor}(TP)$ is the horizontal part of w. These parts are with respect to the decomposition of the tangent space $TP = \text{Vert}(TP) \oplus \text{Hor}(TP)$.

Definition 11.1 *Let $G \hookrightarrow P \xrightarrow{\pi} M$ be a principal bundle with a given Ehresmann connection ω. Then the* curvature $D\omega : T^{\diamond 2}P \to \mathfrak{g}$ *of ω is defined by*

$$D\omega(v_1, v_2) := d\omega(v_1^H, v_2^H)$$

for all $(v_1, v_2) \in T^{\diamond 2}P$, that is, for all $v_1, v_2 \in TP$ with $\tau_P(v_1) = \tau_P(v_2)$.

Recall that $\tau_P : TP \to P$ is the tangent bundle map for the manifold P. Notice that the Ehresmann connection ω enters this formula directly in the expression $d\omega$, but also indirectly in the definitions of v_1^H and v_2^H.

Exercise 11.1 *Show that $D\omega$ is a $\mathfrak{g}$-valued 2-form on P.*

Many texts introduce separate notation for the curvature. We prefer the functional notation $D\omega$ since it directly indicates the connection ω under consideration. Also, D is the covariant derivative of a $\mathfrak{g}$-valued differential form, as discussed in the previous chapter.

The original version of this chapter was revised: typographical error was corrected. The correction to this chapter is available at https://doi.org/10.1007/978-3-319-14765-9_18

© Springer International Publishing Switzerland 2015   145
S.B. Sontz, *Principal Bundles*, Universitext, DOI 10.1007/978-3-319-14765-9_11

The next theorem ties together various threads that we have seen. While the proof, based directly on the Frobenius integrability theorem, is not at all difficult, we prefer just to mention the result for your cultural enrichment.

Theorem 11.1 *Suppose that* $G \hookrightarrow P \xrightarrow{\pi} M$ *is a principal bundle with Ehresmann connection* ω. *Then the associated horizontal distribution*

$$p \mapsto \mathrm{Hor}(T_p P)$$

is integrable if and only if the curvature of ω *is zero,* $D\omega = 0$.

As another cultural aside, we should note that the vertical distribution of any principal bundle is always integrable. You should be able to see this directly without appealing to Frobenius.

11.2   The Structure Theorem

Theorem 11.2 (Cartan structure equation) *Let* $G \hookrightarrow P \xrightarrow{\pi} M$ *be a principal bundle with connection* ω. *Then the curvature of* ω *satisfies*

$$D\omega = d\omega + \frac{1}{2}\omega \wedge \omega.$$

Proof: Here $\omega : TP \to \mathfrak{g}$ is a $\mathfrak{g}$-valued 1-form. So $\omega \wedge \omega : T^{\diamond 2}P \to \mathfrak{g}$ is the $\mathfrak{g}$-valued 2-form given by

$$(\omega \wedge \omega)(v, w) = [\omega(v), \omega(w)] - [\omega(w), \omega(v)] = 2[\omega(v), \omega(w)]$$

for all pairs v, w in the same fiber of TP, where $[\cdot, \cdot]$ is the Lie bracket in $\mathfrak{g}$. Hence, we have to prove that

$$d\omega(v, w) = D\omega(v, w) - [\omega(v), \omega(w)] \tag{11.1}$$

for all $(v, w) \in T^{\diamond 2}P$, that is, for all pairs v, w in the same fiber in TP.

We now use the bilinearity of both sides of equation (11.1) to reduce proving equation (11.1) to the following cases.

- **Case 1:** v, w are horizontal.
  Since ω annihilates horizontal vectors, we have $\omega(v) = 0$ and $\omega(w) = 0$. Thus, the second term on the right side of equation (11.1) vanishes. But the definition of curvature and the fact that $v^H = v$ and $w^H = w$ in this case imply that

$$D\omega(v, w) = d\omega(v^H, w^H) = d\omega(v, w).$$

And so equation (11.1) holds in this case.

- **Case 2:** v, w are vertical.

  In this case, $v^H = 0$ and $w^H = 0$, and so the curvature is zero since we then have $D\omega(v, w) = d\omega(v^H, w^H) = 0$.

  Write $p = \tau_P(v) = \tau_P(w)$ for the common point of tangency of the pair of tangent vectors v, w. (So $p \in P$.) Since v and w are vertical, we have previously shown that there exist elements $A, B \in \mathfrak{g}$ such that $v = A^\sharp(p)$ and $w = B^\sharp(p)$. So we obtain

$$d\omega(v, w) = d\omega(A^\sharp(p), B^\sharp(p)) = (d\omega(A^\sharp, B^\sharp))(p).$$

Now, we know how to evaluate the 2-form $d\omega$ on the pair of vector fields $A^\sharp, B^\sharp$; namely, we have this identity of C^∞-functions on P:

$$d\omega(A^\sharp, B^\sharp) = A^\sharp(\omega(B^\sharp)) - B^\sharp(\omega(A^\sharp)) - \omega([A^\sharp, B^\sharp]).$$

[See the identity (5.9).] But $\omega(B^\sharp) = B$ and $\omega(A^\sharp) = A$ are constant functions, so that the first two terms on the right-hand side of the preceding identity vanish. So we have that

$$
\begin{aligned}
d\omega(A^\sharp, B^\sharp) &= -\omega([A^\sharp, B^\sharp]) \\
&= -\omega([A, B]^\sharp) \\
&= -[A, B] \\
&= -[\omega(A^\sharp), \omega(B^\sharp)],
\end{aligned}
$$

which is also a constant function on P as the third line shows. Then, evaluating the preceding identity (of constant functions) at the point $p \in P$, we get

$$
\begin{aligned}
d\omega(v, w) &= -[\omega(A^\sharp), \omega(B^\sharp)](p) \\
&= -[\omega(A^\sharp(p)), \omega(B^\sharp(p))] \\
&= -[\omega(v), \omega(w)],
\end{aligned}
$$

which shows equation (11.1) in this case.

- **Case 3:** v is vertical and w is horizontal.

  (By the antisymmetry of equation (11.1) in v, w, this is the last case we need to prove.) Now we have $v^H = 0$ and $w^H = w$. So we immediately see that

$$D\omega(v, w) = d\omega(v^H, w^H) = d\omega(0, w) = 0.$$

Thus, the curvature in this case, as in the preceding one, is zero. Also, using $\omega(w) = 0$ since w is horizontal, we get

$$[\omega(v), \omega(w)] = [\omega(v), 0] = 0.$$

So it suffices to show that $d\omega(v, w) = 0$ in this case. (Note that all three terms in equation (11.1) turn out to be zero in this case.) Again, we use the fact that the vertical vector v can be written as $v = A^\sharp(p)$ for some $A \in \mathfrak{g}$.

We would also like some canonical way to write the horizontal vector w, but we only have a weaker result, namely, that a *horizontal* vector field W on P exists such that $W(p) = w$. (This is left to the reader as an exercise. See ahead.) So

$$d\omega(v, w) = d\omega(A^\sharp, W)(p),$$

and then (starting with the identity (5.9)), we get

$$d\omega(A^\sharp, W) = A^\sharp(\omega(W)) - W(\omega(A^\sharp)) - \omega([A^\sharp, W]) = -\omega([A^\sharp, W])$$

since $\omega(W) = 0$ follows from W being horizontal and $\omega(A^\sharp) = A$ is a constant function. Hence, everything reduces to showing that

$$[A^\sharp, W] \in \ker \omega = \mathrm{Hor}(TP);$$

that is, the subspace of fundamental vector fields on P acting via the Lie bracket on vector fields on P leaves invariant the subspace of horizontal vector fields.

Now,

$$[A^\sharp, W]_p = \lim_{t \to 0} \frac{1}{t}\left((\gamma_t^{-1})_* W(\gamma_t(p)) - W(p)\right), \tag{11.2}$$

where γ_t is the flow associated to the vector field $A^\sharp$ in some neighborhood of p. But we have that

$$\gamma_t(q) = q \cdot \exp(tA) \tag{11.3}$$

for all $q \in P$, where

$$\mathbb{R} \ni t \mapsto \exp(tA) \in G$$

is the flow in the Lie group G for the left invariant vector field associated to $A \in \mathfrak{g} = T_e(G)$. (See the following exercise.) It follows that

$$\gamma_t^{-1}(\tilde{p}) = \tilde{p} \cdot \exp(tA)^{-1} = R_{\exp(tA)^{-1}}(\tilde{p}).$$

But this implies that

$$(\gamma_t^{-1})_* W(\gamma_t(p)) = (R_{\exp(tA)^{-1}})_* W(\gamma_t(p)) \in \mathrm{Hor}(TP)$$

for all t by one of the defining properties of the horizontal subspaces of a connection. So the two terms $(\gamma_t^{-1})_* W(\gamma_t(p))$ and $W(p)$ on the right-hand side of equation (11.2) are horizontal vectors, which implies that

$$\frac{1}{t}\left((\gamma_t^{-1})_* W(\gamma_t(p)) - W(p)\right)$$

is also horizontal for every t. But the subspace of horizontal vectors is closed, being finite dimensional, and consequently the limit on the right-hand side of equation (11.2) is a horizontal vector. This says that $[A^\sharp, W]_p \in \mathrm{Hor}(TP)$, which concludes the proof for this case. ∎

Exercise 11.2 *Fill in these gaps in the preceding proof:*

1. *Suppose that w is a horizontal vector at $p \in P$. Show that there exists a horizontal vector field W on P such that $W(p) = w$. Is W unique?*
2. *Show that equation (11.3) is valid.*

11.3 Structure Equation for a Lie Group

The following theorem is usually proved in the context of Lie group theory without any use of connections.

Theorem 11.3 *(Structure equation of a Lie group) Suppose that Ω is the Maurer–Cartan form on a Lie group G. Then*

$$d\Omega + \frac{1}{2}\Omega \wedge \Omega = 0.$$

Proof: We have already seen that Ω is an Ehresmann connection for the principal bundle $G \hookrightarrow G \to \{x\}$ over a one-point space. So it suffices to show by the theorem that its curvature is zero. But its curvature satisfies

$$D\Omega(v_1, v_2) = d\Omega(v_1^H, v_2^H) = 0$$

by the definition of the curvature and the fact that for this principal bundle the only horizontal vector is zero. ∎

11.4 Bianchi's Identity

In this section, we prove a classic geometric result that seems as remote from physics as one could imagine. My gentle reader may have already realized that there is some physics going on here. And we will see that is exactly so in Chapter 14.

Theorem 11.4 *(Bianchi's identity) Let ω be a connection in a principal bundle. Let $\Theta = D\omega$ denote its curvature. Then the covariant derivative of Θ is given by*

$$D\Theta = 0.$$

Remark: Since D^2 generally is not equal to zero, there is something to prove here.

Proof: By Cartan's structure equation, we have

$$\Theta = D\omega = d\omega + \frac{1}{2}\omega \wedge \omega.$$

Recall that ω is a 1-form. So Θ is a 2-form and $D\Theta$ is a 3-form. By the definition of the covariant derivative D, we have

$$D\Theta(v_1, v_2, v_3) = d\Theta(v_1^H, v_2^H, v_3^H)$$

for $v_1, v_2, v_3 \in T^{\diamond 3}P$. But

$$d\Theta = d^2\omega + \frac{1}{2}d\omega \wedge \omega - \frac{1}{2}\omega \wedge d\omega.$$

Beware! The wedge product is not antisymmetric here, so the last two terms do not cancel. However, the first term does vanish, $d^2\omega = 0$. Of course, the connection ω annihilates horizontal vectors since that is the defining property of a horizontal vector. Thus, for the second term we get

$$\frac{1}{2}(d\omega \wedge \omega)(v_1^H, v_2^H, v_3^H) = 0$$

since, when the left side is expanded as a sum, the second factor ω is always evaluated on one of the horizontal vectors v_1^H, v_2^H, v_3^H.

Similarly, the third term vanishes on this triple of horizontal vectors. In short,

$$d\Theta(v_1^H, v_2^H, v_3^H) = 0$$

for all $v_1, v_2, v_3 \in T^{\diamond 3}P$. And this implies that $D\Theta = 0$. ∎

Chapter 12
Classical Electromagnetism

An example of a classical field is the electromagnetic field, which is some sort of combination of the electric and magnetic fields. This also turns out to be a motivational example in the physics of a gauge theory.

12.1 Maxwell's Equations in Vector Calculus

In Section 6.2, we discussed the electric field E and the magnetic field B separately as vector fields on an open set $U \subset \mathbb{R}^3$. It is important to emphasize that there is no mathematical definition of these fields. Rather, they are mathematical models meant to capture experimental properties of many physical systems. And these fields enter into equations of motion of matter with electric charge as well as equations that describe the time evolution of the fields themselves. The first five chapters, about 200 pages, of [28] is devoted to the steady state, that is, when all fields are constant in time and the matter also has a time-independent behavior. We write in vector calculus notation the differential equations for the steady state in a vacuum in some open set U of $\mathbb{R}^3$ in what are called *Gaussian units*:

$$\nabla \cdot E = 4\pi\rho, \quad \text{Coulomb's law} \tag{12.1}$$

$$\nabla \times B = \frac{4\pi}{c} J, \quad \text{Ampère's law} \tag{12.2}$$

$$\nabla \times E = 0, \quad \text{Electric force is conservative} \tag{12.3}$$

$$\nabla \cdot B = 0. \quad \text{No magnetic charges} \tag{12.4}$$

There is a lot to explain here! First off, c is the *speed of light in vacuum*, which is a universal constant of nature. Next, ρ is the *electric charge density*, which is just what the name implies it should be; it is the density of electric charge per unit

© Springer International Publishing Switzerland 2015

S.B. Sontz, *Principal Bundles*, Universitext, DOI 10.1007/978-3-319-14765-9_12

volume of U. For this steady-state setting, this is a time-independent function ρ : $U \to \mathbb{R}$. Notice that electric charge can be positive, negative, or zero, so that the codomain of ρ has to be the full real line $\mathbb{R}$. And J is the electric current density, which also is time-independent, so that $J : U \to \mathbb{R}^3$. Since J is electric charge (scalar) times velocity (vector), J is a vector field. The factor 4π is geometric in origin; it is the surface area of the unit sphere in $\mathbb{R}^3$. The justification, history, and names of these equations are presented in many excellent texts, of which [28] is but one.

This steady-state case admits a decoupling into one set of equations for the electric field E [namely, (12.1) and (12.3)] and another independent set for the magnetic field B [namely, (12.2) and (12.4)]. For many years, this was the state of the science. There were even experiments to demonstrate that magnetism had nothing to do with electricity. The null effects of these experiments were due to other physical factors, such as friction. It was first noted experimentally that an electric current in a wire can produce motion in a nearby magnet. (The story is that this was a demonstration experiment in a physics class that was intended to show the absence of such a phenomenon.) Anyway, this is the experimental origin of Ampère's law.

But then the above steady-state equations were proved to be inadequate as well. It was observed that changing magnetic fields produced electric fields. So Faraday introduced the ideas that led to this replacement for (12.3)):

$$\nabla \times E = -\frac{1}{c}\frac{\partial B}{\partial t} \qquad \text{Faraday's law.} \tag{12.5}$$

Here t denotes time. So now we have a time derivative in the game as well as a coupling of the magnetic to the electric field. This equation involves a time-varying magnetic field. It thus seems reasonable to study the general case where E, ρ, and J all can depend on time t as well as on the spatial variables $(x, y, z) \in U$. By the way, the minus sign in Faraday's law has physical significance. But we leave that to the reader's curiosity.

Now these new equations for the electric and magnetic fields gave Maxwell a headache. It takes a genius to see that something is awry! The problem is that Ampère's law implies that

$$\nabla \cdot J = \nabla \cdot \left(\frac{1}{4\pi}\nabla \times B\right) = 0,$$

using the vector calculus identity $\nabla \cdot (\nabla \times \mathbf{F}) = 0$ for any (smooth) vector field $\mathbf{F}$. (By the way, we will take units from now on so that $c = 1$.) So what is wrong with this? Well, a time-dependent J and a time-dependent ρ will be related by the equation

$$\nabla \cdot J + \frac{\partial \rho}{\partial t} = 0, \tag{12.6}$$

which is the differential equation that says that electric charge is a *locally* conserved quantity. This is called the *continuity equation* of electric charge.

Let's linger a bit over why (12.6) says what I claim it says. As with many of the interpretations of differential equations, the idea is to use vector calculus and, in particular, *Stokes' theorem* to derive an integral version of the differential equation. So we consider an open, bounded domain $\Omega \subset \mathbb{R}^3$ with a "nice" boundary $\partial\Omega$. Then

$$\int_\Omega \frac{\partial \rho}{\partial t} = \frac{\partial Q}{\partial t},$$

where $Q := \int_\Omega \rho$ is interpreted as the total electric charge in Ω as a function of time t. As it should be! For my mathematically strict readers, let me note that, yes, an integral and a derivative have been interchanged and this requires a justification. Since this is important for you, I urge you to find that justification.

For the other term in (12.6), we see by Stokes' theorem that

$$\int_\Omega \nabla \cdot J = \int_{\partial\Omega} J \cdot n, \tag{12.7}$$

where n is the unit outward normal vector at each point in $\partial\Omega$. The existence of this normal vector is a detail that makes the boundary "nice." The integral on the right side of (12.7) is taken with respect to the area measure on $\partial\Omega$ induced by Lebesgue measure on Ω. This special case of Stokes' theorem is also called the *divergence theorem*. For example, see [45] for more details.

The right side of (12.7) is interpreted as the total flux of charge leaving Ω by crossing its boundary. Then when we integrate (12.6) over Ω, we get an integral equation that says the flux of charge leaving Ω plus the change of charge in Ω is always zero:

$$\int_{\partial\Omega} J \cdot n + \frac{\partial Q}{\partial t} = 0.$$

Thinking for the moment that all the charge under consideration is positive, this says that the flux of this charge leaving Ω is compensated (exactly!) by a decrease in the total charge in Ω. Now, "leaving" and "decreasing" are to be taken in an algebraic sense; that is, if the quantity of charge leaving is negative, this means that positive charge is entering. And the negative of "decreasing" is "increasing," as it must be. Similar considerations hold for negative charges and so for any mixture of positive and negative charges. Moreover, this is true for *every* such domain Ω. This is what conservation of charge means in physics; that is, the only way charge can increase or decrease in a given region is when some charge crosses the boundary of that region. Also, the quantity of charge crossing the boundary is exactly the change of charge in the region. For example, if the charge on a chip in a laboratory in Tokyo were to suddenly decrease while at the same time the charge on another chip in Buenos Aires were to increase by exactly the same amount, this would be a violation of the principle of conservation of charge as stated in (12.6). Conservation of charge requires that the charge leaves Tokyo and travels to Buenos Aires.

Now, Maxwell knew that experiment had established that electric charge is a conserved quantity. So he had to accept the continuity equation. But, then, what is wrong with the above equations for the electric and magnetic fields? Well, Ampère's law, while true when ρ is time-independent, must be wrong in the time-dependent case. We are missing a term in Ampère's law, so that we get the continuity equation (12.6) rather than $\nabla \cdot J = 0$. Looking at Coulomb's law (12.1), we see that

$$\rho = \frac{1}{4\pi}\nabla \cdot E,$$

and so

$$\frac{\partial \rho}{\partial t} = \frac{1}{4\pi}\frac{\partial}{\partial t}(\nabla \cdot E) = \nabla \cdot \left(\frac{1}{4\pi}\frac{\partial E}{\partial t}\right).$$

Now it is clear how Ampère's law should be modified. The time-dependent version of Ampère's law is

$$\nabla \times B = 4\pi J + \frac{\partial E}{\partial t}. \qquad \text{Ampére-Maxwell law} \qquad (12.8)$$

The name Maxwell gave to the new term in this equation is the *displacement current*. Notice that it gives another coupling of the electric and magnetic fields. In particular, an electric field that changes in time in the absence of a current will produce a magnetic field. Of course, the Ampère–Maxwell law (12.8) implies the continuity equation (12.6). Maxwell made sure of that! This one contribution to physics alone would have guaranteed Maxwell's fame as one of the greatest physicists of all times. The introduction of the displacement current was a great success of theoretical physics, but not just because of the beautiful argument that produced it. The Ampère–Maxwell equation agrees with all experiments in classical electrodynamics to date. One of the most dramatic theoretical consequences of it together with the other equations is the existence of electromagnetic waves propagating at the speed of light. Those waves can have all frequencies, not just optical frequencies. Read about it in [28], for example. Now we have the following complete set of Maxwell's equations:

$$\nabla \cdot B = 0, \qquad \text{No magnetic charges} \qquad (12.9)$$

$$\nabla \times E + \frac{\partial B}{\partial t} = 0, \qquad \text{Faraday's law} \qquad (12.10)$$

$$\nabla \cdot E = 4\pi\rho, \qquad \text{Coulomb's law} \qquad (12.11)$$

$$\nabla \times B - \frac{\partial E}{\partial t} = 4\pi J. \qquad \text{Ampére–Maxwell's law} \qquad (12.12)$$

We have followed the convention of writing the fields on the left side and their "sources" on the right side and also of writing the homogeneous (source free) equations first. These equations have to be modified for the electromagnetic field in a medium such as air or water, but the essential ideas are mostly in evidence here. So we will not go into that.

There have been conjectures that magnetic charges, and therefore also magnetic currents, exist in nature and so these would enter as nonzero sources on the right side of equations (12.9) and (12.10), respectively. While this would introduce a beautiful symmetry between electricity and magnetism that is not present in the currently accepted Maxwell's equations, there is no experimental evidence to date for the existence of nonzero magnetic charges.

There is a way to get such a symmetry in Maxwell's equations, and that is by studying the special case when $\rho = 0$ and $J = 0$, that is, when there are no electric charges nor currents present. That special case is known as the *source-free Maxwell's equations*. It is an amazing consequence of this theory that time-varying electric and magnetic fields can exist in a region U without the presence of sources in U. And observations confirm this.

If a test particle with electric charge q is placed in the electromagnetic field as given by Maxwell's equations, it will experience a force F on it given by

$$F = q(E + v \times B). \tag{12.13}$$

This is known as *Lorentz's force law* though it seems Heaviside introduced it first in 1889. However, there seems to be some reason to believe that Maxwell knew it, or special cases of it, as early as 1865. We have seen the two terms in (12.13) independently in Section 6.2. This deceptively simple equation is not proved mathematically. It is the encapsulation of an enormous quantity of experimental results in one little equation. It also continues to agree with all experiments in classical electrodynamics to date.

Historically speaking, the vector calculus was devised after Maxwell's equations were put into their final form by Maxwell. In those days, all of the equations were written out with all partial derivatives explicitly given in the usual notation due to Leibniz—or one of those old-timers. In this regard, the four equations (12.9)–(12.12) known today as Maxwell's equations were written as eight scalar equations. This is because (12.10) and (12.12) are vector equations and so give us a total of six scalar equations, while (12.9) and (12.11) are scalar equations and so give us two more scalar equations.

The importance of Maxwell's equations in basic science and technology is enormous. Not only do these equations continue to be verified to ever-higher precision in experiments, but they also are a basic tool in engineering applications such as microwave ovens, fluorescent lights, X-ray generators, radio antennas and receivers, electric motors, electric generators, radar, magnetic resonance imaging, and many more. Just as an example, without Maxwell's equations at our disposal, I do not see how the GPS could ever have been constructed and operated.

And Maxwell's equations have had an amazing influence in theoretical science, only a fraction of which we will see in this book. The first was the recognition that light is an electromagnetic wave, and simultaneously the recognition that electromagnetic waves should exist at all frequencies, not just the optical frequencies. Also, Maxwell's equations led to the development of special relativity since what Einstein did from a modern perspective was to choose between two incompatible symmetry groups: the Lorentz group for Maxwell's equations and the Galilean group for Newtonian mechanics. So we nowadays say that Einstein decided that the Lorentz group was the correct symmetry group for mechanics and so Newtonian mechanics had to be modified. Without his knowing it, Maxwell's work led to an upending of the great Newton! Also, they are the first example in the history of physics of the unification of what were previously thought to be two independent forces. The process of unification had already begun before Maxwell, but it was Maxwell who put the finishing touch by correctly modifying Ampère's law. We may not be too aware of it any more, but Ampère was a true genius. Even so, he only got part of the picture. And the program of the unification of all physical fields continues as an open problem though most researchers expect quantum mechanics as well to play a central role in that.

12.2   Maxwell's Equations in Differential Forms

While physicists for the most part continue to write Maxwell's equations using the notation of vector calculus, those equations can also be written using the notation of differential forms. Yes, they are exactly the same equations, but the new way of writing them helps one to think what other theories might look like. We now look into this.

The first rather uncomfortable change is that one has to think of force as a 1-form instead of as a vector field. This may not be too difficult for a mathematician, but a physicist has heard for innumerable years that force is a vector (meaning, a vector field). This goes back to Newton! Of course, since we always have an inner product available, at least implicitly, we can go back and forth freely between vector fields and 1-forms. So it seems that it really does not matter which way we view force: as either a vector or a 1-form. But why fly in the face of tradition and insist that it is a 1-form when everyone has known since kindergarten that it is a vector? Well, it is because of how we want to use force. We want to use it to calculate work W, and we have also learned that

$$W = \int_\gamma \mathbf{F} \cdot \mathbf{dx}$$

is the formula for the work done on a particle as it moves through a force (vector?) field $\mathbf{F} = (F_1, F_2, F_3)$ along a curve γ. Here $\mathbf{dx} = (dx_1, dx_2, dx_3)$ is some sort of infinitesimal vector object and $\mathbf{F} \cdot \mathbf{dx}$ is some sort of inner product of a finite vector with an infinitesimal vector. What does this all mean? Try this:

$$W = \int_\gamma \omega = \int_\gamma (F_1 dx_1 + F_2 dx_2 + F_3 dx_3),$$

where $\omega := F_1 dx_1 + F_2 dx_2 + F_3 dx_3$ is a 1-form on some open set U in $\mathbb{R}^3$ and $\gamma : (a, b) \to U$ is a curve in U. Here $a < b$ are real numbers. So now it becomes apparent that we should have identified the force from the very beginning as $\mathbf{F} = F_1 dx_1 + F_2 dx_2 + F_3 dx_3$, a 1-form. And the auxiliary notation ω gets dropped. We now adopt the convention of writing differential forms in boldface. So we now have

$$W = \int_\gamma \mathbf{F}.$$

As simple as that! Notice that the infinitesimals are safely encased inside the 1-form $\mathbf{F}$, where they have a well-defined meaning.

Now charge is a scalar quantity (which is an experimental result of quite high precision; see [28] pp. 547–548), and so the simple equation relating the electric force $\mathbf{F}$ and the electric field $\mathbf{E}$, that is, $\mathbf{F} = q\mathbf{E}$, where q is some nonzero electric charge, implies that $\mathbf{E}$ is also a 1-form. Recall now Faraday's law,

$$\nabla \times E + \frac{\partial B}{\partial t} = 0.$$

We know that the curl operation $\nabla \times \cdot$ is the de Rham differential that maps 1-forms into 2-forms. Let's see how that works out for the electric field $E = (E_1, E_2, E_3)$ viewed as a 1-form $\mathbf{E} = E_1 dx_1 + E_2 dx_2 + E_3 dx_3$. Then

$$d\mathbf{E} = \left(\frac{\partial E_3}{\partial x_2} - \frac{\partial E_2}{\partial x_3}\right) dx_2 \wedge dx_3 + \left(\frac{\partial E_1}{\partial x_3} - \frac{\partial E_3}{\partial x_1}\right) dx_3 \wedge dx_1$$
$$+ \left(\frac{\partial E_2}{\partial x_1} - \frac{\partial E_1}{\partial x_2}\right) dx_1 \wedge dx_2.$$

So we see from Faraday's law that the magnetic field B must be a 2-form. If $B = (B_1, B_2, B_3)$ is its representation as a vector, then we must identify it with the 2-form

$$\mathbf{B} = B_1 \, dx_2 \wedge dx_3 + B_2 \, dx_3 \wedge dx_1 + B_3 \, dx_1 \wedge dx_2$$

in order to get Faraday's law, which in this notation is

$$d\mathbf{E} + \frac{\partial \mathbf{B}}{\partial t} = 0.$$

The zero on the right side is a 2-form as well even though we do not put it in boldface.

Since now we have that the magnetic field is a 2-form $\mathbf{B}$, we immediately can write the equation saying that there are no magnetic charges as

$$d\mathbf{B} = 0$$

since d on a 2-form is the divergence operator. Here the right side is the zero 3-form.

Now Coulomb's law (12.11) requires us to take the divergence of a 1-form, which we know is not right. The divergence is defined on 2-forms. But now the Hodge star comes to the rescue. (We use the standard orientation on U, namely, that $dx_1 \wedge dx_2 \wedge dx_3$ is a positive orientation.) We have

$$\mathbf{E} = E_1 dx_1 + E_2 dx_2 + E_3 dx_3,$$

which implies that

$$*\mathbf{E} = E_1\, dx_2 \wedge dx_3 + E_2\, dx_3 \wedge dx_1 + E_3\, dx_1 \wedge dx_2.$$

Now we can take the divergence of this by applying the de Rham differential. So Coulomb's law becomes

$$d(*\mathbf{E}) = (4\pi\rho)\, dx_1 \wedge dx_2 \wedge dx_3,$$

$$\text{or}\quad *d(*\mathbf{E}) = 4\pi\rho.$$

Finally, for the Ampère–Maxwell law, we see through similar reasoning that this becomes

$$*d(*\mathbf{B}) - \frac{\partial \mathbf{E}}{\partial t} = 4\pi\mathbf{J},$$

where $\mathbf{J} = J_1 dx_1 + J_2 dx_2 + J_3 dx_3$ is a 1-form.

We now collect these new versions of Maxwell's equations in one place, but with a slight change in notation. We write d_3 instead of d to indicate that this is the de Rham differential associated to three-dimensional space. Here they are:

$$d_3\mathbf{B} = 0, \qquad \text{no magnetic charges}$$

$$d_3\mathbf{E} + \frac{\partial \mathbf{B}}{\partial t} = 0, \qquad \text{Faraday's law}$$

$$*d_3(*\mathbf{E}) = 4\pi\rho, \qquad \text{Coulomb's law}$$

$$*d_3(*\mathbf{B}) - \frac{\partial \mathbf{E}}{\partial t} = 4\pi\mathbf{J}. \qquad \text{Ampére–Maxwell's law}$$

Recall that all of these objects are functions of (t, x_1, x_2, x_3), which is a point in the *Minkowski space* $\mathbb{R}^4$, also called *spacetime*, used in special relativity theory.

However, ρ is a scalar, $\mathbf{J}$ is a 1-form, and the spatial derivatives are hidden in the notation d_3 while the time derivatives are visible.

In short, we do not have everything in the notation of special relativity. This is an additional condition that arises from considering the physics of the situation. While this is a quite natural consideration for a physicist, for a mathematician this might seem like a bolt from the blue. It turns out that the scalar ρ and the vector J fit together nicely to give a map $j : U \to \mathbb{R}^4$ defined by

$$U \ni p \xrightarrow{\ j\ } (\rho(p), J(p)) \in \mathbb{R} \times \mathbb{R}^3 = \mathbb{R}^4.$$

One says that j is a 4-*vector*, while J is a 3-*vector*.

Now the electric and magnetic field give six independent scalar fields, and it seems that this just cannot fit into the scheme of relativistic quantities such as 4-vectors. Well, quite so as far as 4-vectors are concerned. But there are also tensors in $V \otimes V$, where $V = \mathbb{R}^4$ is the Minkowski space. But these have 16 independent entries! Where can we find six dimensions in special relativity? Well, $\Lambda^2(V^*) = V^* \wedge V^*$ has dimension

$$\binom{4}{2} = \frac{4!}{2!\,2!} = 6.$$

And a 2-form on an open set $\tilde{U} \subset \mathbb{R}^4$ is a smooth function $\mathbf{F} : \tilde{U} \to \Lambda^2(V^*)$. (Apologies! This 2-form will be called the field strength and should not be confused with the force 1-form used earlier.) And the electric and magnetic fields are functions on such an open set $\tilde{U}$. Let's use the standard coordinates (t, x_1, x_2, x_3) on $\tilde{U}$. We will not use the more common notation $x_0 = t$ as used in physics. But we mention it since many other books have formulas in terms of the coordinates (x_0, x_1, x_2, x_3). Using our notation, we see that the standard associated basis of $\Lambda^2(V^*)$ has three elements with the time infinitesimal, namely,

$$dt \wedge dx_1, \quad dt \wedge dx_2, \quad dt \wedge dx_3.$$

The remaining three basis elements contain only spacial infinitesimals, namely,

$$dx_2 \wedge dx_3, \quad dx_3 \wedge dx_1, \quad dx_1 \wedge dx_2.$$

Since we already have the magnetic field 2-form written in terms of these spacial infinitesimals, it makes sense to use the same expression here. So we will define the field strength 2-form as something like

$$\mathbf{F} = B_1\, dx_2 \wedge dx_3 + B_2\, dx_3 \wedge dx_1 + B_3\, dx_1 \wedge dx_2 + \text{more stuff}$$

$$= \mathbf{B} + \text{more stuff},$$

where we have yet to fill in the mixed temporal-spatial infinitesimals. Clearly, that is where we have to put in the electric field **E**, a 1-form. But we are looking for a 2-form! But there is not much choice in the matter since $dt \wedge \mathbf{E}$ is a 2-form with the mixed temporal-spatial infinitesimals. So we define

$$\mathbf{F} := \mathbf{B} - dt \wedge \mathbf{E}$$

$$B_1\,dx_2 \wedge dx_3 + B_2\,dx_3 \wedge dx_1 + B_3\,dx_1 \wedge dx_2$$

$$-\Big(E_1\,dt \wedge dx_1 + E_2\,dt \wedge dx_2 + E_3\,dt \wedge dx_3\Big). \tag{12.14}$$

The only "mystery" about this definition is the minus sign. But if I had put $+\mathbf{E} \wedge dt$ for the second term, no one would have asked, "Why a plus sign?" Let's see what the exterior derivative d (which is not d_3) of this 2-form is in this four-dimensional setting. Then we calculate directly

$$d\mathbf{F} = \Big(\frac{\partial B_1}{\partial t}dt + \frac{\partial B_1}{\partial x_1}dx_1\Big) \wedge dx_2 \wedge dx_3 + \Big(\frac{\partial B_2}{\partial t}dt + \frac{\partial B_2}{\partial x_2}dx_2\Big) \wedge dx_3 \wedge dx_1$$

$$+\Big(\frac{\partial B_3}{\partial t}dt + \frac{\partial B_3}{\partial x_3}dx_3\Big) \wedge dx_1 \wedge dx_2 - \Big(\frac{\partial E_1}{\partial x_2}dx_2 + \frac{\partial E_1}{\partial x_3}dx_3\Big) \wedge dt \wedge dx_1$$

$$-\Big(\frac{\partial E_2}{\partial x_1}dx_1 + \frac{\partial E_2}{\partial x_3}dx_3\Big) \wedge dt \wedge dx_2 - \Big(\frac{\partial E_3}{\partial x_1}dx_1 + \frac{\partial E_3}{\partial x_2}dx_2\Big) \wedge dt \wedge dx_3.$$

Now we have 12 terms here, each of which is a multiple of a standard basis element in a four-dimensional space. So we combine terms, getting

$$d\mathbf{F} = \Big(\frac{\partial B_1}{\partial x_1} + \frac{\partial B_2}{\partial x_2} + \frac{\partial B_3}{\partial x_3}\Big)dx_1 \wedge dx_2 \wedge dx_3$$

$$+\Big(\frac{\partial B_1}{\partial t} - \frac{\partial E_2}{\partial x_3} + \frac{\partial E_3}{\partial x_2}\Big)dt \wedge dx_2 \wedge dx_3$$

$$+\Big(\frac{\partial B_2}{\partial t} + \frac{\partial E_1}{\partial x_3} - \frac{\partial E_3}{\partial x_1}\Big)dt \wedge dx_3 \wedge dx_1$$

$$+\Big(\frac{\partial B_3}{\partial t} - \frac{\partial E_1}{\partial x_2} + \frac{\partial E_2}{\partial x_1}\Big)dt \wedge dx_1 \wedge dx_2.$$

So the equation $d\mathbf{F} = 0$ encodes Faraday's law (12.10) in the last three terms and the absence of magnetic charge (12.9) in the first term. These are the homogeneous equations in Maxwell's equations. Notice that this worked out nicely because of the choice of the mysterious minus sign in (12.14).

We still have to write the nonhomogeneous Maxwell's equations in this formalism. To do this, we will compute the Hodge star of **F** with respect to the oriented, orthonormal basis dt, dx_1, dx_2, dx_3. As we remarked earlier, the Minkowski space $\mathbb{R}^4$ has this nondegenerate, but not positive definite, inner product:

$$g(v, w) = v_0 w_0 - v_1 w_1 - v_2 w_2 - v_3 w_3,$$

where $v = (v_0, v_1, v_3, v_4)$ and $w = (w_0, w_1, w_3, w_4)$ are in $\mathbb{R}^4$. In fact, the more conventional notation for Minkowski space is $\mathbb{R}^{(1,3)}$, which indicates that there is one plus sign and three minus signs in the inner product. As vector spaces and differentiable manifolds, $\mathbb{R}^{(1,3)}$ and $\mathbb{R}^4$ are identical. The difference between them resides in the inner product. But the invariant definition of the Hodge star in $\mathbb{R}^{(1,3)}$, denoted $\star$, involves its inner product g, and so is not the Hodge star $*$ in $\mathbb{R}^4$, with its standard positive definite inner product.

For now we note these identities, all of which follow from Corollary 5.1:

$$\star(dx_1 \wedge dx_2) = -dt \wedge dx_3,$$

$$\star(dx_3 \wedge dx_1) = -dt \wedge dx_2,$$

$$\star(dx_2 \wedge dx_3) = -dt \wedge dx_1,$$

$$\star(dt \wedge dx_1) = dx_2 \wedge dx_3,$$

$$\star(dt \wedge dx_2) = dx_3 \wedge dx_1,$$

$$\star(dt \wedge dx_3) = dx_1 \wedge dx_2.$$

Here are some more Hodge stars, also calculated with Corollary 5.1:

$$\star(dx_1 \wedge dx_2 \wedge dx_3) = -dt,$$

$$\star(dt \wedge dx_2 \wedge dx_3) = -dx_1,$$

$$\star(dt \wedge dx_3 \wedge dx_1) = -dx_2,$$

$$\star(dt \wedge dx_1 \wedge dx_2) = -dx_3.$$

So we see that

$$\star\mathbf{F} = B_1 \star(dx_2 \wedge dx_3) + B_2 \star(dx_3 \wedge dx_1) + B_3 \star(dx_1 \wedge dx_2)$$
$$- E_1 \star(dt \wedge dx_1) - E_2 \star(dt \wedge dx_2) - E_3 \star(dt \wedge dx_3)$$
$$= -B_1 (dt \wedge dx_1) - B_2 (dt \wedge dx_2) - B_3 (dt \wedge dx_3)$$
$$- E_1 (dx_2 \wedge dx_3) - E_2 (dx_3 \wedge dx_1) - E_3 (dx_1 \wedge dx_2).$$

Next, we evaluate the exterior derivative of this 2-form, getting

$$d(\star\mathbf{F}) = dB_1 \wedge dt \wedge dx_1 + dB_2 \wedge dt \wedge dx_2 + dB_3 \wedge dt \wedge dx_3$$
$$+ dE_1 \wedge dx_2 \wedge dx_3 + dE_2 \wedge dx_3 \wedge dx_1 + dE_3 \wedge dx_1 \wedge dx_2$$
$$= \left(\frac{\partial B_1}{\partial x_2} dx_2 + \frac{\partial B_1}{\partial x_3} dx_3 \right) \wedge dt \wedge dx_1 + \left(\frac{\partial B_2}{\partial x_1} dx_1 + \frac{\partial B_2}{\partial x_3} dx_3 \right) \wedge dt \wedge dx_2$$

$$+ \left(\frac{\partial B_3}{\partial x_1} dx_1 + \frac{\partial B_3}{\partial x_2} dx_2 \right) \wedge dt \wedge dx_3 + \left(\frac{\partial E_1}{\partial t} dt + \frac{\partial E_1}{\partial x_1} dx_1 \right) \wedge dx_2 \wedge dx_3$$

$$+ \left(\frac{\partial E_2}{\partial t} dt + \frac{\partial E_2}{\partial x_2} dx_2 \right) \wedge dx_3 \wedge dx_1 + \left(\frac{\partial E_3}{\partial t} dt + \frac{\partial E_3}{\partial x_3} dx_3 \right) \wedge dx_1 \wedge dx_2.$$

Now, combining terms, we see that

$$-d(\star \mathbf{F}) = \left(\frac{\partial E_1}{\partial x_1} + \frac{\partial E_2}{\partial x_2} + \frac{\partial E_3}{\partial x_3} \right) dx_1 \wedge dx_2 \wedge dx_3$$

$$+ \left(\frac{\partial E_1}{\partial t} + \frac{\partial B_2}{\partial x_3} - \frac{\partial B_3}{\partial x_2} \right) dx_2 \wedge dx_3 \wedge dt$$

$$+ \left(\frac{\partial E_2}{\partial t} - \frac{\partial B_1}{\partial x_3} + \frac{\partial B_3}{\partial x_1} \right) dt \wedge dx_3 \wedge dx_1$$

$$+ \left(\frac{\partial E_3}{\partial t} + \frac{\partial B_1}{\partial x_2} - \frac{\partial B_2}{\partial x_1} \right) dt \wedge dx_1 \wedge dx_2.$$

Taking the Hodge star of this gives us

$$\star d(\star \mathbf{F}) = \left(\frac{\partial E_1}{\partial x_1} + \frac{\partial E_2}{\partial x_2} + \frac{\partial E_3}{\partial x_3} \right) dt + \left(\frac{\partial E_1}{\partial t} + \frac{\partial B_2}{\partial x_3} - \frac{\partial B_3}{\partial x_2} \right) dx_1$$

$$+ \left(\frac{\partial E_2}{\partial t} - \frac{\partial B_1}{\partial x_3} + \frac{\partial B_3}{\partial x_1} \right) dx_2 + \left(\frac{\partial E_3}{\partial t} + \frac{\partial B_1}{\partial x_2} - \frac{\partial B_2}{\partial x_1} \right) dx_3.$$

So

$$\star d(\star \mathbf{F}) = 4\pi \mathbf{J},$$

where we define $\mathbf{J} := \rho \, dt - J_1 \, dx_1 - J_2 \, dx_2 - J_3 \, dx_3$, which is called the *density* 1-*form* and is equivalent to Coulomb's law (12.11) and Ampère–Maxwell's law (12.12) together. These are the inhomogeneous equations in Maxwell's equations. The minus signs in this definition of $\mathbf{J}$ give us a Lorentz covariant 1-form. Think of it as the Minkowski inner product of the 4-vector (ρ, J) with the infinitesimal 4-vector (dt, dx) if you like.

Resuming, here are Maxwell's equations written in terms of differential forms:

$$d\mathbf{F} = 0, \tag{12.15}$$

$$\star d(\star \mathbf{F}) = 4\pi \mathbf{J}, \tag{12.16}$$

where $\mathbf{F}$ is the field strength 2-form of the electromagnetic field and $\mathbf{J}$ is the density 1-form, as defined above. It is worth emphasizing once more that d here is the exterior derivative in $\mathbb{R}^4$.

12.3 Gauge Theory of Maxwell's Equations

Equation (12.15), $d\mathbf{F} = 0$, says that the field strength 2-form is closed. So if the domain $\tilde{U}$ under consideration is contractible (for example), then we can find a globally defined 1-form $\mathbf{A}$ such that

$$\mathbf{F} = d\mathbf{A}.$$

Here the word "globally" means that the domain of the section $\mathbf{A}$ is $\tilde{U}$. We call $\mathbf{A}$ a *potential* or *gauge field*.

Even when the topology of $\tilde{U}$ obstructs the existence of such a globally defined potential, we can always find potentials for $\mathbf{F}$ that are defined in small enough open, contractible subsets of $\tilde{U}$.

One of the most important properties is that the potential $\mathbf{A}$ is not unique. To see this, take any $f \in \Omega^0(\tilde{U}) = C^\infty(\tilde{U})$ and any potential $\mathbf{A}$ of $\mathbf{F}$. Then define $\mathbf{A}' := \mathbf{A} + df$. Clearly, $\mathbf{A}'$ is also a potential of $\mathbf{F}$. This change from $\mathbf{A}$ to $\mathbf{A}'$ is called a *gauge transformation*. Since in classical electrodynamics the dynamics depends only on $\mathbf{F}$, a given potential in and of itself has no innate physical significance, but rather the class of potentials determined by it under all possible gauge transformations.

Exercise 12.1 *We say that the 1-forms $\mathbf{A}$ and $\mathbf{A}'$ are gauge equivalent if there is a gauge transformation changing $\mathbf{A}$ into $\mathbf{A}'$. Show that this is an equivalence relation on the space $\Omega^1(\tilde{U})$ of all 1-forms on $\tilde{U}$.*

Definition 12.1 *We say that a gauge field $\mathbf{A}$ is pure gauge if its associated field strength $\mathbf{F} = d\mathbf{A}$ satisfies $\mathbf{F} = 0$.*

So the gauge fields that are pure gauge form the gauge equivalence class of the gauge field $\mathbf{A} = 0$. And these are exactly the gauge fields that have zero field strength, that is, those with zero electric field and zero magnetic field.

The physics (meaning the time evolution of physical systems) is found by solving Maxwell's equations for the fields E and B and then using Lorentz's force law and Newton's second law of motion to understand how test charged particles move in the electromagnetic field. And for this we only need the gauge equivalence class that gives us E and B. However, just a single potential in the gauge equivalence class is a useful mathematical tool because if we can find one, then we can immediately calculate the electromagnetic field using $\mathbf{F} = d\mathbf{A}$ and proceed as described above. So we are interested in studying potentials.

We write $\mathbf{A} = -\phi \, dt + A_1 \, dx_1 + A_2 \, dx_2 + A_3 \, dx_3$ (Lorentz covariance again) and calculate

$$d\mathbf{A} = -\left(\frac{\partial \phi}{\partial x_1} dx_1 + \frac{\partial \phi}{\partial x_2} dx_2 + \frac{\partial \phi}{\partial x_3} dx_3\right) \wedge dt$$

$$+ \left(\frac{\partial A_1}{\partial t} dt + \frac{\partial A_1}{\partial x_2} dx_2 + \frac{\partial A_1}{\partial x_3} dx_3\right) \wedge dx_1$$

$$+ \left(\frac{\partial A_2}{\partial t} dt + \frac{\partial A_2}{\partial x_1} dx_1 + \frac{\partial A_2}{\partial x_3} dx_3 \right) \wedge dx_2$$

$$+ \left(\frac{\partial A_3}{\partial t} dt + \frac{\partial A_3}{\partial x_1} dx_1 + \frac{\partial A_3}{\partial x_2} dx_2 \right) \wedge dx_3.$$

This gives 12 terms, each of which is a multiple of a standard basis element in the six-dimensional space. Combining, we get these six terms:

$$d\mathbf{A} = \left(\frac{\partial \phi}{\partial x_1} + \frac{\partial A_1}{\partial t} \right) dt \wedge dx_1 + \left(\frac{\partial \phi}{\partial x_2} + \frac{\partial A_2}{\partial t} \right) dt \wedge dx_2$$

$$+ \left(\frac{\partial \phi}{\partial x_3} + \frac{\partial A_3}{\partial t} \right) dt \wedge dx_3 + \left(\frac{\partial A_2}{\partial x_1} - \frac{\partial A_1}{\partial x_2} \right) dx_1 \wedge dx_2$$

$$+ \left(\frac{\partial A_3}{\partial x_2} - \frac{\partial A_2}{\partial x_3} \right) dx_2 \wedge dx_3 + \left(\frac{\partial A_1}{\partial x_3} - \frac{\partial A_3}{\partial x_1} \right) dx_3 \wedge dx_1.$$

When we use $\mathbf{F} = d\mathbf{A}$ and the identification of $\mathbf{F}$ with $B - dt \wedge E$, this means in vector calculus notation that

$$B = \nabla \times A \quad \text{and} \quad E = -\nabla \phi - \frac{\partial A}{\partial t}, \tag{12.17}$$

where $A = (A_1, A_2, A_3)$ is called the *vector potential* and ϕ is called the *scalar potential*. It is important to emphasize that for any (smooth) A and ϕ (or equivalently for any $\mathbf{A}$), this gives us a solution of the homogeneous Maxwell's equations. That is a lot of solutions!

Exercise 12.2 *Check by substitution that any E and B as given in (12.17) is a solution of the homogeneous Maxwell's equations.*

Then $\mathbf{A}$ can be substituted into Maxwell's nonhomogeneous equations to give the remaining equations for $\mathbf{A}$. Notice that the original Maxwell's equations have six unknown scalar fields, which are the components of the vectors E and B. These have now been reduced to four unknown scalar fields that have to satisfy these four scalar equations:

$$\Delta \phi + \frac{\partial}{\partial t} \left(\nabla \cdot A \right) = -4\pi\rho, \quad \text{Coulomb's law} \tag{12.18}$$

$$\Delta A - \frac{\partial^2 A}{\partial t^2} - \nabla \left(\nabla \cdot A + \frac{\partial \phi}{\partial t} \right) = -4\pi J, \quad \text{Ampére–Maxwell's law} \tag{12.19}$$

where

$$\Delta = \frac{\partial^2}{\partial x_1^2} + \frac{\partial^2}{\partial x_2^2} + \frac{\partial^2}{\partial x_3^2}$$

is the famous *Laplacian operator*. These are coupled scalar equations since all of the unknown scalar functions ϕ, A_1, A_2, A_3 appear in each one of these scalar equations. But we still have some free play to address this question, as we will shortly see.

And we already have seen that **A** is not uniquely determined but can be changed within its gauge equivalence class. So it is often convenient in physics to impose some extra condition on **A**. For example, if we require the *Lorenz condition*

$$\nabla \cdot A + \frac{\partial \phi}{\partial t} = 0,$$

then we say that **A** is in the *Lorenz gauge*. The motivation for doing this is that in the Lorenz gauge the nonhomogeneous Maxwell's equations (12.18) and (12.19) become

$$\Delta \phi - \frac{\partial^2 \phi}{\partial t^2} = -4\pi\rho, \quad \text{Coulomb's law}$$

$$\Delta A - \frac{\partial^2 A}{\partial t^2} = -4\pi J, \quad \text{Ampére–Maxwell's law}$$

Now we have four decoupled scalar equations since the first is for ϕ only, the second for A_1 only, and so forth. Moreover, these equations all have the same form, which is well known in mathematical physics; they are the wave equation with source term. The only difference among these four equations is that in general we have four different source terms on the right side. These are still, in general, nonhomogeneous equations. And we remind the reader that they are still highly nontrivial since the source terms ρ and J depend in general on time t and all the spatial coordinates x_1, x_2, and x_3.

And, even so, there are nontrivial gauge transformations that preserve the Lorenz condition.

Exercise 12.3 *Show that the gauge transformation* **A** $\mapsto$ **A** $+ df$ *preserves the Lorenz condition if f satisfies the* Klein–Gordon equation

$$\frac{\partial^2 f}{\partial t^2} - \Delta f = 0.$$

As the reader familiar with the theory of special relativity might have already recognized, all these extra equations in the Lorenz gauge are invariant under Lorentz transformations. And the reader will also notice that the name of the Danish physicist Ludvig Lorenz (1829–1891) is used for the gauge, while the name of the Dutch physicist Hendrik Lorentz (1853–1928) is used for the transformations of relativity theory. Just to keep things complicated, the Lorenz gauge was first used by the Irish physicist George Francis FitzGerald (1851–1901), who is most famous for the Lorentz–FitzGerald contraction.

Another class of potentials is determined by the condition

$$\nabla \cdot A = 0.$$

This condition is not invariant under Lorentz transformations. It is known by (at least!) three different names: Coulomb gauge, radiation gauge, and transverse gauge. Again, this is a class of gauge potentials that are related by gauge transformations among themselves. This gauge is handy when there is a particular reference frame that we wish to use in a particular problem. In this gauge, (12.18) becomes Poisson's equation for ϕ and is independent of A. So we can solve (12.18) for ϕ and then substitute this into (12.19), which then becomes three decoupled equations in the three scalar unknowns A_1, A_2, and A_3.

In the static (or steady-state) case, all quantities are independent of time. And so $E = -\nabla\phi$. Combining this with Coulomb's law $\nabla \cdot E = 4\pi\rho$, we arrive at *Poisson's equation*

$$\Delta\phi = -4\pi\rho.$$

Just this one special case of Maxwell's equations has been intensely studied. It gives rise to a whole area of mathematics known as *potential theory*. In the ever more particular case when $\rho = 0$, we have *Laplace's equation*

$$\Delta\phi = 0,$$

whose solutions are called *harmonic functions*. This paragraph so far is about a whole world unto itself known as *electrostatics*. There is also a world of mathematics that comes from *magnetostatics*, which is the case of time-independent magnetic fields. This is a lot of classical mathematics motivated by physics.

12.4   Gauge Transformations and Quantum Mechanics

We now see how the world of classical electrodymanics relates to one basic aspect of quantum mechanical theory, the Schrödinger equation. This falls short of quantum electrodynamics, but it will give us some idea of how and why gauge transformations are important also in quantum theories. As we have seen on other occasions, the basic equations of physics are not proved in a mathematical sense but rather are an encapsulation of many experimental results in a formula. The Schrödinger equation is no exception. It is much like Newton's second law of motion, $F = ma$, in that it is really a family of equations that is meant to describe a wide variety of physical systems. Here is one form of it that describes a one-particle quantum system in $\mathbb{R}^3$ that is under the influence of a scalar potential energy function, denoted by $V : \mathbb{R}^3 \to \mathbb{R}$ and zero magnetic field:

$$i\hbar \frac{\partial \psi}{\partial t} = -\frac{\hbar^2}{2m} \Delta \psi + V\psi. \qquad (12.20)$$

As expected, a lot of explanation is in order. There are three quite different types of constants in this equation. The first is i, the square root of -1, a mathematical constant. The second is Planck's constant $\hbar > 0$, a universal constant of physics. And the third is the mass $m > 0$ of the particle under consideration, a case-specific physical constant.

The Schrödinger equation is a second-order partial differential equation because of the Laplacian operator Δ in $\mathbb{R}^3$. It is a linear equation in the unknown $\psi : \mathbb{R}^3 \to \mathbb{C}$ since the differential operators are linear and so is the operator $\psi \mapsto V\psi$, which is pointwise multiplication by the function V. It also has a first-order partial derivative with respect to time t and so leads to a Cauchy problem in the theory of partial differential equations. This problem is to find the solution $\psi(t, x_1, x_2, x_3)$ for all $t \geq 0$ given the initial condition

$$\psi(0, x_1, x_2, x_3) = \phi(x_1, x_2, x_3)$$

for some given function ϕ. A mathematical framework within the theory of Hilbert spaces has been established where this Cauchy problem has a unique solution under quite general hypotheses, and in that setting this is a deterministic model with no probabilistic aspects at all.

The most difficult thing to explain about the Schrödinger equation is its solution, the so-called wave function ψ. This is where both probability and nonlinearity enter into the theory. One restricts attention to those solutions of the Schrödinger equation that also satisfy the *normalization condition*

$$\int_{\mathbb{R}^3} |\psi|^2 = 1, \qquad (12.21)$$

which is a nonlinear condition on the solution ψ. Then one can and does interpret the nonnegative real-valued function

$$|\psi|^2 : \mathbb{R}^3 \to \mathbb{R}$$

as a probability density with respect to the Lebesgue measure on $\mathbb{R}^3$. How this probability density should be related to experimental measurements is a long, complicated, and controversial issue. But the interpretation of many experimental results has something to do with this probability density (or with some other secondary expression that involves the solution wave function ψ in a nonlinear way). Fortunately, we need not go into any of that beyond emphasizing that the wave function ψ itself does not have any known direct physical interpretation although $|\psi|^2$ does. So if we replace a solution ψ with $e^{ir}\psi$, where $r \in \mathbb{R}$ is an arbitrary real number, we will again have a solution of the Schrödinger equation satisfying the normalization condition (12.21). [The fact that the initial condition changes by the factor e^{ir} is not important.] One puts this into words by saying that the overall phase factor of the wave function has no physical significance. Notice that this phase factor does not depend on the coordinates of space and time.

But then a physicist is led to ask if there would be a change in physical significance due to a change of phase factor that *did* depend on space and time coordinates. And the answer has to be no. The idea is that if one changes the phase at some point in spacetime in one way and at other points in spacetime in other ways, then the wave function should be describing the same physical system since the normalization condition (12.21), which is the principal aspect of the physical interpretation, remains unchanged. But what about the Schrödinger equation? It will change, and shortly we will see exactly how, but its new form must then be interpreted as describing the same physical system. Let's now see what this all means in mathematical detail.

We are interested in studying how the transformation

$$\psi \mapsto \tilde{\psi} = e^{i\Phi} \psi, \tag{12.22}$$

where $\Phi : \mathbb{R}^4 \to \mathbb{R}$ is an arbitrary smooth function, changes the Schrödinger equation (12.20). The codomain of the so-called phase Φ is important since we want to preserve the normalization condition (12.21). Actually, something stronger holds. Since Φ is real-valued, $|\tilde{\psi}|^2 = |\psi|^2$, which tells us that the associated probability density does not change.

Inverting this transformation yields

$$\psi = e^{-i\Phi} \tilde{\psi}. \tag{12.23}$$

Now given that ψ satisfies (12.20), we want to find the transformed equation that $\tilde{\psi}$ satisfies. To do this, we start by calculating the partial derivative of (12.23) with respect to time t:

$$\frac{\partial \psi}{\partial t} = -ie^{-i\Phi} \frac{\partial \Phi}{\partial t} \tilde{\psi} + e^{-i\Phi} \frac{\partial \tilde{\psi}}{\partial t}. \tag{12.24}$$

This is enough for the left side of (12.20). But for the right side of that equation, we have to calculate the Laplacian $\Delta \psi$ in terms of $\tilde{\psi}$. So we start by taking the first partial derivative with respect to the spatial coordinate x_j and get (12.24) with x_j in place of t:

$$\frac{\partial \psi}{\partial x_j} = -ie^{-i\Phi} \frac{\partial \Phi}{\partial x_j} \tilde{\psi} + e^{-i\Phi} \frac{\partial \tilde{\psi}}{\partial x_j}. \tag{12.25}$$

Now we continue by taking the partial of (12.25) with respect to x_j again, thereby getting

$$\frac{\partial^2 \psi}{\partial x_j^2} = -e^{-i\Phi} \left(\frac{\partial \Phi}{\partial x_j} \right)^2 \tilde{\psi} - ie^{-i\Phi} \frac{\partial^2 \Phi}{\partial x_j^2} \tilde{\psi} - ie^{-i\Phi} \frac{\partial \Phi}{\partial x_j} \frac{\partial \tilde{\psi}}{\partial x_j}$$

$$- ie^{-i\Phi} \frac{\partial \Phi}{\partial x_j} \frac{\partial \tilde{\psi}}{\partial x_j} + e^{-i\Phi} \frac{\partial^2 \tilde{\psi}}{\partial x_j^2}. \qquad (12.26)$$

Next, we sum on $j = 1, 2, 3$ and factor out $e^{-i\Phi}$ to get

$$\Delta \psi = e^{-i\Phi} \left(-(\nabla \Phi)^2 \tilde{\psi} - i(\Delta \Phi) \tilde{\psi} - 2i (\nabla \Phi) \cdot (\nabla \tilde{\psi}) + \Delta \tilde{\psi} \right). \qquad (12.27)$$

So applying (12.23) to (12.20), we get this equation for $\tilde{\psi}$:

$$i\hbar \left(-ie^{-i\Phi} \frac{\partial \Phi}{\partial t} \tilde{\psi} + e^{-i\Phi} \frac{\partial \tilde{\psi}}{\partial t} \right) =$$

$$-\frac{\hbar^2}{2m} e^{-i\Phi} \left(-(\nabla \Phi)^2 \tilde{\psi} - i(\Delta \Phi) \tilde{\psi} - 2i (\nabla \Phi) \cdot (\nabla \tilde{\psi}) + \Delta \tilde{\psi} \right) + Ve^{-i\Phi} \tilde{\psi}.$$

Canceling out the everywhere nonzero factor $e^{-i\Phi}$ and isolating the time derivative on the right side, we simplify this to the following:

$$i\hbar \frac{\partial \tilde{\psi}}{\partial t} = \qquad (12.28)$$

$$-\frac{\hbar^2}{2m} \left(-(\nabla \Phi)^2 \tilde{\psi} - i(\Delta \Phi) \tilde{\psi} - 2i (\nabla \Phi) \cdot (\nabla \tilde{\psi}) + \Delta \tilde{\psi} \right) + \left(V - \hbar \frac{\partial \Phi}{\partial t} \right) \tilde{\psi}.$$

Hence, we have changed the potential energy term in a rather simple way by introducing another energy term. (My readers who are physicists will have already recognized that the new term, $\hbar \, \partial \Phi / \partial t$, has dimensions of energy.) However, it seems that we have three incomprehensible terms as well as a Laplacian, for a total of four terms in place of just one simple term: the Laplacian. What a mess! Those who see order in chaos are to be greatly honored. And there is order in this expression. Without any motivation whatsoever, we think of the operators

$$\frac{\partial}{\partial x_j} - i \frac{\partial \Phi}{\partial x_j}$$

for $j = 1, 2, 3$, where the second term means the operator of multiplying by $i \, \partial \Phi / \partial x_j$. Next, we compute the square of this operator acting on $\tilde{\psi}$, which for the time being could be just any smooth function. So we obtain

$$\left(\frac{\partial}{\partial x_j} - i \frac{\partial \Phi}{\partial x_j} \right)^2 \tilde{\psi} = \left(\frac{\partial}{\partial x_j} - i \frac{\partial \Phi}{\partial x_j} \right) \left(\frac{\partial \tilde{\psi}}{\partial x_j} - i \frac{\partial \Phi}{\partial x_j} \tilde{\psi} \right)$$

$$= \frac{\partial^2 \tilde{\psi}}{\partial x_j^2} - i \frac{\partial^2 \Phi}{\partial x_j^2} \tilde{\psi} - i \frac{\partial \Phi}{\partial x_j} \frac{\partial \tilde{\psi}}{\partial x_j} - i \frac{\partial \Phi}{\partial x_j} \frac{\partial \tilde{\psi}}{\partial x_j} - \left(\frac{\partial \Phi}{\partial x_j}\right)^2 \tilde{\psi}$$

$$= \frac{\partial^2 \tilde{\psi}}{\partial x_j^2} - i \frac{\partial^2 \Phi}{\partial x_j^2} \tilde{\psi} - 2i \frac{\partial \Phi}{\partial x_j} \frac{\partial \tilde{\psi}}{\partial x_j} - \left(\frac{\partial \Phi}{\partial x_j}\right)^2 \tilde{\psi}.$$

Taking the summation on j then gives

$$\sum_{j=1}^{3} \left(\frac{\partial}{\partial x_j} - i \frac{\partial \Phi}{\partial x_j}\right)^2 \tilde{\psi} = \Delta \tilde{\psi} - i (\Delta \Phi) \tilde{\psi} - 2i (\nabla \Phi) \cdot (\nabla \tilde{\psi}) - (\nabla \Phi)^2 \tilde{\psi},$$

which are exactly the four messy terms in (12.28). Having now arrived at this expression, the reader can surely appreciate that the above unmotivated considerations were well worth the effort.

So we rewrite (12.28) as

$$i\hbar \frac{\partial \tilde{\psi}}{\partial t} = -\frac{\hbar^2}{2m} \sum_{j=1}^{3} \left(\frac{\partial}{\partial x_j} - i \frac{\partial \Phi}{\partial x_j}\right)^2 \tilde{\psi} + \left(V - \hbar \frac{\partial \Phi}{\partial t}\right) \tilde{\psi}.$$

Then we see that the Laplacian in (12.20),

$$\Delta \psi = \sum_{j=1}^{3} \frac{\partial^2 \psi}{\partial x_j^2},$$

has been replaced by

$$\sum_{j=1}^{3} \left(\frac{\partial}{\partial x_j} - i \frac{\partial \Phi}{\partial x_j}\right)^2 \tilde{\psi}.$$

One expresses this by saying that the transformation $\psi \mapsto \tilde{\psi} = e^{i\Phi} \psi$ must be done simultaneously with the transformations

$$V \mapsto V - \hbar \frac{\partial \Phi}{\partial t} \quad \text{and} \quad \frac{\partial}{\partial x_j} \mapsto \frac{\partial}{\partial x_j} - i \frac{\partial \Phi}{\partial x_j}.$$

At this point, it is much more convenient to make two changes that will clean up these formulas. First, we pick physical units so that $\hbar = 1$, a dimensionless constant. Second, we replace Φ with $-q\Phi$, where q is the electric charge of the particle being described all along here. So now $\psi \mapsto \tilde{\psi} = e^{-iq\Phi} \psi$. (This has the upshot that the new Φ has nontrivial dimensions, those of inverse electric charge.) Then the above replacements read as

$$V \mapsto V + q\frac{\partial \Phi}{\partial t} \quad \text{and} \quad \frac{\partial}{\partial x_j} \mapsto \frac{\partial}{\partial x_j} + iq\frac{\partial \Phi}{\partial x_j}. \tag{12.29}$$

So $\partial\Phi/\partial t$ when multiplied by electric charge is an energy. This means that $\partial\Phi/\partial t$ is an *electric potential* function, which is also known as a *voltage*. (It is measured in the *Système International (SI) d'Unités* in *volts*. But beware, since $\hbar = 1$ is not part of the SI.) This leads us to interpret the remaining three spatial partials of Φ as the components of the magnetic potential. So we have seen (up to a factor of electric charge) that Φ corresponds to f in a classical gauge transformation $\mathbf{A} \mapsto \mathbf{A}' = \mathbf{A} + df$.

The reader might be aware that the imaginary unit $i = \sqrt{-1}$ enters (12.29) in a seemingly asymmetrical way, that is, only in the transformation of the three spatial derivatives. However, if we write those substitutions equivalently as

$$\frac{1}{i}\frac{\partial}{\partial x_j} \mapsto \frac{1}{i}\frac{\partial}{\partial x_j} + q\frac{\partial \Phi}{\partial x_j},$$

then we see that we are transforming symmetric operators into yet other symmetric operators since both q and Φ are real. And the transformation of V also takes a symmetric operator to a symmetric operator. These symmetric operators will be *self-adjoint* (or *essentially self-adjoint*) under certain rather general hypotheses and so will have an interpretation in quantum mechanics in terms of real-valued quantities that are measurable experimentally. This is a whole theory in and of itself within functional analysis, which we will not discuss further, but we refer the reader to the four volumes of Reed and Simon.

Another asymmetry concerns the transformation of the potential energy, a real-valued function, in the first transformation given in (12.29), while the remaining transformations in (12.29) change first-order partial differential operators. However, the potential energy gauge transformation is equivalent to this transformation of a first-order partial differential operator:

$$\frac{1}{i}\frac{\partial}{\partial t} \mapsto \frac{1}{i}\frac{\partial}{\partial t} + q\frac{\partial \Phi}{\partial t}.$$

Exercise 12.4 *Prove this last statement.*

In general, if $\mathbf{A} = (\phi, A_1, A_2, A_3)$ is the 4-potential of an electromagnetic field, the Schrödinger equation of a single particle of mass m and electric charge q in this electromagnetic field will be

$$i\hbar\frac{\partial \psi}{\partial t} = -\frac{\hbar^2}{2m}\sum_{j=1}^{3}\left(\frac{\partial}{\partial x_j} + iq A_j\right)^2 \psi + (q\phi)\,\psi. \tag{12.30}$$

Please note we have not proved anything in the rigorous mathematical sense, but rather we have given some physics motivation for the previous sentence.

Notice that the electric potential ϕ can depend on time and so the energy of the particle need not be constant in time. This makes sense physically since the electric field is interacting with the particle and thereby possibly augmenting or diminishing its energy. (As an aside, let's note that since the classical magnetic field produces a force orthogonal to the velocity, it does zero work on a test particle.) The replacement that introduced the magnetic potential into Schrödinger's equation, namely,

$$\frac{\partial}{\partial x_j} \mapsto \frac{\partial}{\partial x_j} + iq \, A_j$$

for $j = 1, 2, 3$, is known in the physics literature as the *minimal coupling* of the magnetic field to the wave function. The history behind the choice of this strange terminology is unknown to this author.

Now, if we start with Schrödinger's equation (12.30) based on a given 4-vector potential $\mathbf{A} = (\phi, A_1, A_2, A_3)$ and apply a *local gauge transformation* $\psi \mapsto \tilde{\psi} = e^{-iq\Phi} \psi$, then we get Schrödinger's equation (12.30) again, but now associated to the gauge-transformed 4-vector $\mathbf{A}' = \mathbf{A} + q \, d\Phi$. This is essentially a repetition of the material of this section where we worked out the special case $\mathbf{A} = (V/q, 0, 0, 0)$. We leave the details to the reader.

Exercise 12.5 *Indeed, there are a lot of details. But it may be useful to go through them to make sure you understand this section.*

The upshot of this section is that a local gauge transformation of the wave function satisfying Schrödinger's equation with minimal coupling corresponds to a classical gauge transformation of the associated classical electromagnetic field. But in the quantum mechanics setting, the local gauge transformation is a function $\mathbb{R}^4 \to S^1$, the unit circle in the complex plane. It turns out that S^1 is a Lie group where the product is that of complex numbers. For historical reasons (which remain totally unclear to this author), this particular Lie group in this particular context is always written as $U(1)$, the group of 1×1 unitary matrices. This is isomorphic to S^1 since the unique entry in such a matrix is exactly an element in S^1. In short, one says that electromagnetism in quantum mechanics is a $U(1)$ gauge theory. And $U(1)$ is an abelian group. The importance of the seemingly trivial comments in this paragraph only became apparent with the introduction of general gauge theory.

Chapter 13
Yang–Mills Theory

In this chapter, we try to present the theory as Yang and Mills saw it in [54] some 60 years ago. Rather than develop gauge theory in all its generality and then remark that the Yang–Mills theory is just a special case, we want to show the ideas and equations that these physicists worked with, and see later how the generalization came about. So we will use much of the notation of [54], including a lot of indices, as is the tradition in the physics literature to this day. This will require the reader to understand the operations of raising and lowering indices, sometimes not too lovingly known as "index mechanics." Even though [54] is only five pages long, we only discuss in detail some of its topics. Nonetheless, our approach is somewhat anachronistic since it is our current understanding of gauge theory that illuminates the present reading of [54]. Also, we use a bit of the language of differential forms to prepare the reader for the next chapter. However, this language is inessential for the strict purpose of this chapter.

13.1 Motivation

This theory was introduced by Yang and Mills in [54] in 1954, and we use the phrase "Yang–Mills theory" to refer to that specific theory with Lie group $SU(2)$, rather than its generalization as gauge theories with an arbitrary Lie group. Those authors had no idea at that time of its relation with differential geometry (see Yang's comments in [55]). Rather, they were motivated by Heisenberg's idea that the proton p and the neutron n were essentially the same particle, the *nucleon*, at least as far as the strong (nuclear) interaction is concerned. Heisenberg thought of p and n as something like two spin states of a spin $1/2$ particle, typically named the "up" state and the "down" state with respect to some arbitrary oriented line in Euclidean space $\mathbb{R}^3$. But this was a new type of spin eventually called *isospin*, a shortened version of the original *isotopic spin*. Again, isospin is not a special case of spin but rather an analog of spin.

© Springer International Publishing Switzerland 2015

S.B. Sontz, *Principal Bundles*, Universitext, DOI 10.1007/978-3-319-14765-9_13

But analogs have something in common with the original. And in this case, the element in common is the symmetry group $SU(2)$, the universal covering space of $SO(3)$, the rotation group in the Euclidean space $\mathbb{R}^3$. In the case of p and n, Heisenberg conjectured that the electromagnetic interaction did not respect this new $SU(2)$ symmetry (and so one says that the symmetry is *broken*), and this accounted for the known differences between p and n—basically that the mass of n is slightly larger than that of p. [It seems that Heisenberg only referred to the group $SO(3)$. But this is a small matter since it has the same Lie algebra as $SU(2)$ and only the Lie algebra enters into his discussion.]

The states p and n are represented by the two standard basis elements in the complex vector space $\mathbb{C}^2$. The action of the group $SU(2)$ on $\mathbb{C}^2$ is just the standard multiplication of a 2×2 matrix (on the left) and a 2×1 column vector (on the right). This action provides an *irreducible representation* of $SU(2)$ on $\mathbb{C}^2$. And this particular irreducible representation is called the *spin-1/2 representation*. One says that p and n are states of an *isospin doublet* of isospin-1/2 states much as spin up and spin down form a *spin doublet* of spin-1/2 states.

Exercise 13.1 *If this terminology is unfamiliar, you can consult any of the many excellent books on the representation theory of Lie groups. One of my favorites is [36] though [17, 20] and [49] are among many other outstanding choices. Or you can grin and bear it since for now it is just a lot of jargon for the matrix multiplication of 2×2 matrices and 2×1 column vectors.*

While this analogy might satisfy a physicist, it might be inadequate for a mathematician. After all, how did $SU(2)$ get into physics in the first place? And what does it have to do with spin? And spin has never been defined! The short answer is that certain binary results were observed in experiment (either one way or another, sort of like bits in computer science) when a continuum of results was expected. In the laboratory experiment of Stern and Gerlach, performed in 1922 just a few years before the modern theory of quantum mechanics, particles either went up along one path or down along another, instead of following those two paths as well as all of the paths at the intermediate angles. Here "up" and "down" describe the two directions that we all know about here on planet Earth, including physics laboratories. In modern terms, we would say that Stern and Gerlach had discovered the *qubit*, the quantum bit.

That class of phenomena eventually was dubbed "spin" but only later, after the advent of quantum mechanics. Spin is a form of angular momentum, but not its only form. Angular momentum is already seen in quantum mechanics in the solution of Schrödinger's equation for the hydrogen atom (which is the quantum mechanics two-body problem, where the particles are an electron and a proton). This is worked out in any elementary quantum mechanics text. What one uses is a separation-of-variables argument for the eigenvalue Schrödinger equation for the bound states in the Hilbert space $L^2(\mathbb{R}^3, d^3x)$. One finds that the invariance of this equation under rotations in $\mathbb{R}^3$, that is, under the action of elements of $SO(3)$, leads to subspaces of solutions, traditionally labeled by the integer $l \geq 0$, of dimension $2l + 1$. Each such subspace, denoted as $V_l \subset L^2(\mathbb{R}^3, d^3x)$, carries an irreducible representation of the Lie algebra $so(3)$ of $SO(3)$ given by three first-order linear partial differential operators, denoted by L_x, L_y, L_z. These are

$$L_x = y\frac{\partial}{\partial z} - z\frac{\partial}{\partial y}, \quad L_y = z\frac{\partial}{\partial x} - x\frac{\partial}{\partial z}, \quad L_z = x\frac{\partial}{\partial y} - y\frac{\partial}{\partial x}.$$

Each unit length vector in V_l has angular momentum $l\hbar$, where the integer l is called the *orbital angular momentum quantum number* and $\hbar$ is Planck's constant, the basic constant of quantum theory. By analyzing the equation $E = \hbar\omega$ (essentially from Planck's famous 1900 paper that started quantum theory), where E is the energy of a photon and ω its angular frequency, one easily shows that the dimensions of $\hbar$ are the dimensions of angular momentum.

The physics intuition is that the same symmetry group, $SO(3)$, gives rise to angular momentum solutions in the quantum mechanics setting as it was known to do in the classical mechanics two-body (or Kepler) problem of two "point" particles, such as the Sun and any one of its planets, interacting via the gravitational interaction. The conservation of angular momentum in this planetary situation is equivalent to Kepler's law that a planet sweeps out equal areas (with respect to a radial vector emanating from the Sun) in equal periods of time. So the quantum picture looked complete. It seemed that angular momentum in quantum mechanics is always "quantized"; that is, it is an integer multiple of $\hbar$.

As an aside, we note that for any representation of a compact Lie group on a vector space V, we can find a positive definite inner product on V that is invariant under the action of the representation. We also note that $SO(3)$ and $SU(2)$ are compact. So when we refer to unit vectors in a representation space of a compact Lie group, it is with respect to the norm associated to an invariant inner product.

But it turns out that the fundamental group $\pi_1(SO(3)) = \mathbb{Z}_2$, the group with two elements. This topological fact (which you may or may not know) combines with the theory of representations of Lie groups. More explicitly, this implies that there is a universal covering space, in this case a two-to-one group morphism $p : SU(2) \to SO(3)$, which induces an isomorphism of the corresponding Lie algebras, $su(2) \to so(3)$. Also, each irreducible representation of the Lie algebra $su(2) \cong so(3)$ can be realized as the infinitesimal version (essentially, the derivative at the identity element) of a Lie group irreducible representation of the universal cover $SU(2)$. And there is one such irreducible representation of $SU(2)$ for each positive integer. By tradition, this positive integer is written as $2s + 1$, with the values $s = 0, 1/2, 1, 3/2, 2, 5/2, \ldots$, which is called the *spin quantum number*. The dimension of the space of the representation associated with the spin quantum number s is $2s + 1$. When $s = 0, 1, 2, \ldots$, the irreducible representation of $SU(2)$ factors through the quotient map $p : SU(2) \to SO(3)$ to give the irreducible representation of $SO(3)$ described above, that is, $s = l$.

When $s = 1/2, 3/2, 5/2, \ldots$, the irreducible representation of $SU(2)$ does *not* factor through the quotient map $p : SU(2) \to SO(3)$. Some people call these two-to-one representations of $SO(3)$, but I find this to be a rather confusing way of saying a representation of $SU(2)$ that does not pass to the quotient space $SO(3)$. It is a extrapolation of theoretical ideas to suppose that these "half-integer" representations of $SU(2)$ have a physical meaning, let alone a meaning related to angular momentum. But this is what happened, and it worked! Next, we let V_{2s+1} denote a vector space carrying an irreducible representation of $SU(2)$ with half-integer spin $s = m/2$, where m is an odd, positive integer. Then it turns out that the unit vectors in V_{2s+1} have *spin angular momentum* $s\hbar = m\hbar/2$. So with spin in quantum mechanics, angular momentum is still quantized but now as integer multiples of $\hbar/2$.

A basic aspect of this theory is the basis of matrices for the Lie algebra $\mathfrak{su}(2)$ of $SU(2)$. These three matrices are named the *Pauli matrices* for W. Pauli, one of the first generation of quantum physicists. However, Dirac arrived at these matrices independently. The idea that a variety of experimental results could be explained by three explicit 2×2 matrices is rather astounding. It took a genius to do it. It was not obvious in the slightest. But it explained a lot of known data as well as a slew of data taken in subsequent experiments. Nothing succeeds like success, as the saying goes. And Pauli took one more giant step forward by introducing what we nowadays call the Pauli exclusion principle to describe a collection of spin-$1/2$ particles in quantum mechanics. Among many other accomplishments, this in turn led to the founding of modern solid-state physics. Dirac got the Nobel Prize in Physics in 1933, and Pauli got the Nobel Prize in Physics in 1945. Good choices.

The whole point of all this chit-chat about spin is to show how angular momentum arises from the rotation group $SO(3)$ and then passes to $SU(2)$. Both of these groups have dimension 3, as does their common Lie algebra. There is no motivation whatsoever to consider the group $U(2)$, which has dimension 4, in the study of angular momentum. And according to what they say in [54], Yang and Mills chose $SU(2)$ for the same reason Heisenberg did. And that is the analogy with spin that we have presented here.

But is there an intrinsic reason for this choice? Consider a basic wave function for the nucleon, say

$$\psi = \begin{pmatrix} \psi_p \\ \psi_n \end{pmatrix},$$

where the components $\psi_p, \psi_n \in \mathcal{H}$, a Hilbert space, represent the proton part and the neutron part, respectively, of the nucleon. So the normalization condition for ψ in the Hilbert space $\mathcal{H} \oplus \mathcal{H}$ is

$$||\psi_p||^2_{\mathcal{H}} + ||\psi_n||^2_{\mathcal{H}} = 1,$$

where $|| \cdot ||_{\mathcal{H}}$ denotes the norm in $\mathcal{H}$. Compare this with (12.21). The group of 2×2 matrices that preserve this condition is $U(2)$, all the unitary 2×2 matrices. Analogies can be nice and helpful, but it seems that the physics in this setting dictates that $U(2)$ should be the Lie group of interest, not $SU(2)$, which comes from nothing more than an analogy with spin. But for any $U \in U(2)$, we have a factorization

$$U = (\det U)^{1/2}((\det U)^{-1/2}U)$$

with $(\det U)^{1/2} \in S^1$ and $(\det U)^{-1/2}U \in SU(2)$. The particular choice of the complex square root is not an essential aspect of this argument. The point as far as physics is concerned is that the phase factor $(\det U)^{1/2}$ lying on the unit circle in $\mathbb{C}$ corresponds to a known gauge theory (electromagnetism, as we have seen), while the second factor in $SU(2)$ is quite new for us and might have interesting physics consequences. So this is a physics justification for using $SU(2)$ rather than $U(2)$ in this setting.

13.2 Basics

So the starting point of Yang and Mills was to formulate a suitable theory of the $SU(2)$ symmetry of isospin, where strong nuclear forces would be mediated by some new particle (or particles), as had already been conjectured by Yukawa. They were also motivated, as was Yukawa, by electromagnetic interactions in quantum theory which are mediated by a particle known as the photon, the quantum version of light.

Following the theory of spin-$1/2$ particles, we assume that isospin-$1/2$ states are described by a wave function

$$\psi(t, x_1, x_2, x_3) = \begin{pmatrix} \psi_1(t, x_1, x_2, x_3) \\ \psi_2(t, x_1, x_2, x_3) \end{pmatrix} \in \mathbb{C}^2, \tag{13.1}$$

where each component $\psi_j(t, x_1, x_2, x_3) \in \mathbb{C}$ is a standard wave function in some Hilbert space $\mathcal{H}$ over the field of complex numbers. Then ψ is in the Hilbert space $\mathcal{H} \oplus \mathcal{H}$. The Hilbert space $\mathcal{H}$ need not be specified further for now though $L^2(\mathbb{R}^4, d^4x)$ is the usual choice.

Then in this setting, a *gauge transformation* is

$$\psi \to \tilde{\psi} = U^{-1}\psi,$$

where U is a unitary 2×2 matrix with determinant 1. All such matrices form the Lie group $SU(2)$. [Unlike Heisenberg, we cannot get by only using $SO(3)$.] Notice that the important aspect of the theory here is that $SU(2)$ acts by matrix multiplication on the 2×1 column vectors $\psi \in \mathcal{H} \oplus \mathcal{H}$, and this structure does not depend on the explicit choice of $\mathcal{H}$.

There are two important cases of a gauge transformation:

1. Global gauge transformations where $U \in SU(2)$ is exactly one fixed matrix that depends on no parameters.
2. Local gauge transformations where $U : \mathbb{R}^4 \to SU(2)$; that is, the matrix U depends on the time and space coordinates.

178 13 Yang–Mills Theory

Of course, a global gauge transformation is a particular case of a local gauge transformation. An argument based on physics ideas leads us to the nontrivial conclusion that a physical theory invariant under global gauge transformations should also be invariant under local gauge transformations. This is an amazingly powerful scientific statement. It is not, nor could it be, a mathematical theorem. This argument is given in [54], and it is the reason Yang and Mills were interested in local gauge transformations. It is the same argument we presented leading up to the local gauge transformation (12.22) in the setting of electrodynamics.

At this stage, we do not have partial differential equations to describe the time evolution of this isospin-1/2 system. This is totally different from our discussion of the gauge transformations of electromagnetism in quantum mechanics, where from the start we had the Schrödinger equation that even gave us a Cauchy problem. We follow Yang and Mills by looking for a theory whose equations of time evolution are invariant under local gauge transformations.

Since this involves replacing all the partial derivatives by transformed expressions, it behooves us to understand what is a partial derivative of a two-component isospin wave function (13.1). We define it to be the partial derivative acting in each component:

$$
\frac{\partial \psi}{\partial x_j}(t, x_1, x_2, x_3) := \begin{pmatrix} \dfrac{\partial \psi_1}{\partial x_j}(t, x_1, x_2, x_3) \\ \dfrac{\partial \psi_2}{\partial x_j}(t, x_1, x_2, x_3) \end{pmatrix}. \tag{13.2}
$$

And a similar expression defines the partial with respect to time t. At this point, we require that the wave functions in $\mathcal{H}$ have derivatives. So the choice $\mathcal{H} = L^2(\mathbb{R}^4, d^4x)$ is now more natural. Though, as is well known and well studied in functional analysis, these are unbounded operators only defined on a dense linear subspace of $L^2(\mathbb{R}^4, d^4x)$.

A much more compact way of doing this is to let the matrix $(\partial/\partial x_j)I$, where I is the 2×2 identity matrix, act on the 2×1 column vectors ψ in the usual way. Then taking this one step further, we simply drop the I and let $\partial/\partial x_j$ denote this matrix. This is a mild abuse of notation, but context should always clarify what is meant.

So in analogy with the situation with electrodynamics, we want

$$
\frac{1}{i}\frac{\partial}{\partial t} \mapsto \frac{1}{i}\frac{\partial}{\partial t} + gB_0 \quad \text{and} \quad \frac{1}{i}\frac{\partial}{\partial x_j} \mapsto \frac{1}{i}\frac{\partial}{\partial x_j} + gB_j, \tag{13.3}
$$

where the so-called coupling constant $g \neq 0$ measures the strength of the strong interaction much as the electric charge q measures the strength of the electromagnetic interaction. We take g to be a real number. However, this time we do not have an equation to transform under these mappings. Rather, the symbol $\mapsto$ here means that instead of using the expressions to its left, in the theory we will always use the expressions to the right of it. Now the objects B_μ for $\mu = 0, 1, 2, 3$ act on the two-component isospin wave functions, so the natural choice is that they should be 2×2 matrices. As before, we want to impose a "reality" condition on them. In the context of 2×2 matrices with complex entries, that condition is $B_\mu^* = B_\mu$, where B_μ^* is the adjoint matrix of B_μ. One says that B_μ is a *hermitian matrix*. (Those with some background in functional analysis might say *self-adjoint matrix*.) And the partial derivatives act on these two-component wave functions as explained above.

The four objects B_μ all together correspond to the potential (or gauge field) in electrodynamics

$$A = A_0\, dt + A_1\, dx_1 + A_2\, dx_2 + A_3\, dx_3,$$

where $A_0 = \phi$ in our previous notation. But the objects B_μ are matrix-valued functions, rather than complex-valued functions, of the time and space variables:

$$B_\mu : \mathbb{R}^4 \to M_H(2, \mathbb{C}).$$

Here $M_H(2, \mathbb{C})$ is the set of hermitian 2×2 matrices with complex entries. $M_H(2, \mathbb{C})$ is a four-dimensional vector space over the field of real numbers. Consequently, we think of B_μ for $\mu = 0, 1, 2, 3$ as the coefficients of a 1-form on the spacetime $\mathbb{R}^4$, namely,

$$B := B_0\, dt + B_1\, dx_1 + B_2\, dx_2 + B_3\, dx_3.$$

Strictly speaking, this is a not a usual differential 1-form, whose coefficients are scalar-valued C^∞-functions. But the notation pretty much tells you what is happening, so we will not discuss this issue further for the time being.

Now we want to impose the gauge invariance for the first-order differential operators in (13.3) under *local* gauge transformations. To facilitate matters, we at last introduce the notation $x_0 \equiv t$. And we use Greek letters as indices varying over the values $0, 1, 2, 3$. So we want

$$U\left(\frac{1}{i}\frac{\partial}{\partial x_\mu} + gB_\mu'\right)\tilde\psi = \left(\frac{1}{i}\frac{\partial}{\partial x_\mu} + gB_\mu\right)\psi, \tag{13.4}$$

where B_μ' are the still-unknown gauge transforms of the matrices B_μ. This corresponds exactly to equation (2) in [54], provided that we take $g = -\epsilon$.

Then, expanding the left side of (13.4) gives

$$U\left(\frac{1}{i}\frac{\partial}{\partial x_\mu} + gB_\mu'\right)\tilde\psi = U\left(\frac{1}{i}\frac{\partial}{\partial x_\mu} + gB_\mu'\right)\left(U^{-1}\psi\right)$$

$$= U\frac{1}{i}\frac{\partial U^{-1}}{\partial x_\mu}\psi + U\frac{1}{i}U^{-1}\frac{\partial \psi}{\partial x_\mu} + gUB_\mu'U^{-1}\psi$$

$$= \frac{1}{i}U\frac{\partial U^{-1}}{\partial x_\mu}\psi + \frac{1}{i}\frac{\partial \psi}{\partial x_\mu} + gUB_\mu'U^{-1}\psi.$$

To have this equal to the right side of (13.4) for all ψ, we must have

$$\frac{1}{i} U \frac{\partial U^{-1}}{\partial x_\mu} + gUB'_\mu U^{-1} = gB_\mu.$$

Since $g \neq 0$, this can be rearranged as

$$B'_\mu = U^{-1} B_\mu U - \frac{1}{ig} \frac{\partial U^{-1}}{\partial x_\mu} U.$$

Differentiating the identity $U^{-1}U = I$, the identity matrix, we see for each μ that

$$\frac{\partial U^{-1}}{\partial x_\mu} U + U^{-1} \frac{\partial U}{\partial x_\mu} = 0,$$

where the right side is the 2×2 zero matrix. As an aside, let's also note that this implies this well-known identity, which we will also be needing:

$$\frac{\partial U^{-1}}{\partial x_\mu} = -U^{-1} \frac{\partial U}{\partial x_\mu} U^{-1}. \tag{13.5}$$

So we get this gauge transformation for the matrices B_μ:

$$B'_\mu = U^{-1} B_\mu U + \frac{1}{ig} U^{-1} \frac{\partial U}{\partial x_\mu}. \tag{13.6}$$

It is most likely completely clear that this is equivalent to

$$B' = U^{-1} BU + \frac{1}{ig} U^{-1} dU.$$

However, we have not defined the expressions and operations in this equation!

We compare this with a local gauge transformation of the gauge potential A_μ in electromagnetism. Taking the local gauge transformation

$$U = e^{iq\Phi} : \mathbb{R}^4 \to S^1 \cong U(1),$$

where $\Phi : \mathbb{R}^4 \to \mathbb{R}$, we calculate

$$\frac{1}{iq} U^{-1} \frac{\partial U}{\partial x_\mu} = \frac{1}{iq} e^{-iq\Phi} (iq) \left(\frac{\partial \Phi}{\partial x_\mu} \right) e^{iq\Phi} = \frac{\partial \Phi}{\partial x_\mu}.$$

So we have the gauge transformation

$$A'_\mu = A_\mu + \frac{\partial \Phi}{\partial x_\mu} = U^{-1} A_\mu U + \frac{1}{iq} U^{-1} \frac{\partial U}{\partial x_\mu} \qquad (13.7)$$

since $U^{-1} A_\mu U = A_\mu$ holds because multiplication of complex numbers is commutative. Consequently, we see that the electromagnetic gauge field transforms in the same way as the isospin gauge field B_μ does, except that (13.6) is an equation relating 2×2 matrices that do not necessarily commute among themselves, while (13.7) is an equation involving only scalars.

Note that (13.6) shows that the isospin gauge field B_μ is invariant under *global* gauge transformations since then the second term vanishes. (We say "invariant" here because that is what is commonly said. It is more correct to say "covariant.") But it is not invariant under local gauge transformations. And we want to find expressions that are invariant under local gauge transformations in order to build a theory that is also invariant under local gauge transformations. As we remarked earlier, this consideration is motivated by physics ideas.

Here are some definitions in analogy with the electromagnetic theory.

Definition 13.1 *Suppose that B_μ and B'_μ are gauge fields. We say that B_μ is gauge equivalent to B'_μ if there exists a local gauge transformation $U : \mathbb{R}^4 \to SU(2)$ so that*

$$B'_\mu = U^{-1} B_\mu U + \frac{1}{ig} U^{-1} \frac{\partial U}{\partial x_\mu}.$$

Exercise 13.2 *Show that gauge equivalence of gauge fields is an equivalence relation, as it should be to deserve its name!*

Definition 13.2 *We say that a gauge field B_μ is* pure gauge *if*

$$B_\mu = \frac{1}{ig} U^{-1} \frac{\partial U}{\partial x_\mu}$$

for some local gauge transformation $U : \mathbb{R}^4 \to SU(2)$. Equivalently, B_μ is pure gauge *if it is gauge equivalent to the zero gauge field.*

Some elementary results are often overlooked in this sort of elementary introduction. These are introduced in the next exercise.

Exercise 13.3 *Let $U : \mathbb{R}^4 \to SU(2)$ be a local gauge transformation.*

1. Show that $(ig)^{-1} U^{-1} (\partial U / \partial x_\mu)$ is a hermitian matrix whose trace is zero; that is, the pure gauge term satisfies

$$\frac{1}{ig} U^{-1} \frac{\partial U}{\partial x_\mu} \in \mathfrak{su}(2),$$

the Lie algebra of $SU(2)$. (Here we are using the physics definition of the Lie algebra of a matrix group, to be explained in the next chapter. In short, the physics and math definitions differ by a factor of i.)

2. Suppose that B_μ is a hermitian matrix. Show that $U^{-1} B_\mu U$ is also a hermitian matrix.

Therefore, the two results of this exercise combine to show that B'_μ is a hermitian matrix given that B_μ is a hermitian matrix; that is, local gauge transformations preserve the property of being a hermitian matrix. In other words, if $B_\mu \in \mathfrak{u}(2)$, then $B'_\mu \in \mathfrak{u}(2)$.

Because of the results in this exercise, the two terms on the right side of (13.6) have different properties since the first is in $\mathfrak{u}(2)$ while the second is in $\mathfrak{su}(2)$. So it seems reasonable to change the theory a bit so that the first term on the right side of (13.6) is in $\mathfrak{su}(2)$ as well. And the change is simple enough; one takes B_μ with values in $\mathfrak{su}(2)$ for all μ instead of B_μ with values in $\mathfrak{u}(2)$.

Yang and Mills also give an argument in [54] for this change, based on the following considerations. The Lie algebra $\mathfrak{su}(2)$ of dimension 3 carries the irreducible representation of $SU(2)$ with $l = 2s + 1 = 3$. So this is the isospin space with isospin $s = 1$. Hence, if B_μ had terms from other isospin irreducible representations, that is, $s \neq 1$, then those terms under local gauge transformations would have to transform as $B_\mu \to B'_\mu = U^{-1} B_\mu U$ without the pure gauge term, which only appears for the $s = 1$ case. Such additional terms are described in [54] as "extraneous" and "irrelevant" as far as one is interested in isospin gauge theory. What this means is that such extra terms could be included in the theory with no serious impact on the way to construct a field strength that is locally gauge invariant. The tricky bit is that pure gauge term in the isospin-1 case. Momentarily, we will address this nontrivial issue.

Exercise 13.4 Suppose that $B_\mu : \mathbb{R}^4 \to \mathfrak{su}(2)$ is a gauge field and that $U : \mathbb{R}^4 \to SU(2)$ is a local gauge transformation. Prove that $U^{-1} B_\mu U$ is also an $\mathfrak{su}(2)$-valued gauge field, that is, $U^{-1} B_\mu U : \mathbb{R}^4 \to \mathfrak{su}(2)$.

In short, the moral of this little story is that the Lie algebra $\mathfrak{su}(2)$ of the "structure" group $SU(2)$ of this theory is where the gauge fields B_μ have to "live." Even though the Yang–Mills theory was based on an analogy with spin, the Lie algebra $\mathfrak{u}(2)$ was introduced at first, only to be replaced by $\mathfrak{su}(2)$ later on. This is unlike the theory of spin, which starts with $SO(3)$ and continues on to its universal covering space $SU(2)$ without any reference at all to $U(2)$ or to its Lie algebra $\mathfrak{u}(2)$.

We will see this relation as well in the general gauge theory presented in the next chapter. In that setting, we will have a "structure" Lie group G and its Lie algebra $\mathfrak{g}$.

The field strength in analogy with electrodynamics should be the 2-form $F := dB$ or, equivalently,

$$F_{\mu\nu} := \frac{\partial B_\mu}{\partial x_\nu} - \frac{\partial B_\nu}{\partial x_\mu}.$$

Notice that $F_{\mu\nu} : \mathbb{R}^4 \to M_H(2, \mathbb{C})$. Let's see if this is invariant under local gauge transformations. So we are going to compute

$$F'_{\mu\nu} = \frac{\partial B'_\mu}{\partial x_\nu} - \frac{\partial B'_\nu}{\partial x_\mu}.$$

To keep the formulas reasonably manageable, we introduce the standard notation $\partial_\mu = \partial/\partial x_\mu$. First, we compute

$$\partial_\nu B'_\mu = (\partial_\nu U^{-1}) B_\mu U + U^{-1}(\partial_\nu B_\mu)U + U^{-1} B_\mu (\partial_\nu U)$$

$$+ \frac{1}{ig}(\partial_\nu U^{-1})\partial_\mu U + \frac{1}{ig}U^{-1}(\partial_\nu \partial_\mu U)$$

$$= (-U^{-1}(\partial_\nu U)U^{-1}) B_\mu U + U^{-1}(\partial_\nu B_\mu)U + U^{-1} B_\mu (\partial_\nu U)$$

$$- \frac{1}{ig}(U^{-1}(\partial_\nu U)U^{-1})\partial_\mu U + \frac{1}{ig}U^{-1}(\partial_\nu \partial_\mu U),$$

where we used (13.5). Now we see that

$$F'_{\mu\nu} = \partial_\nu B'_\mu - \partial_\mu B'_\nu$$

$$= (-U^{-1}(\partial_\nu U)U^{-1}) B_\mu U + U^{-1}(\partial_\nu B_\mu)U + U^{-1} B_\mu (\partial_\nu U)$$

$$- \frac{1}{ig}(U^{-1}(\partial_\nu U)U^{-1})\partial_\mu U + \frac{1}{ig}U^{-1}(\partial_\nu \partial_\mu U)$$

$$+ (U^{-1}(\partial_\mu U)U^{-1}) B_\nu U - U^{-1}(\partial_\mu B_\nu)U - U^{-1} B_\nu (\partial_\mu U)$$

$$+ \frac{1}{ig}(U^{-1}(\partial_\mu U)U^{-1})\partial_\nu U - \frac{1}{ig}U^{-1}(\partial_\mu \partial_\nu U).$$

The two terms with second-order derivatives cancel. Then rearranging terms, we get more than we bargained for, namely,

$$F'_{\mu\nu} = U^{-1}\left(\partial_\nu B_\mu - \partial_\mu B_\nu\right)U$$

$$+ U^{-1}\left((-(\partial_\nu U)U^{-1}) B_\mu + (\partial_\mu U)U^{-1}) B_\nu\right)U$$

$$- U^{-1}(\frac{1}{ig}(\partial_\nu U)U^{-1} + B_\nu)\partial_\mu U$$

$$+ U^{-1}(\frac{1}{ig}(\partial_\mu U)U^{-1} + B_\mu)\partial_\nu U.$$

The first line gives what we want for local gauge invariance, but the remaining three lines need not be zero. Thus, the proposed definition for the field strength is not invariant under local gauge transforms, and we therefore must keep looking for another expression. The immediate hope would be to modify the definition of the field strength so that the undesired terms all cancel out. At this stage, it seems that Yang and Mills played around with various expressions, always hoping that something would work out. They do not specify any method for arriving at the desired modified field strength. It seems that they just kept trying one expression after another until something worked out. Here is one of the things that they probably did.

First, we consider a product of two of these gauge-transformed fields:

$$B'_\mu B'_\nu = \left(U^{-1}B_\mu U + \frac{1}{ig}U^{-1}\partial_\mu U\right)\left(U^{-1}B_\nu U + \frac{1}{ig}U^{-1}\partial_\nu U\right)$$

$$= U^{-1}B_\mu B_\nu U + \frac{1}{ig}U^{-1}B_\mu(\partial_\nu U) + \frac{1}{ig}U^{-1}(\partial_\mu U)U^{-1}B_\nu U$$

$$+ \frac{1}{(ig)^2}U^{-1}(\partial_\mu U)U^{-1}(\partial_\nu U).$$

Then let's form the antisymmetrization of this since the field strength $F_{\mu\nu}$ is antisymmetric in the electromagnetic case, and it might be antisymmetric in this case, too. But who knows? This is just a hunch. Anyway, this is what we get:

$$B'_\mu B'_\nu - B'_\nu B'_\mu = U^{-1}B_\mu B_\nu U + \frac{1}{ig}U^{-1}B_\mu(\partial_\nu U) + \frac{1}{ig}U^{-1}(\partial_\mu U)U^{-1}B_\nu U$$

$$+ \frac{1}{(ig)^2}U^{-1}(\partial_\mu U)U^{-1}(\partial_\nu U) - U^{-1}B_\nu B_\mu U - \frac{1}{ig}U^{-1}B_\nu(\partial_\mu U)$$

$$- \frac{1}{ig}U^{-1}(\partial_\nu U)U^{-1}B_\mu U - \frac{1}{(ig)^2}U^{-1}(\partial_\nu U)U^{-1}(\partial_\mu U).$$

Grouping like terms then yields

$$B'_\mu B'_\nu - B'_\nu B'_\mu = U^{-1}(B_\mu B_\nu - B_\nu B_\mu)U$$

$$+ \frac{1}{ig}U^{-1}\left((\partial_\mu U)U^{-1}B_\nu - (\partial_\nu U)U^{-1}B_\mu\right)U + \frac{1}{ig}U^{-1}B_\mu(\partial_\nu U)$$

$$- \frac{1}{ig}U^{-1}B_\nu(\partial_\mu U) + \frac{1}{(ig)^2}U^{-1}(\partial_\mu U)U^{-1}(\partial_\nu U) - \frac{1}{(ig)^2}U^{-1}(\partial_\nu U)U^{-1}(\partial_\mu U).$$

Again, the first term is what we want, but it is followed by more terms, which in general will not be zero. Now when we combine these calculations, we get an enormous number of cancellations, giving us

$$\partial_\nu B'_\mu - \partial_\mu B'_\nu - ig\left(B'_\mu B'_\nu - B'_\nu B'_\mu\right) = U^{-1}\left(\partial_\nu B_\mu - \partial_\mu B_\nu - ig\left(B_\mu B_\nu - B_\nu B_\mu\right)\right)U.$$

We have found the golden expression! So we redefine the field strength to be the local gauge-invariant expression

$$F_{\mu\nu} := \partial_\nu B_\mu - \partial_\mu B_\nu - \mathrm{ig}\big(B_\mu B_\nu - B_\nu B_\mu\big). \tag{13.8}$$

From a modern perspective, this definition is the crucial step in the paper [54] of Yang and Mills. [Note that we are using $g = -\epsilon$, and so our definition agrees with equation (4) in [54].] Then what we have just shown can be ever so briefly written as

$$F'_{\mu\nu} = U^{-1} F_{\mu\nu} U, \tag{13.9}$$

which is equation (5) in [54]. This equation says how the isospin field strength changes under local gauge transformations. One says the isospin field strength is *invariant* under local gauge transformations though *covariant* would be better terminology. The simple equation (13.9) is the key result in the paper by Yang and Mills, at least from this author's decidedly 21st-century point of view.

The derivation given here of (13.8) and (13.9) is long and messy. In [54], the authors simply write down (13.8), and immediately after that they write, "One easily shows ..." (13.9). They give no details. In particular, they do not present a quick way to verify (13.9), which leads me to suspect that they were not aware of any such quick way and that they did something similar to what we have presented. But there is a quick way, which is as follows. First, for each μ one defines

$$D_\mu := \partial_\mu + \mathrm{ig} B_\mu. \tag{13.10}$$

Then we note that we easily compute this commutator of matrices:

$$\begin{aligned}
[D_\mu, D_\nu] &= [\partial_\mu + \mathrm{ig} B_\mu, \partial_\nu + \mathrm{ig} B_\nu] \\
&= [\mathrm{ig} B_\mu, \partial_\nu] + [\partial_\mu, \mathrm{ig} B_\nu] + [\mathrm{ig} B_\mu, \mathrm{ig} B_\nu] \\
&= (-\mathrm{ig})\big(\partial_\nu B_\mu - \partial_\mu B_\nu - \mathrm{ig}[B\mu, B_\nu]\big) \\
&= (-\mathrm{ig}) F_{\mu\nu}.
\end{aligned}$$

Second, equation (13.4) can be written as $D'_\mu = U^{-1} D_\mu U$. Then one immediately has

$$\begin{aligned}
F'_{\mu\nu} &= \mathrm{ig}^{-1} [D'_\mu, D'_\nu] = \mathrm{ig}^{-1} [U^{-1} D_\mu U, U^{-1} D_\mu U] \\
&= \mathrm{ig}^{-1} U^{-1} [D_\mu, D_\nu] U = U^{-1} F_{\mu\nu} U,
\end{aligned}$$

which proves (13.9) in record time! Also, this proof underlines the importance of the expression (13.10), which will eventually be identified as a *covariant derivative*.

For the electromagnetic field, we have seen that the fields are given by

$$F^{\mathrm{em}}_{\mu\nu} = \partial_\nu A_\mu - \partial_\mu A_\nu,$$

where A_μ are the four *scalar* coefficients of the potential 1-form. But the corresponding Yang–Mills potential is defined as the 1-form on spacetime with *matrix* coefficients B_μ; this is a noteworthy difference between the two theories. However, there are some other rather minor differences between these two theories. First of all, we can see that the electromagnetic field transforms under local gauge transformations according to

$$(F_{\mu\nu}^{\text{em}})' = F_{\mu\nu}^{\text{em}}.$$

So this field is not only gauge invariant; it is out-and-out invariant. Of course, this relation can also be written as

$$(F_{\mu\nu}^{\text{em}})' = U^{-1} F_{\mu\nu}^{\text{em}} U,$$

where U is the $U(1)$-valued local gauge transformation. This is so since $U(1)$ is simply S^1 and the multiplication on the right side is the commutative multiplication of complex numbers. Another common way of quickly saying this is that $U(1)$ is an abelian group though that is inaccurate since the middle factor $F_{\mu\nu}^{\text{em}}$ is real-valued.

On the other hand, in the Yang–Mills setting we have (13.9), which has the product of three 2×2 matrices on the right side. Since matrix multiplication is not commutative, we cannot in general cancel the factors U^{-1} and U in (13.9) as we can in the abelian case of electromagnetism.

Another difference is the new term in the definition (13.8) of the field strength. This term contains the commutator of 2×2 matrices

$$[B_\mu, B_\nu] = B_\mu B_\nu - B_\nu B_\mu$$

and consequently need not be zero. But we can also write

$$F_{\mu\nu}^{\text{em}} = \partial_\nu A_\mu - \partial_\mu A_\nu - iq[A_\mu, A_\nu]$$

since the commutator $[A_\mu, A_\nu]$ of 1×1 matrices is identically equal to zero. (Equivalently, the multiplication of complex numbers is commutative.)

The addition by Yang and Mills of the new term in (13.8) has something in common with the addition by Maxwell of the displacement current term to Ampère's law. The point is that a general theoretical principle—local gauge invariance for Yang and Mills and charge conservation for Maxwell—led to the addition of a new nontrivial term. It should be noted that Yang and Mills had experimental evidence in favor of their principle of local gauge invariance of isospin. (See [54] for the discussion of that evidence.) And this new term has important consequences. It is also rather curious that Yang and Mills actually used the Maxwell theory of electromagnetism in order to motivate in part their theory of isotopic spin. Again, I wish to emphasize that the motivation for Yang and Mills came from physics, and not from differential geometry.

13.3 Yang–Mills Equations

We still do not have the dynamics of Yang–Mills theory. In the theory of electromagnetism, we had Maxwell's equations from the start, but Yang and Mills did not have anything of the sort when they started. Rather, they looked for the simplest dynamics consistent with the principle of local gauge invariance. What they came up with has became known as the Yang–Mills equations. They had to work "backward" from general physical principles to specific equations for this setting. This procedure is based on conjecture since one is never quite sure which physical principles are the general ones! Of course, Yang and Mills had taken local gauge invariance as a general physical principle just to get their research going. So they continued using it for this situation.

Also, they knew from any number of examples that the equations of motion for a physical system can often be obtained from a *Lagrangian formalism*. This is the basis of *Lagrangian mechanics*. For example, see [2]. But it also can be applied in other contexts, such as electrodynamics. (See Chapter 12 of Jackson's classic text [28].) In particular, for the electromagnetic field alone (that is, without the presence of sources), the Lagrangian density is defined in [28] to be

$$\mathcal{L}_{\text{free}} = -\frac{1}{16\pi} \sum_{\alpha,\beta=0}^{3} F_{\alpha\beta}^{\text{em}} (F^{\text{em}})^{\alpha\beta}.$$

The second factor is $F_{\alpha\beta}^{\text{em}}$, with its indices raised by using the *Minkowski inner product (or metric)* $g^{\alpha\beta}$ on $\mathbb{R}$. In Jackson's (and our) convention, $g^{00} = +1$, while $g^{11} = g^{22} = g^{33} = -1$ and $g^{\alpha\beta} = 0$ for $\alpha \neq \beta$. (Warning: The Minkowski inner product $g^{\alpha\beta}$ should not be confused with the strong coupling constant g.) If this density is integrated over a volume in the space $\mathbb{R}^3$, the result is called the *Lagrangian* for that volume and has dimensions of energy. If the Lagrangian density is integrated over a volume in the spacetime $\mathbb{R}^4$, the result is called the *action* and has dimensions of angular momentum, which by no accident whatsoever are also the dimensions of Planck's constant $\hbar$. For example, the *action integral* in this setting is the functional

$$S := \int_{\mathbb{R}^4} \mathcal{L}_{\text{free}}.$$

We simply assume that this integral converges in order to be able to continue the discussion. Of course, this depends on the behavior of the integrand near infinity, an issue we do not wish to consider further here.

The *least action principle* of the Lagrangian formalism is now applied to this functional. By taking the *variational derivatives* of the action integral with respect to each of the *generalized coordinates* (here, A_μ and $\partial_\mu A_\nu$) and then setting them equal to zero, one obtains the *Euler–Lagrange equations*, or for short the *E–L equations*. This is part of a theory in mathematics known as the *calculus of variations*. The E–L equations are differential equations for the potential A_μ. If this exercise is carried out (again, see [28]), one obtains the nonhomogeneous Maxwell's equations, but with the source terms set to zero because in this particular example the variational derivatives with respect to the A_μs are zero. All this depends on the initial choice of the Lagrangian density. On the other hand, starting with the Lagrangian density

$$\mathcal{L}_{\text{int}} = -\frac{1}{16\pi} \sum_{\alpha,\beta=0}^{3} F_{\alpha\beta}^{\text{em}} (F^{\text{em}})^{\alpha\beta} - \sum_{\alpha=0}^{3} J^\alpha A_\alpha,$$

one gets the nonhomogeneous Maxwell's equations but with the source terms present, since now the variational derivatives with respect to the A_μs give the components J^α of the current.

This should give the reader some flavor of how physical theories are built out of Lagrangian densities. One changes the definition of the Lagrangian density, and varying theories are produced. But in this particular case, the homogeneous Maxwell's equations *never* come out as a consequence of a particular choice of Lagrangian density for the potential A_μ. Rather, the A_μs were constructed (though only modulo gauge transformations) from the homogeneous Maxwell's equations and so automatically satisfy them. Also, the choice of constants and signs in the Lagrangian density were "tuned" just so the E–L equations became the nonhomogeneous Maxwell's equations. A lot of experience is required to play this game well. Some might even call it physical intuition.

So Yang and Mills had what it takes to do this sort of theory building. And they wanted a theory that would give E–L equations that do not change when one applies a local gauge transformation. Having constructed the field strength

$$F_{\mu\nu} = \partial_\nu B_\mu - \partial_\mu B_\nu - ig[B_\mu, B_\nu],$$

it is trivial to raise indices to define $F^{\mu\nu}$ as

$$F^{\mu\nu} := \sum_{\alpha,\beta=0}^{3} g^{\mu\alpha} g^{\nu\beta} F_{\alpha\beta},$$

where $g^{\alpha\beta}$ is Minkowski's metric on $\mathbb{R}^4$, and then form

$$\mathcal{L}_{\text{YMfree}} = -\frac{1}{4} \sum_{\mu,\nu=0}^{3} F_{\mu\nu} F^{\mu\nu}.$$

Actually, this is not what appears in the paper [54], but it is not far from what they did. Also, the change in the negative constant is unimportant though the sign is important since eventually one wants to *minimize* the action integral defined below. Under a gauge transformation $U : \mathbb{R}^4 \to SU(2)$, this Lagrangian density can change since

$$\mathcal{L}'_{\text{YMfree}} = U^{-1} \mathcal{L}_{\text{YMfree}} U$$

follows from the gauge transformation identity for $F_{\alpha\beta}$. In electromagnetism, the corresponding product on the right side is the commutative product of scalars. But now we have the product of three 2×2 matrices, so in general there is no way that the U^{-1} and the U will cancel. Now we want a gauge-independent action integral so that we will finally arrive at gauge-independent equations of time evolution. And $\mathcal{L}_{\text{YMfree}}$ does not do the job! But if we take the trace, denoted Tr, of the previous equation, we find that

$$\text{Tr}(\mathcal{L}'_{\text{YMfree}}) = \text{Tr}(U^{-1}\mathcal{L}_{\text{YMfree}}\,U) = \text{Tr}(\mathcal{L}_{\text{YMfree}}),$$

where in the last equality we used the well-known property $\text{Tr}(AB) = \text{Tr}(BA)$ of the trace of matrices. Next, we define the action integral in the obvious way as

$$S_{\text{YMfree}} := \int_{\mathbb{R}^4} \text{Tr}(\mathcal{L}_{\text{YMfree}}),$$

thus getting a locally gauge-independent action integral functional. But again, we do not discuss the delicate issue of the convergence of this integral. Rather, we assume that it does converge and carry on from there. Much as we have seen before, this definition of the Yang–Mills action integral is compatible with the motivating example of the commutative electromagnetic theory since the trace of a 1×1 matrix is simply its unique matrix entry.

We then apply the calculus of variations in order to get the corresponding E–L equations for this setting. These E–L equations are called *Yang–Mills equations* for the free theory.

The game continues in [54] with the addition of more terms to account for sources of Yang–Mills fields. This is an ever more technical theory whose many details are not of interest to us now. Suffice it to say that the fields themselves are part of the source term; that is, the fields carry isospin "charge" and isospin "current." This is unlike the electromagnetic case, where the electromagnetic field (read: photon) itself does not have an electric charge. So the physics in Yang–Mills theory does have marked differences from that of the electromagnetic theory. For those readers who understand these words, let me say that there are Feynman diagrams for Yang–Mills theory that are not present for the electromagnetism theory.

Here are a few words on what Feynman has to do with this theory. The key idea of Feynman is to think about an integral of the form

$$\int_{\mathcal{P}} d\mu_{\mathcal{P}}(\gamma)\, e^{iS(\gamma,\dot{\gamma})/\hbar}, \tag{13.11}$$

where γ is a path in the configuration space of a physical system whose action $S(\gamma, \dot{\gamma})$ along the path γ is a physical property of the system. Here $\dot{\gamma}$ is the derivative (tangent vector) of the path, $\mathcal{P}$ is the space of all possible paths under consideration, and $d\mu_{\mathcal{P}}(\gamma)$ is a measure on this path space. The expression in (13.11) is called the *Feynman path integral*.

There is no need to go into details about a mathematical definition of the measure space $\mathcal{P}$ with its measure $d\mu_{\mathcal{P}}(\gamma)$ because such details do not exist! It has been mathematically proved that in most cases of physical interest (where one has a good idea what $\mathcal{P}$ has to be), this measure does not exist. Nevertheless, one can consider (13.11) with an adequate choice of S as the unitary group $e^{-iHt/\hbar}$, where H is a Schrödinger operator. Again, the details are completely formal without any mathematical justification, so we do not go into that. However, the unitary group applied to a wave function ψ_0, that is, $\psi = e^{-iHt/\hbar}\psi_0$, is the solution of the quantum mechanical problem, namely, the Schrödinger equation

$$i\hbar\frac{\partial\psi}{\partial t} = H\psi$$

with initial condition ψ_0. For more details, see [19].

Exercise 13.5 *Check by using a formal (= nonrigorous) calculation that $\psi = e^{-iHt/\hbar}\psi_0$ satisfies the Schrödinger equation and the initial condition.*

And for self-adjoint operators, that is, $H^* = H$ in a Hilbert space, the unitary group *does* exist by spectral theory. For those who do not know what these words mean, the message for you is that learning spectral theory is important. A nice, quick introduction for this with applications in analysis and mathematical physics is [7].

So Feynman starting in the 1940s and generations of physicists since have taken the integral (13.11) quite seriously. And a multitude of experimentally verified results have been obtained in this way. The basic method used to study (13.11) is called the *method of stationary phase*. This is a rigorously established method for integrals over Euclidean spaces. So what one does is to take the equation that results from the rigorous theory, and one applies it here. This is called *perturbation theory*. Without going into the details of perturbation theory, let's simply note that this method is based on finding the critical points of the phase $e^{iS/\hbar}$, which means finding the stationary points of the action S. And this is *grosso modo* how the Yang–Mills equations come into the quantum theory. Feynman in [13] made the first step in quantizing Yang–Mills theory and arrived at the one-loop Feynman diagrams in perturbation theory. It is curious that Feynman was really interested in the quantization of gravitation, which is still an unsolved problem, and was using Yang–Mills theory as a warmup exercise before addressing that more difficult problem. Then De Witt and independently Faddeev and Popov got the complete quantization (from a physicist's point of view) of Yang–Mills theory in 1966. See Faddeev's review paper [12] for these and other interesting details. The rigorous quantization from a mathematician's point of view remains an open problem.

This has been a bird's-eye view of how one gets Yang–Mills equations without ever doing an experiment. To go through all the rigorous details might try the patience of even my most mathematically inclined readers. At one level, the whole discussion up to now in this section can be considered, if one wishes, to be nothing other than a nonrigorous but heuristic motivation for the Yang–Mills equations, which then are written down and taken as the true starting point of a rigorous discussion. For someone learning this material for the first time, that can be an adequate approach. But at another level we would like to know that the solutions of the Yang–Mills equations really and truly minimize the action integral. And so to achieve that level of understanding, one is well advised to work through all the mathematical details rigorously. For example, see [18] for the theory of the calculus of variations.

The exact equations that one gets are important, of course. Here are the Yang–Mills equations with zero source terms:

$$\sum_{\alpha=0}^{3} \left(\partial_\alpha F_{\alpha\beta} + ig[B_\alpha, F_{\alpha\beta}] \right) = 0, \qquad (13.12)$$

for $\beta = 0, 1, 2, 3$, where we remind the reader that

$$F_{\alpha\beta} = \partial_\beta B_\alpha - \partial_\alpha B_\beta - ig[B_\alpha, B_\beta].$$

These are four coupled partial differential equations in the four unknowns (B_0, B_1, B_2, B_3). They are of second order and moreover, due to the presence of the nontrivial commutator terms, are nonlinear in the unknowns, which themselves are 2×2 traceless, hermitian matrix-valued functions. In the electromagnetic theory, the corresponding commutators are of 1×1 matrices and so are zero. So in that theory the corresponding nonlinear terms drop out, giving us four of Maxwell's equations, which are linear. It seems that in mathematical studies, one is mainly interested in equations (13.12), where the source terms have been put equal to zero. But in physics studies, starting with the paper [54] itself, nonzero sources coming from Dirac matter fields are included in the Lagrangian density, and these then lead to nonzero current terms being added to (13.12). We will not discuss these important details further. See equations (11)–(13) in [54].

Again, we remind the reader that the four Yang–Mills equations (13.12) correspond to the four nonhomogeneous Maxwell's equations with source terms put equal to zero. What corresponds in the present setting to the homogeneous Maxwell's equations is an interesting and important part of Yang–Mills theory. This is more easily understood in the general setting of gauge theory rather than in the particular case of Yang–Mills theory. So we will get back to this topic in the next chapter.

13.4 Mass of the Mesons

One consideration that was left unresolved in the paper [54] of Yang and Mills was the determination of the masses of the mediating mesons that they had conjectured to explain the strong interactions. They called these mesons the *b quanta.* They tried to come up with some way of showing that these masses were not zero. But they found no convincing argument either in favor of or against nonzero mass.

However, Pauli soon showed in the original setting of Yang–Mills theory that the Yang–Mills mesons have zero mass, just as the photon does. The point is that by introducing into the Lagrangian density a standard mass term, $m^2 \sum_\mu B_\mu B^\mu$ with nonzero mass m, one would thereby break the gauge invariance. And finding a gauge-invariant theory was the original idea! This was probably communicated in one of Pauli's famous unpublished letters. Anyway, this greatly discouraged the physics community since such particles would have already been produced in abundance in experiment. And no such particle had been detected. Later, it was realized that the quantization of a Yang–Mills theory with a nonabelian Lie group and with a mass term could not be renormalizable. It is no easy matter to explain what "quantization" and "renormalization" are, but suffice it to say for now that in physics these are essential aspects needed to get a "good" quantum field theory of the elementary particles.

At that point in time, the experimentally observed charged particles π^+ and π^-, nowadays called *pions*, were thought to be the mesons that mediated the strong interaction. The original meaning of "meson" is a particle of intermediate mass, which in turn means a mass between the electron mass and the proton mass. (The modern meaning of "meson" is a particle that interacts via the strong interaction and has integer spin $0, 1, 2$, etc.) And the measured mass of these pions was in that region. The experimental discovery of the neutral pion π^0 with almost the same mass completed the picture of a triplet of pions with isospin 1. The pions were also measured as having spin 0. But the **b** quanta of Yang–Mills theory were supposed to have spin 1. So it seemed that Yang–Mills theory was not correct.

Well, Yang–Mills theory to this day is thought to be an incorrect theory of strong interactions. That role is currently being played by *quantum chromodynamics, QCD*. But the story of Yang–Mills is part of the story of QCD. While Yang and Mills could not argue successfully for a nonzero mass of the mesons in the setting of their theory, the addition of a new theoretical consideration, called *spontaneous symmetry breaking*, allowed one to do Yang–Mills theory and have nonzero masses for the mesons. (See the 1966 article [22] by Peter Higgs, who won the Nobel Prize in Physics in 2013 for this work, which anticipated the discovery of the Higgs boson in 2012.) This led to renewed interest in the physics community in Yang–Mills theory and in its generalization, *gauge theory*, the subject of the next chapter.

And QCD is a gauge theory, though now the Lie group is $SU(3)$, the symmetry group of "color." In this case, there are eight mass-zero particles, called *gluons*, which mediate the strong interaction. The electromagnetic and *weak (nuclear) interaction* have been unified into what is called the *electroweak theory*, which is a gauge theory with Lie group $U(1) \times SU(2)$. In this theory for the weak part of the interaction, there are three spin- 1 particles, denoted W^+, Z^0, W^-, with masses much larger than that of a proton. And for the electromagnetic part of the interaction, there is the spin-1, mass-zero photon. The *standard model* of particle physics is the gauge theory obtained by combining QCD and the electroweak theory. It is a gauge theory whose Lie group is $U(1) \times SU(2) \times SU(3)$. In all of these theories, we have compact Lie groups.

The remaining known interaction in nature is gravitation, which is best described using the theory of *general relativity (GR)*, which is a classical, not a quantum, theory. However, GR can also be viewed as a sort of gauge theory. However, the group is all transformations of spacetime coordinates, and this is not a compact group. Whether GR can be incorporated into quantum theory remains an overriding challenge in contemporary physics. Even if that is not the correct way of understanding gravitation, there is the problem of unifying all the known interactions of nature into one theory. Somehow gauge theory has got to be an aspect of that, given its success in the standard model.

Chapter 14
Gauge Theory

If this were a novel, then this chapter would be the climax of the story, because here we present the now-legendary result identifying the gauge fields of physics theory with the principal bundles that have a connection of mathematical theory. But in science the story continues and continues. And to this day we do not know how it will end.

The very first mention of the basic idea of gauge invariance seems to be in the paper [51] written almost 100 years ago by H. Weyl. Unfortunately, the idea did not work in the setting that Weyl had in mind, which was the unification of electromagnetism with general relativity using a scaling of lengths, and apparently this led to his general idea being abandoned by physicists. Although the realization that Yang–Mills theory could be generalized to gauge theory came rather quickly after the publication of [54] (see [48]), it took more time to realize that there might be some physics in all of this. So only then did gauge theories come into their own in the physics literature. By the way, what we call a *gauge theory* here is often called a *nonabelian gauge theory*. The extra adjective is superfluous and misleading. The Lie group in a gauge theory can well be abelian, as the example of electrodynamics shows. Also, what we call a gauge theory is often called a Yang–Mills theory, but I prefer to reserve the latter expression for the specific $SU(2)$ theory introduced in [54] by Yang and Mills. Any way one classifies these theoretical developments, the published scientific literature concerning Yang–Mills and gauge theories is enormous. In round numbers, our bibliography omits 100% of this literature.

Before the publication of [54], Ehresmann in [11] had already developed the theory of connections without any motivation from physics. This was a continuation of trends in classical differential geometry that go back to T. Levi-Civita and E. Cartan. The relation between these two independent developments was worked out later. According to Mayer in [10], the paper [34] by Lubkin in 1963 was an early paper to present a bundle interpretation of gauge theories. The introduction by Mayer in [10] contains a lot of references to papers published in the period 1963–1977 on the relation of principal bundles and gauge theory. Rather curiously, based on his reading of [54], Mayer asserts in [10] that it must have been clear

© Springer International Publishing Switzerland 2015   195
S.B. Sontz, *Principal Bundles*, Universitext, DOI 10.1007/978-3-319-14765-9_14

to Yang and Mills that their theory had a "geometric meaning" because "... they use the gauge-covariant derivative and the curvature form of the connection ...." However, Yang contradicts this in his interview published in [55].

14.1   Gauge Theory in Physics

An outstanding reference for learning how a Nobel laureate physicist views gauge theories is Weinberg's text [50], which is just chock-full of wonderful information. However, my mathematically inclined readers are warned that Weinberg does not follow the modern accepted criteria of mathematical rigor. Another view of the theory up to 1980 is given for a general reader in [24] by 't Hooft, almost two decades before he himself received the Nobel Prize in Physics. For the first 50 years of the theory from a physics viewpoint, [25] has an ample collection of review articles. You also might enjoy the history of the early years of gauge theory in [41] by O'Raifeartaigh. The enormous tome [42] by Penrose is supposedly written for a general public, but scientists can gain some insights from it too. You might find Chapter 15 of [42] helpful for more intuition on gauge theory.

A fact, though not a theorem, seems to be that all Lie groups used in physics are matrix groups. (See [20], Appendix C.3 for examples of Lie groups that are not matrix groups.) Even if this is false, it is almost true. So the situation is that every "physics" Lie group G is explicitly given as a closed subgroup $G \subset GL(n, \mathbb{C})$, the Lie group of all invertible $n \times n$ matrices with complex entries. Even if the entries in the matrices of G are real numbers, we consider $G \subset GL(n, \mathbb{C})$. The reason for doing this will soon be explained.

It follows that the tangent bundle $T(G)$ of G is a subbundle of the tangent bundle $T(GL(n, \mathbb{C}))$ of $GL(n, \mathbb{C})$. But we can identify $GL(n, \mathbb{C})$ with an open set in $M(n, \mathbb{C}) \cong \mathbb{R}^{2n^2}$, where $M(n, \mathbb{C})$ is the vector space of all $n \times n$ matrices with complex entries. This is true since the determinant function det $: M(n, \mathbb{C}) \to \mathbb{C}$ is a continuous function and $GL(n, \mathbb{C}) = \det^{-1}(\mathbb{C} \setminus \{0\})$ exhibits $GL(n, \mathbb{C})$ as the inverse image under a continuous function of an open subset of $\mathbb{C}$. So the tangent bundle of $GL(n, \mathbb{C})$ is trivial, with each fiber being a copy of $M(n, \mathbb{C})$. Recalling that the Lie algebra of a Lie group is its tangent space at its identity element e, we see that

$$\mathfrak{g} := T_e(G) \subset T_e(GL(n, \mathbb{C})) =: \mathfrak{gl}(n, \mathbb{C}) = M(n, \mathbb{C}).$$

This is an inclusion of Lie algebras, so the Lie bracket on $\mathfrak{g}$ is the restriction of the Lie bracket on $M(n, \mathbb{C})$. But the Lie bracket on the latter Lie algebra is the commutator of matrices. So everything looks as nice as one could possibly want. Except for physicists! In practice, in most specific cases the elements of $\mathfrak{g}$ are *antihermitian matrices* A with respect to the adjoint operation on $M(n, \mathbb{C})$, that is, $A^* = -A$.

Exercise 14.1 *As a quick interlude, prove that if A_1, A_2 are antihermitian matrices, then their commutator $[A_1, A_2] := A_1 A_2 - A_2 A_1$ is also an antihermitian matrix.*

While mathematicians have no problem dealing with a mere minus sign, physicists have an attitude problem. And that attitude is that in quantum mechanics we want to study *hermitian matrices B*, that is, those satisfying $B^* = B$. The *spectrum*, namely, the set of *eigenvalues*, of a hermitian matrix consists of real numbers and can be interpreted directly as possible values measured in experiment. So a physicist takes A, the mathematician's antihermitian matrix, and says that $B = iA$ is an element in the Lie algebra of G. So even if G has only matrices with real entries, the matrix B will have complex entries. This is why we are taking the group G to always be a subgroup of $GL(n, \mathbb{C})$. In short, the physicist replaces $\mathfrak{g} = T_e(G)$ with $i\mathfrak{g} = i T_e(G)$. Of course, the physicist uses the symbol $\mathfrak{g}$ to mean $i T_e(G)$. Unfortunately, what should be a minor point can end up being an obstacle to communication.

The physicist's convention wreaks havoc as far as a mathematician is concerned. For example, the Lie bracket in the physicist's Lie algebra is *not* the commutator but rather is given by

$$[B_1, B_2]_{\text{phys}} := -ig\,(B_1 B_2 - B_1 B_2) = -ig\,[B_1, B_2].$$

In the nicest assignment of dimensions to the basic terms in gauge theory, the coupling constant $g \in \mathbb{R}$ is dimensionless, and so its presence here is not explained by a dimension argument. On the other hand, neither can it be excluded by such an argument. However, it will give a pretty result, which we cannot see right now. Many texts take $g = 1$, and so this detail is invisible.

Exercise 14.2 *As a quicker interlude, prove that if B_1 and B_2 are hermitian matrices, then their Lie bracket $[B_1, B_2]_{\text{phys}}$ is also a hermitian matrix.*

Also, the one-parameter subgroup $\exp(sA)$ becomes $\exp(-isB)$ for $s \in \mathbb{R}$, where exp is the exponential map of matrices $\exp : M(n, \mathbb{C}) \to GL(n, \mathbb{C})$ as defined for a matrix T by the usual infinite series

$$\exp T := \sum_{j=0}^{\infty} \frac{1}{j!} T^j,$$

which *is* convergent. Here the matrix A is antihermitian and B is hermitian.

This may seem like a tempest in a teapot, which I guess it should be, but it is best to be forewarned about these varying conventions. From the vigor of my comments, I imagine that many readers feel that I am against the physics conventions. Not so! I grew up with them and feel uncomfortable with the mathematics conventions.

We want to see how general Yang–Mills theory can be made. Well, not quite that since there never seems to be a limit to generalization and, worse than that, generalization for generalization's sake. But just for now we are thinking about the simplest sort of generalization. So we take an arbitrary "physics" Lie group

G with its Lie algebra $\mathfrak{g}$. We consider a local coordinate system x_μ in some chart on an open subset V of some differentiable manifold M with $\dim_\mathbb{R} M = m$. One can consider that $\mu = 1, 2, \ldots, m$ or that $\mu = 0, 1, \ldots, m - 1$. The reader can choose (consistently!) either one of these two options. Indices with Greek letters are assumed to range over the chosen index set.

We let $B_\mu : V \to \mathfrak{g}$ be a smooth function for each μ. This is the *gauge field*. As with Yang–Mills theory, we consider matter fields ψ in some Hilbert space. As before, the exact Hilbert space is not very important. However, there is a representation (or action) of the group G on the Hilbert space, and this is important. This is denoted as $U\psi$ for $U \in G$ and ψ in the Hilbert space. Some texts use $g \in G$ to denote the typical group element, but this conflicts with our use of g as the coupling constant.

We now run through the basic results from the previous chapter in this new setting. The *local gauge transformation* for a matter field is now given by

$$\psi \to \tilde{\psi} = U^{-1}\psi,$$

where $U : V \to G$ is a smooth function. Then the condition for the gauge-transformed gauge field is the same formula as in Yang–Mills theory, except now μ ranges over a possibly different index set. Here it is again:

$$U\left(\frac{1}{i}\frac{\partial}{\partial x_\mu} + gB'_\mu\right)\tilde{\psi} = \left(\frac{1}{i}\frac{\partial}{\partial x_\mu} + gB_\mu\right)\psi,$$

where B'_μ are the still-unknown gauge transforms of the matrices B_μ. As before, the Hilbert space of wave functions ψ must admit operations of partial differentiation. And exactly the same argument leads to the same solution:

$$B'_\mu = U^{-1}B_\mu U + \frac{1}{ig}U^{-1}\frac{\partial U}{\partial x_\mu}.$$

Since we are dealing with "physics" Lie groups, all the factors in this equation are $n \times n$ matrices for some integer $n \geq 1$. So all the products are simply the products of such matrices. It is important to understand that $B'_\mu : V \to \mathfrak{g}$; that is, we have transformed the gauge field B_μ with values in the Lie algebra $\mathfrak{g}$ into another gauge field B'_μ with values in the *same* Lie algebra $\mathfrak{g}$. We leave the proof of this detail to the interested and motivated reader.

Exercise 14.3 *Finding a proof of this should not be so trivial. But sometimes the best way to learn to swim is to jump into the deep end of the pool. For this reason, there are no hints for this exercise.*

We define the gauge equivalence of gauge fields just as in the Yang–Mills setting. We also define pure gauge exactly as before. We next define the *field strength* by the same formula as in Yang–Mills theory:

$$F_{\mu\nu} := \partial_\nu B_\mu - \partial_\mu B_\nu - ig[B_\mu, B_\nu].$$

Then the reader is kindly requested to verify that the derivation of the local gauge transformation of $F_{\mu\nu}$ in the Yang–Mills setting is also valid in this new setting. This is a question of rereading the argument for the Yang–Mills field strength and checking that everything works in this more general setting. So we have

$$F'_{\mu\nu} = U^{-1}F_{\mu\nu}U.$$

Recall that $U : V \to G$ here. This is much, much stronger than saying that the previous formula is valid for all $U \in G$.

Next, we define the Lagrangian density for this theory just as before. But we have to raise indices as before. To do this, we assume that we also have a nondegenerate metric given on the manifold M. Then we define the Lagrangian density by

$$\mathcal{L}_{\text{gauge}} := -\frac{1}{4}\sum_{\mu\nu} F_{\mu\nu}F^{\mu\nu}.$$

The action functional is also defined as before,

$$S_{\text{gauge}} := \int_V \text{Tr}(\mathcal{L}_{\text{gauge}}),$$

where we integrate with respect to the volume form on M defined by its nondegenerate metric. As before, we will assume that the integral converges and continue the discussion.

We apply the calculus of variations to the action functional in order to derive the associated Euler–Lagrange equations. These have the same form as in the Yang–Mills setting, only now the indices vary over a possibly different index set. The resulting Euler–Lagrange equations are called the *Yang–Mills equations* for a general gauge theory. Here they are:

$$\sum_\alpha \left(\partial_\alpha F_{\alpha\beta} - ig[B_\alpha, F_{\alpha\beta}]\right) = 0.$$

So there are m equations here, one for each value of β. As we saw before in the Yang–Mills setting, these are second-order partial differential equations in the unknown gauge fields B_α. They are also nonlinear if the Lie algebra $\mathfrak{g}$ is noncommutative.

In the physics approach, one focuses on equations written in terms of the local coordinates of a specific chart. Actually, all this section is equivalent to a discussion of a theory for an open set V in $\mathbb{R}^m$ with its standard coordinates.

14.2   Gauge Fields in Geometry

The global approach of considering a gauge theory on all of a manifold M is part of the geometric interpretation of gauge theory. The major characteristic of the geometric approach is to define everything globally in terms of manifolds and bundles. The physics approach in the last section was dominated by a lot of local coordinates. Of course, each of these approaches is both mathematical and physical. And they are logically consistent approaches. But the geometric approach wins the day, and not just for its elegance, but also because it leads to a deeper understanding based on global constructs, such as characteristic classes. Even though we will not see those, it is time to view this theory in a global setting. But that requires us to express local information in global language. This will take a while. And even though the purpose of this section is to give the global generalization of the previous section, to facilitate matters we choose to start with the primordial example of the nonabelian Yang–Mills theory of [54].

First, let's consider again the undefined object B that contains the gauge field $B_\mu = B_\mu(x)$, where $x \in \mathbb{R}^4$, of Yang–Mills theory encoded in its smooth hermitian matrix-valued coefficients:

$$B = B_0\, dx_0 + B_1\, dx_1 + B_2\, dx_2 + B_3\, dx_3 = \sum_\mu B_\mu\, dx_\mu. \tag{14.1}$$

This certainly *looks* like a 1-form since it is some sort of linear combination of 1-forms. If we were to evaluate this on a vector field

$$V = V_0\, \partial_0 + V_1\, \partial_1 + V_2\, \partial_2 + V_3\, \partial_3 = \sum_\mu V_\mu\, \partial_\mu,$$

where each $V_\mu = V_\mu(x)$ is a C^∞ *real-valued* function of $x \in \mathbb{R}^4$, then by naïvely using bilinearity plus the duality of the two bases of the vector fields $\partial_\mu = \partial/\partial x_\mu$ and the 1-forms dx_μ, we would expect to get

$$V_0 B_0 + V_1 B_1 + V_2 B_2 + V_3 B_3 = \sum_\mu V_\mu B_\mu = \sum_\mu V_\mu(x) B_\mu(x),$$

a smooth $M_H(2,\mathbb{C})$-valued function, since the coefficients $V_\mu(x)$ are real. Here $M_H(2,\mathbb{C})$ is the real vector space of the 2×2 hermitian matrices.

So we could define B to be the linear map $V \mapsto \sum_\mu V_\mu B_\mu$. Equivalently, we prefer to *define* B as the smooth function

$$B : T(\mathbb{R}^4) \to M_H(2,\mathbb{C}), \tag{14.2}$$

given by the formula

$$B(v) := \sum_{\mu} v_{\mu} B_{\mu}(x),$$

where $v = (v_{\mu}) \in T_x(\mathbb{R}^4)$ and $x \in \mathbb{R}^4$, which is a linear map on each fiber $T_x(\mathbb{R}^4)$. This motivates the next definition.

Definition 14.1 *Let M be a differential manifold and W a vector space over the real numbers. Then a smooth function $A : T(M) \to W$ that is linear on every fiber of $T(M)$ is called a* vector- (space-) valued (differential) 1-form *on M. If W is moreover a Lie algebra, then A is called a* Lie algebra-valued (differential) 1-form *on M.*

If Definition 14.1 looks like déjà vu all over again, recall Definition 10.2. They are the same! We have arrived at the same definition twice, before by going down a pure mathematics path and now by going down a physics path.

Letting x_{μ} be local coordinates on an open subset $V \subset M$ in a chart and A be as in Definition 14.1, then in that chart we have

$$A = \sum_{\mu} A_{\mu} \, dx_{\mu},$$

where $A_{\mu} := A \circ \partial_{\mu} : V \to W$. Here $\partial_{\mu} : V \to T(V) \subset T(M)$ denotes the standard vector field associated to the coordinate x_{μ}. Conversely, every such expression defines a vector space-valued 1-form locally.

It turns out that in the physics convention, the Lie algebra $\mathfrak{u}(2)$ of the group $U(2)$ of all 2×2 unitary matrices is exactly $M_H(2, \mathbb{C})$. Hence, B as in (14.1) or (14.2) is a Lie algebra-valued 1-form on $\mathbb{R}^4$ with values in $\mathfrak{u}(2)$. Even though Yang and Mills present their theory as being based on the group $SU(2)$, the gauge field that they first present has matrices in $\mathfrak{u}(2)$, not $\mathfrak{su}(2)$. Already we have seen that the way to generalize their construction is by using the Lie algebra of any Lie group.

By the way, the condition on the time component of the gauge field, denoted B_4, in [54] is that it should be antihermitian. But Yang and Mills also use $x_4 = it$, where t is the time coordinate and dimensions are used so that the speed of light c is dimensionless and $c = 1$. So this is consistent with our condition that the time component B_0 should be hermitian since for us $x_0 = t$ as well as $c = 1$.

Comparing Definition 5.3, where we defined differential k-forms, with Definition 14.1, we see that there are two differences. First, in Definition 14.1, we have only the special case of 1-forms. And, second, we have W instead of $\mathbb{R}$ as the codomain of the form. This leads us to a rather natural definition of a vector-valued k-form.

Definition 14.2 *Let M be a differential manifold and W a vector space over the real numbers. Then a smooth function $A : T^{\diamond k}(M) \to W$ that is* antisymmetric *and k-multilinear on every fiber of $T^{\diamond k}(M)$ is called a* vector- (space-) valued (differential) k-form *on M. If, moreover, W is a Lie algebra, then A is called a* Lie algebra-valued (differential) k-form *on M.*

If x_μ are local coordinates on an open set $V \subset M$ in a chart, then in that chart we have that the k-form A with values in W is given by

$$A = \sum_I A_I \, dx_I,$$ (14.3)

where $A_I := A \circ \partial_I : V \to W$. Here I is a k-*index*, that is, a strictly increasing sequence of k indices $1 \le i_1 < i_2 < \cdots < i_k \le m = \dim M$, and

$$\partial_I := \partial_{i_1, i_2, \cdots, i_k} : V \to T^{\diamond k}(V) \subset T^{\diamond k}(M)$$

is defined for $p \in V$ by

$$\partial_I(p) = \partial_{i_1, i_2, \cdots, i_k}(p) := (\partial_{i_1}(p), \partial_{i_2}(p), \cdots, \partial_{i_k}(p)),$$

where each entry is a standard basis element in the vector space $T_p(M)$. Just in case the reader has not already guessed, here we are also defining

$$dx_I = dx_{i_1, i_2, \cdots, i_k} := dx_{i_1} \wedge dx_{i_2} \wedge \cdots \wedge dx_{i_k}.$$

Conversely, every such expression (14.3) defines a vector space-valued 1-form locally. The exterior derivative of this k-form A is by definition given by

$$dA := \sum_I \sum_{\alpha=1}^m \partial_\alpha A_I \, dx_\alpha \wedge dx_I.$$ (14.4)

Note that $\partial_\alpha A_I : V \to W$, so that the coefficients in this last expression, like the coefficients A_I, are W-valued smooth functions.

Then the standard game starts up. First, one notes that a given W-valued k-form will have local representations in charts with a certain compatibility condition between any pair of charts. Second, one notes that any system of W-valued k-forms, each being defined locally in an atlas of charts and each pair of them also satisfying the compatibility condition, defines a unique global W-valued k-form that restricts in the open set of each chart in the given atlas to the given W-valued k-form defined locally in that chart.

In physics parlance, one expresses these facts somewhat cryptically by saying that

"a k-form is something that transforms like a k-form."

Third, one proves that the formula (14.4) given for each chart satisfies the compatibility condition for a W-valued $(k + 1)$-form. So (14.4) does define a W-valued $(k + 1)$-form dA on M.

For example, for the 1-form $B = \sum_\mu B_\mu dx_\mu$, this yields

$$dB = \sum_{\mu\nu} \partial_\nu B_\mu \, dx_\nu \wedge dx_\mu.$$

By a standard argument using the anticommutativity of the expressions $dx_\nu \wedge dx_\mu$, one can show that

$$dB = \sum_{\mu<\nu} (\partial_\nu B_\mu - \partial_\mu B_\nu) \, dx_\nu \wedge dx_\mu.$$

This captures the information in our first, but unsuccessful, definition of the field strength $F_{\mu\nu}$. However, the definition that we eventually adapted does include this term. But it also had a commutator, $[B_\mu, B_\nu]$. We can just stick this as is to get

$$F = dB - ig \sum_{\mu<\nu} [B_\mu, B_\nu] \, dx_\nu \wedge dx_\mu.$$

But we would rather write the second term directly in terms of the Lie algebra-valued 1-form B itself rather than its coefficients in local coordinates. Who knows? Just because you want something is no guarantee that you can get it. However, there might be a window of opportunity here. The expression $[B_\mu, B_\nu]$ is antisymmetric and bilinear. And so is $B \wedge B$. But the wedge of something with itself gives zero, right? So this looks hopeless. But wait! The coefficients of B are matrices, not scalars. In the usual argument showing that $\omega \wedge \omega = 0$, one assumes that ω is a *real-valued* 1-form. Let's see what happens when we evaluate $B \wedge B$. We do this explicitly for the case when $\dim M = m = 4$, but the proof for the case of general m is the same. So we see that

$$B \wedge B = \tag{14.5}$$

$$(B_0 \, dx_0 + B_1 \, dx_1 + B_2 \, dx_2 + B_3 \, dx_3) \wedge (B_0 \, dx_0 + B_1 \, dx_1 + B_2 \, dx_2 + B_3 \, dx_3).$$

Recall that the product of the matrix coefficients on the right side is their Lie bracket in $\mathfrak{g}$, which is

$$[M, N]_{\text{phys}} = -ig \, (MN - NM) = -ig \, [M, N],$$

where M, N are square matrices.

This expression (14.5) gives us a grand total of 16 terms. But four of these terms are given by $[B_\mu, B_\mu]_{\text{phys}} (dx_\mu \wedge dx_\mu) = 0$. So we have to deal with the remaining 12 terms. Here they are:

$$(-ig)^{-1}(B \wedge B) =$$

$$[B_0, B_1] \, dx_0 \wedge dx_1 + [B_0, B_2] \, dx_0 \wedge dx_2 + [B_0, B_3] \, dx_0 \wedge dx_3 +$$

$$[B_1, B_0] \, dx_1 \wedge dx_0 + [B_1, B_2] \, dx_1 \wedge dx_2 + [B_1, B_3] \, dx_1 \wedge dx_3 +$$
$$[B_2, B_0] \, dx_2 \wedge dx_0 + [B_2, B_1] \, dx_2 \wedge dx_1 + [B_2, B_3] \, dx_2 \wedge dx_3 +$$
$$[B_3, B_0] \, dx_3 \wedge dx_0 + [B_3, B_1] \, dx_3 \wedge dx_1 + [B_3, B_2] \, dx_3 \wedge dx_2.$$

Writing these 12 terms as linear combinations of the 6 standard basis elements $dx_\mu \wedge dx_\nu$ with $\mu < \nu$ reduces this to 6 terms, each of which is a combination of 2 terms in the previous expression. So we have

$$(-\mathrm{i}\, g)^{-1} (B \wedge B) =$$
$$([B_0, B_1] - [B_1, B_0]) \, dx_0 \wedge dx_1 + ([B_0, B_2] - [B_2, B_0]) \, dx_0 \wedge dx_2$$
$$+ ([B_0, B_3] - [B_3, B_0]) \, dx_0 \wedge dx_3 + ([B_1, B_2] - [B_2, B_1]) \, dx_1 \wedge dx_2$$
$$+ ([B_1, B_3] - [B_3, B_1]) \, dx_1 \wedge dx_3 + ([B_2, B_3] - [B_2, B_3]) \, dx_2 \wedge dx_3$$
$$= \sum_{\mu < \nu} 2 [B_\mu, B_\nu] \, dx_\mu \wedge dx_\nu.$$

In this setting, each B_μ is a matrix and so the coefficients of $B \wedge B$ need not be zero. We can write this result as

$$B \wedge B = -2\mathrm{i} g \sum_{\mu < \nu} [B_\mu, B_\nu] \, dx_\mu \wedge dx_\nu.$$

Substituting back into the equation for the field strength, we see that the two factors $\mathrm{i} g$ cancel out, this being the promised pretty result of the definition of $[\cdot, \cdot]_{\text{phys}}$. So the field strength is

$$F = dB + \frac{1}{2} B \wedge B. \tag{14.6}$$

We have seen this equation before! It is the equation for the curvature of an Ehresmann connection, but so far we only know that B is a $\mathfrak{g}$-valued 1-form. So it is reasonable to ask whether B is actually an Ehresmann connection as well.

But in order to address this question, we need to define a principal bundle. Since we want to work with a chart defined on an open subset of a manifold M, it seems reasonable to take the open set to be M and consider a trivial principal bundle $\pi : P \to M$. For the structure Lie group of the principal bundle, the only candidate is G, the Lie group of the gauge theory. So for the explicitly trivial case, we take the total space to be $P := M \times G$, where G acts on the right of P by right multiplication on the second factor of P. In other words,

$$(x, h_1) \cdot h_2 := (x, h_1 h_2)$$

for all $x \in M$ and $h_1, h_2 \in G$. Then with the projection map $\pi_1 : M \times G \rightarrow M$ onto the first coordinate, that is, $\pi_1(x, h) = x$ for $h \in G$, we have a principal bundle with structure Lie group G.

Since the coordinates x_μ make sense on both of the spaces P and M, we can consider the $\mathfrak{g}$-valued 1-form B as being defined on either of these spaces. But to get an Ehresmann connection, we need a $\mathfrak{g}$-valued 1-form defined on the total space P. So it is better to think of B as being defined on P. A more precise way of saying this is that given the $\mathfrak{g}$-valued 1-form on the base space M, we can also consider the $\mathfrak{g}$-valued 1-form $\pi_1^* B$ on the total space P. But is $\pi_1^* B$ an Ehresmann connection on P? The answer is no.

Here's what's happening. Let Y be any nonzero vector field on G. Then we can define a vector field W on P for all $p = (x, h) \in P = M \times G$ by

$$W(p) = W(x, h) := (0, Y(h)) \in T_x(M) \oplus T_h(G) \cong T_{(x,h)}(M \times G) = T_p(P).$$

Since $\ker \pi_{1*} = T(G)$, we see that $(\pi_1^* B)(W) = B(\pi_{1*} W) = 0$. In particular, if we take $A \in \mathfrak{g}$, then this defines fundamental vertical vector fields $A^{\sharp G}$ on G and $A^{\sharp P}$ on P. These are related by $A^{\sharp P} = (0, A^{\sharp G})$.

Exercise 14.4 *Prove the last equation.*

So by the above argument, we obtain

$$(\pi_1^* B)(A^{\sharp P}) = 0.$$

By definition, an Ehresmann connection ω on P satisfies $\omega(A^{\sharp P}) = A$. So if $\mathfrak{g} \neq 0$, which is the case of interest, the $\mathfrak{g}$-valued 1-form $\pi_1^* B$ on P is not an Ehresmann connection on P.

Exercise 14.5 *It is even worse! An Ehresmann connection ω on P also must satisfy the other defining property,*

$$\mathrm{Ad}_{g^{-1}} \omega = R_g^* \omega$$

for all $g \in G$. Prove that $\pi_1^* B$ always fails to have this property also, except when $G = \{e\}$, the trivial group.

So it is not such a simple matter to find an Ehresmann connection on P that is somehow related to the given $\mathfrak{g}$-valued 1-form B on the base space M. Also, we want the curvature of this still-to-be-found Ehresmann connection to be related to the field strength 2-form on M. We have found a tantalizing formula in (14.6). The challenge is to get it integrated into a geometric theory of gauge theories.

So here is the challenge explicitly stated: to find an Ehresmann connection ω on the total space P of a trivial principal bundle $\pi : P \rightarrow M$ with a given section $\sigma : M \rightarrow P$ such that $\sigma^* \omega = B$, a given $\mathfrak{g}$-valued 1-form on M. The new condition $\sigma^* \omega = B$ tells us exactly how the desired Ehresmann connection ω should be related to B.

To make the formulas more transparent, we first consider the case where the bundle is $\pi_1 : M \times G \to M$ with its canonical section $\sigma_1(x) = (x, e)$ for $x \in M$. Since $\sigma_1 : M \to M \times G$ embeds M onto its range $M \times \{e\}$, the requirement $\sigma_1^* \omega = B$ only forces the definition of the sought-for ω on the slice $M \times \{e\}$ of $M \times G$. Specifically, restricting to $M \times \{e\}$, we see that

$$\pi_1^* B = \pi_1^* \sigma_1^* \omega = (\sigma_1 \pi_1)^* \omega = \omega$$

since $\sigma_1 \pi_1$ restricted to $M \times \{e\}$ is the identity map.

Using the second defining property of an Ehresmann connection, we must have for all $h \in G$ that

$$\text{Ad}_{h^{-1}} \omega = R_h^* \omega = \omega R_{h*}. \tag{14.7}$$

To see what ω must look like over the slice $M \times \{h\}$ for arbitrary $h \in G$, we start with the assertion that

$$M \times \{e\} \xrightarrow{R_h} M \times \{h\}$$

identifies these two slices. Moreover, we have $\pi_1 R_h = \pi_1 : M \times G \to M$.

Condition (14.7) gives us the commutativity of this diagram:

$$\begin{array}{ccc} T(M \times \{e\}) & \xrightarrow{R_{h*}} & T(M \times \{h\}) \\ \omega = \pi_1^* B \downarrow & & \downarrow \omega \\ \mathfrak{g} & \xrightarrow{\text{Ad}_{h^{-1}}} & \mathfrak{g} \end{array}.$$

Then to find what ω must look like over $M \times \{h\}$, we consider the vertical arrow on the right of this diagram. So ω over $M \times \{h\}$ must be

$$\omega = \text{Ad}_{h^{-1}} B \, \pi_{1*} \, R_{(h^{-1})*} = \text{Ad}_{h^{-1}} B \, \pi_{1*} = \text{Ad}_{h^{-1}} \pi_1^*(B).$$

To convert this into a definition over $M \times G$, we define $g : P = M \times G \to G$ by $g(x, h) := h$. Of course, this is sometimes denoted as π_2, the projection onto the second factor. But for now we want to use both notations. Anyway, we define a $\mathfrak{g}$-valued 1-form on $P = M \times G$ by

$$\omega_1 := \text{Ad}_{g^{-1}} \pi_1^*(B).$$

And this definition gives us an object with nice properties. For example, $\sigma_1^* \omega_1 = B$ since we constructed ω_1 so that it would be equal to $\pi_1^* B$ over $M \times \{e\}$.

Warning: The notation g^{-1} is used here since it is quite common. But it is *not* the inverse of the function $g : M \times G \to G$. Rather, it is that function g composed with the map $G \to G$ that maps each element in G to its inverse element in G. Explicitly, $g^{-1}(x, h) = h^{-1}$ for $x \in M$ and $h \in G$.

But not so obvious is the next property. Here we take $h \in G$ and $v \in T_p(P)$. We will use the identity $g(ph) = g(p)h$ in the equivalent form $g^{-1}(ph) = h^{-1}g^{-1}(p)$. Then we compute as follows:

$$(R_h^* \omega_1)(v) = \omega_1(R_{h*}v)$$

$$= \text{Ad}_{g^{-1}(ph)}\Big((\pi_1^* B)(R_{h*}v)\Big)$$

$$= \text{Ad}_{h^{-1}g^{-1}(p)}\big(B(\pi_{1*}R_{h*}v)\big)$$

$$= \text{Ad}_{h^{-1}}\text{Ad}_{g^{-1}(p)}\big(B(\pi_{1*}v)\big)$$

$$= \text{Ad}_{h^{-1}}\text{Ad}_{g^{-1}(p)}(\pi_1^* B)(v)$$

$$= (\text{Ad}_{h^{-1}}\omega_1)(v).$$

So ω_1 has one of the defining properties of an Ehresmann connection. Does it have the other? Maybe the reader has guessed that the subscript on ω_1 means that this is not the Ehresmann connection on P that we are looking for. And this is exactly the moment when we see why not. This is really a repeat of the argument we used with $\pi_1^*(B)$. So again we take a nonzero element $A \in \mathfrak{g}$. Then, letting $A^\sharp$ denote its fundamental vertical vector field on P, we have

$$\omega_1(A^\sharp) = \text{Ad}_{g^{-1}}(\pi_1^* B)(A^\sharp) = \text{Ad}_{g^{-1}} B(\pi_{1*}A^\sharp) = 0$$

since $\pi_{1*}A^\sharp = 0$, as noted earlier.

So we have good news and bad news. And that news, both good and bad at the same time, is that ω_1 satisfies almost all the desired properties. Actually, the space of all connections on a principal bundle $\pi : P \to M$ with structure Lie group G is an affine space whose associated vector space consists of those $\mathfrak{g}$-valued 1-forms η on P that satisfy these two conditions that are linear in η:

1. $\text{Ad}_{h^{-1}}\eta = R_h^*\eta$.
2. $\eta(A^\sharp) = 0$ for all $A \in \mathfrak{g}$.

Any such η is called an *Ehresmann connection displacement*.

So the $\mathfrak{g}$-valued 1-form ω_1 is an Ehresmann connection displacement. The usual results about affine spaces hold. First, the difference of any two Ehresmann connections is an Ehresmann connection displacement. Second, the sum of an Ehresmann connection and an Ehresmann connection displacement is an Ehresmann connection. Third, the set of all Ehresmann connections is obtained by taking any one specific Ehresmann connection and adding to it all possible Ehresmann connection displacements.

Exercise 14.6 *The reader is asked to reflect on the relation of the last sentence with this familiar chant learned in any (good!) introductory course in linear differential equations. "The general solution of a linear inhomogeneous equation is given by the sum of any particular solution of the inhomogeneous equation with the general solution of the corresponding linear homogeneous equation."*

So we want to add an Ehresmann connection α to our known Ehresmann connection displacement, which contains the information about the physics of the situation, namely, the $\mathfrak{g}$-valued 1-form ω_1. So we want to define

$$\omega := \omega_1 + \alpha,$$

where $\alpha : T(P) \to \mathfrak{g}$ is some Ehresmann connection on P. But at the same time, we want $\sigma_1^*(\omega) = B$. But we already have $\sigma_1^*(\omega_1) = B$. So the unknown Ehresmann connection α must satisfy $\sigma_1^*(\alpha) = 0$, that is, $\alpha\,\sigma_{1*} = 0$. Now

$$\sigma_{1*} : T(M) \to T(P) = T(M \times G) \cong T(M) \oplus T(G)$$

is the derivative of $\sigma_1(x) = (x, e)$. So for $v \in T(M)$, we see that

$$\sigma_{1*}(v) = (v, 0).$$

This implies that

$$\alpha(v, 0) = \alpha\,\sigma_{1*}(v) = 0$$

for all $v \in T(M)$, that is, $\ker \alpha \supset T(M)$. So the sought-for Ehresmann connection α is completely determined by its values on $T(G) \subset T(M) \oplus T(G)$. The inclusion map $j : T(G) \to T(M) \oplus T(G)$ of this inclusion is $j : w \mapsto (0, w)$ for all $w \in T(G)$. Define $\beta := \alpha\,j$. We would like to say that β is a pullback of α by some map $G \to M$, so that we could conclude that β is an Ehresmann connection. This depends on being able to write j as the derivative of some smooth function $G \to M \times G$.

Take any point $x \in M$ and define $\iota_x : G \to M \times G = P$ by $\iota_x(h) := (x, h)$ for all $h \in G$. Then its derivative

$$\iota_{x*} : T(G) \to T(M \times G) \cong T(M) \oplus T(G)$$

is given by $\iota_{x*}(w) = (0, w) = j(w)$ for all $w \in T(G)$. So $j = \iota_{x*}$, the derivative of a function $G \to M \times G$, as desired.

Then we have $\beta = \alpha\,j = \alpha\,\iota_{x*} = \iota_x^*\alpha$, an Ehresmann connection on G. But there is exactly one Ehresmann connection on a Lie group G, as we have seen. And that Ehresmann connection is the Maurer–Cartan form Ω. So we must have

$\beta = \Omega = \iota_x^* \alpha$. We now want to "solve" this equation for α in terms of Ω. Applying π_2^*, we get

$$\pi_2^* \Omega = \pi_2^* \iota_x^* \alpha = (\iota_x \pi_2)^* \alpha.$$

Now $\iota_x \pi_2(y, h) = \iota_x(h) = (x, h)$, so that its derivative is $(\iota_x \pi_2)_*(v, w) = (0, w)$ for $v \in T(M)$ and $w \in T(G)$. Now $(\iota_x \pi_2)^*$ is certainly not the identity map. Nonetheless, here is its action on α, where we take $v \in T(M)$ and $w \in T(G)$:

$$((\iota_x \pi_2)^* \alpha)(v, w) = \alpha(\iota_x \pi_2)_*(v, w) = \alpha(0, w) = \alpha(v, w).$$

The last equality holds since $\alpha(v, 0) = 0$, as we have already seen. Hence,

$$\alpha = (\iota_x \pi_2)^* \alpha = \pi_2^* \Omega.$$

So at last we arrive at the definition:

$$\omega := \mathrm{Ad}_{g^{-1}} \pi_1^*(B) + \pi_2^*(\Omega).$$

We have proved most of the following proposition.

Proposition 14.1 *Consider the explicitly trivial principal bundle*

$$\pi_1 : P = M \times G \to M,$$

with its canonical section $\sigma_1(x) = (x, e)$ for $x \in M$. Suppose that B is a $\mathfrak{g}$-valued 1-form on M, where $\mathfrak{g}$ is the Lie algebra of G. Then there exists a unique Ehresmann connection ω on P such that $\sigma_1^*(\omega) = B$.

The formula for this Ehresmann connection is

$$\omega = \mathrm{Ad}_{\pi_2^{-1}} \pi_1^*(B) + \pi_2^*(\Omega) = \mathrm{Ad}_{g^{-1}} \pi_1^*(B) + g^*(\Omega),$$

where Ω is the Maurer–Cartan form on G and $\pi_2 = g : P \to G$, the projection onto the second factor of $P = M \times G$.

Proof: We have shown above that ω is an Ehresmann connection on P with $\sigma_1^*(\omega) = B$. The uniqueness statement remains to be proved.

So let ω_{any} be any Ehresmann connection on P satisfying $\sigma_1^*(\omega_{any}) = B$. We define $\alpha := \omega_{any} - \mathrm{Ad}_{g^{-1}} \pi_1^*(B)$. Then the same argument as above shows that $\alpha = g^*(\Omega)$. Thus, $\omega_{any} = \mathrm{Ad}_{g^{-1}} \pi_1^*(B) + g^*(\Omega) = \omega$. ∎

As indicated earlier, we studied the explicitly trivial bundle in preparation for the case of an arbitrary trivial bundle. Here is that result.

Theorem 14.1 *Suppose that $\pi : P \to M$ is a principal bundle with structure group G, $\sigma : M \to P$ is a section of this bundle, and $B : T(M) \to \mathfrak{g}$ is a $\mathfrak{g}$-valued*

1-*form on* M, *where* $\mathfrak{g}$ *is the Lie algebra of* G. *Then there exists a unique Ehresmann connection* $\omega : T(P) \to \mathfrak{g}$ *such that* $B = \sigma^*\omega$.

Moreover, the explicit formula for ω *is*

$$\omega = \mathrm{Ad}_{g^{-1}}\pi^*(B) + g^*(\Omega),$$

where Ω *is the Maurer–Cartan form on* G *and* $g : P \to G$ *will be defined in the proof. More explicitly, if we take* $v \in T_p(P)$, *where* $p \in P$, *then*

$$\omega(v) = Ad_{g^{-1}(p)}\,(\pi^*B)v + g^*(\Omega_{g(p)})v$$
$$= Ad_{g^{-1}(p)}\,(B\pi_*)v + (\Omega_{g(p)}g_*)v.$$

Proof: So the idea is to pull back everything from an explicitly trivial bundle where we have the result from the previous proposition. As we have seen, the existence of the section σ not only implies that P is a trivial bundle, but it also gives us an explicit trivialization of P:

$$\phi_\sigma : P \to M \times G.$$

Let's recall how. We take an arbitrary point $p \in P$ and let $x = \pi(p) \in M$. Then $\sigma(x)$ and p lie in the same fiber $\pi^{-1}(x)$ of P. By the definition of principal bundle, there is a *unique* element $g(p) \in G$ satisfying

$$p = \sigma(x)g(p).$$

So we have defined a function $g : P \to G$, which actually depends on σ. Even though g_σ would be better notation for g, we will use the traditional notation g. Actually, in terms of the affine operation, this is

$$g(p) = (\sigma(x))^{-1}p = \big(\sigma(\pi(p))\big)^{-1}p.$$

Finally, we define the trivialization ϕ_σ by

$$\phi_\sigma(p) := (\pi(p), g(p)) \in M \times G.$$

We let ω_0 denote the Ehresmann connection on $M \times G$ constructed in the previous proposition. So

$$\omega_0 = \mathrm{Ad}_{\pi_2^{-1}}\pi_1^*(B) + \pi_2^*(\Omega).$$

Since Ehresmann connections are preserved under pullback, we pull back with ϕ_σ^* to get automatically an Ehresmann connection on P:

$$\omega := \phi_\sigma^*(\omega_0) = \phi_\sigma^*(\mathrm{Ad}_{\pi_2^{-1}}\pi_1^*(B)) + \phi_\sigma^*(\pi_2^*(\Omega)).$$

Now $\pi_1\phi_\sigma = \pi$ and $\pi_2\phi_\sigma = g$, so that $\phi_\sigma^*\pi_1^* = \pi^*$ and $\phi_\sigma^*\pi_2^* = g^*$. Then using these as well as $\phi_\sigma^*(\pi_2) = \pi_2\phi_\sigma = g$, we immediately get the formula

$$\omega = \mathrm{Ad}_{g^{-1}}\pi^*(B) + g^*(\Omega),$$

as desired. (Be warned that ϕ_σ^* has been given two distinct meanings here!) Finally, we use $\phi_\sigma \sigma = \sigma_1$ to see that

$$\sigma^*(\omega) = \sigma^*\phi_\sigma^*(\omega_0) = \sigma_1^*(\omega_0) = B.$$

The uniqueness of ω follows from the fact that ϕ_σ is an isomorphism of principal bundles. Briefly, if there were another, different Ehresmann connection $\hat{\omega}$ on P satisfying the condition $\sigma^*(\hat{\omega}) = B$, then the uniqueness statement in the previous proposition would be contradicted.

And that finishes the proof. ∎

We now consider in more detail the trivialization $\phi_\sigma : P \to M \times G$ of the principal bundle $\pi : P \to M$. The first thing we would like to note is that ϕ_σ is an isomorphism of principal bundles, where the codomain of this isomorphism is the principal bundle $\pi_1 : M \times G \to M$. The bundle map π_1 is the projection onto the first factor of the product space.

The section $\sigma : M \to P$ in this trivialization, $\phi_\sigma \circ \sigma : M \to M \times G$, is

$$(\phi_\sigma \circ \sigma)(x) = \phi_\sigma(\sigma(x)) = (\pi(\sigma(x)), g(\sigma(x))) = (x, e)$$

for $x \in M$. Here we used the definition of ϕ_σ and the identity $g(\sigma(x)) = e$ as used in the previous theorem. So the range in this trivialization is

$$\mathrm{Ran}\,(\phi_\sigma \circ \sigma) = M \times \{e\} \subset M \times G.$$

This is a slice at $e \in G$ in the Cartesian product $M \times G$. But this particularly pretty representation of σ holds for *any* section σ of $\pi : P \to M$. Each one looks pretty, but only with respect to its own trivialization ϕ_σ. If we are just given a trivial bundle $\pi : P \to M$, but without a particular trivialization, then there are in general many trivializations $\phi : P \to M \times G$—in fact, exactly one for each section of P.

This is a special case of a very general, well-known fact from category theory. That fact is that given isomorphic objects A and B in a category, in general there is more than one isomorphism $\phi : A \to B$. What we want to emphasize is that a trivialization of a principal bundle $\pi : P \to M$ is an isomorphism in the category of principal bundles between $\pi : P \to M$ and $\pi_1 : M \times G \to M$, and conversely. One gets accustomed to saying "trivialization" instead of "isomorphism with the product principal bundle" even though these are equivalent. So when we say that a principal bundle is trivial, we mean that it is *isomorphic* to the product bundle, no more and no less. In particular, we are not speaking of a particular isomorphism.

So the question arises of how two trivializations of a principal bundle can differ. Or equivalently, how can two sections of a trivial principal bundle differ? This is a natural question arising from category theory. It is equivalent to asking what is the automorphism group $\text{Aut}(P)$ of P in the category of principal bundles for a trivial principal bundle $\pi : P \to M$, another general fact from category theory. (This comment is left for the reader's further consideration. The idea for one of the implications is that given trivializations $\phi_1, \phi_2 : P \to M \times G$, then $\phi_2^{-1}\phi_1 : P \to P$ is an automorphism of P.)

So we consider two sections $\sigma_1, \sigma_2 : M \to P$ of a given trivial bundle $\pi :$ $P \to M$. For each $x \in M$, we have $\sigma_1(x) \in \pi^{-1}(x)$ and $\sigma_2(x) \in \pi^{-1}(x)$. And having two points in the same fiber of P implies that there exists a unique element $g(x) \in G$ such that we have

$$\sigma_2(x) = \sigma_1(x)g(x).$$

This holds for all $x \in M$, and so we have found a smooth function $g : M \to G$. Conversely, given a smooth function $g : M \to G$ and a particular section $\sigma_1 : M \to P$ of $\pi : P \to M$, then $\sigma_2(x) = \sigma_1(x)g(x)$ defines a section of P. So the sections of P can be parameterized by the set

$$\{g : M \to G \mid g \text{ is smooth}\}.$$

Of course, we have seen these functions before. A function $g : M \to G$ is a local gauge transformation of the principal bundle $\pi : P \to M$. The general idea from category theory leads directly to an important concept in physics!

A crucial point here is that the parameterization of the set of sections by the set of local gauge transformations depended on the choice of a particular section $\sigma_1 : M \to P$ of $\pi : P \to M$. However, when we are speaking of the trivial bundle $\pi_1 : M \times G \to M$, we take σ_1 to be the "canonical" slice at $e \in G$,

$$\sigma_1(x) := (x, e).$$

So in this setting choosing any other section $\sigma : M \to M \times G$ is the same as choosing a local gauge transformation, and conversely. In this case, we say more briefly that any section *is* a local gauge transformation, a slight abuse of language. And for other reasons, as we shall see later, sometimes we say that a section is a choice of gauge.

Now the reader may be wondering how the theory of the special case of trivial principal bundles is going to help us understand the general case of a principal bundle $\pi : P \to M$, which is not necessarily trivial. The point is that any such bundle is locally trivial. So we have an open cover $\{V_\alpha \mid \alpha \in A\}$ of M with each V_α being homeomorphic to an open subset in some Euclidean space that does not depend on α. (Being an open cover means that each V_α is an open subset of M and $\cup_\alpha V_\alpha = M$.) Second, for each index α, we have a principal bundle isomorphism $\phi_\alpha : \pi^{-1}(V_\alpha) \to V_\alpha \times G$ covering the identity map of V_α. In particular, we have this commutative diagram:

$$
\begin{array}{ccc}
\pi^{-1}(V_\alpha) & \xrightarrow{\phi_\alpha} & V_\alpha \times G \\
{\scriptstyle \pi}\downarrow & & \downarrow{\scriptstyle \pi_1} \\
V_\alpha & \xrightarrow{id_{V_\alpha}} & V_\alpha
\end{array}
$$

Exercise 14.7 *These two properties are not exactly what Theorem 9.1 says that we need to have. But these two properties are enough for us to apply Theorem 9.1, as you should verify.*

Notice as you continue reading this discussion that the fact that each V_α is homeomorphic to an open subset in Euclidean space is never used. This is why we have dropped the notation for that homeomorphism and work directly with each V_α instead.

So we have in general an abundance of trivial bundles $\pi^{-1}(V_\alpha) \xrightarrow{\pi} V_\alpha$ with explicit trivializations ϕ_α and so each with its own canonical section $\sigma_\alpha : V_\alpha \to \pi^{-1}(V_\alpha)$ associated to ϕ_α by

$$
\sigma_\alpha(x) := \phi_\alpha^{-1}(x, e) \tag{14.8}
$$

for $x \in V_\alpha$. Moreover, in each nonempty intersection $V_\alpha \cap V_\beta$, we have two sections, namely, σ_α and σ_β, of the trivial bundle

$$
\pi^{-1}(V_\alpha \cap V_\beta) \xrightarrow{\pi} V_\alpha \cap V_\beta.
$$

So we know that for all $x \in V_\alpha \cap V_\beta$, we have

$$
\sigma_\beta(x) = \sigma_\alpha(x) g_{\alpha\beta}(x), \tag{14.9}
$$

where $g_{\alpha\beta} : V_\alpha \cap V_\beta \to G$. These functions $g_{\alpha\beta}$ must somehow be related to the structure of the principal bundle $\pi : P \to M$. Actually, the very notation alone suggests the conjecture that $g_{\alpha\beta}$ is the cocycle used to define the principal bundle. As we will shortly see, this is true. Hence, in physics terminology, a cocycle is a family of local gauge transformations that satisfy certain "compatibility" conditions.

Theorem 14.2 *With the above notation, $g_{\alpha\beta}$ is the cocycle associated to the collection $\{(V_\alpha, \phi_\alpha)\}$.*

Proof: To see if this is so, we compute the transition functions for the total space P. We take $x \in V_\alpha \cap V_\beta$. Rewriting (14.9) using the definition in (14.8) of the canonical sections, we see that

$$
\phi_\beta^{-1}(x, e) = \phi_\alpha^{-1}(x, e) g_{\alpha\beta}(x) = \phi_\alpha^{-1}(x, g_{\alpha\beta}(x))
$$

since ϕ_α^{-1} is an isomorphism of principal bundles. This immediately implies

$$\phi_\alpha \phi_\beta^{-1}(x, e) = (x, g_{\alpha\beta}(x)).$$

Now acting on the right by an element $g \in G$, we find that

$$\phi_\alpha \phi_\beta^{-1}(x, g) = (x, g_{\alpha\beta}(x)g).$$

Here we used that $\phi_\alpha \phi_\beta^{-1}$ is an isomorphism of principal bundles on the left side, and we used the definition of the right G action on both sides. And this corresponds to (9.1) given our conventions here. So the functions $g_{\alpha\beta} : V_\alpha \cap V_\beta \to G$ defined in (14.9) are the cocycle of the principal bundle for the trivialization given by the collection $\{(V_\alpha, \phi_\alpha)\}$. ∎

We next continue this line of inquiry by asking how a given Ehresmann connection looks when pulled back by two sections ($\equiv$ trivializations) of the same trivial principal bundle.

Theorem 14.3 *Let $\pi : P \to M$ be a principal bundle with Lie group G, whose Lie algebra is $\mathfrak{g}$. Let ω be an Ehresmann connection on P. Suppose that $\sigma_1, \sigma_2 : M \to P$ are sections of P and define $B_1 := \sigma_1^* \omega$ and $B_2 := \sigma_2^* \omega$. So B_1 and B_2 are $\mathfrak{g}$-valued 1-forms on the base space. (Or in physics terms, B_1 and B_2 are gauge fields on M.) Also, for all $x \in M$, we have*

$$\sigma_2(x) = \sigma_1(x)g(x),$$

where $g = g_{12} : M \to G$ is a local gauge transformation.
Then we have this relation between B_1 and B_2:

$$B_2 = Ad_{g^{-1}} B_1 + g^* \Omega, \qquad (14.10)$$

where Ω denotes the Maurer–Cartan form on G. If we take $v \in T_x(M)$ for some $x \in M$, then this relation is more explicitly written as

$$B_2(v) = Ad_{g^{-1}(x)}(B_1(v)) + (g^* \Omega)(v).$$

Remarks: The relation between B_1 and B_2 in physics parlance says that B_2 is the local gauge transformation by $g : M \to G$ of B_1. The proof of this theorem uses Theorem 14.1, but these are quite different results. In Theorem 14.1, we start with a $\mathfrak{g}$-valued 1-form in the base space and find the unique corresponding Ehresmann connection on the total space. On the other hand, in this theorem we start with an Ehresmann connection on the total space and then look at what happens when it is pulled back to the base space by two sections. By the way, you should understand what happens to (14.10) in the particular case when $\sigma_1 = \sigma_2$.

Proof: By the uniqueness result of Theorem 14.1, we must have that

$$\omega = Ad_{g_1^{-1}} \pi^* B_1 + g_1^*(\Omega),$$

where $g_1 : P \to G$ is given for $p \in P$ by

$$g_1(p) = (\sigma_1 \circ \pi(p))^{-1} p.$$

Using the definition of $g(x)$, we calculate for $x \in M$ that

$$g_1(\sigma_2(x)) = \big(\sigma_1(\pi\sigma_2(x))\big)^{-1} \sigma_2(x) = \big(\sigma_1(x)\big)^{-1} \sigma_2(x) = g(x).$$

Thus, $g_1 \sigma_2 = g$. Then it immediately follows that

$$B_2 = \sigma_2^* \omega$$
$$= \sigma_2^* \big(Ad_{g_1^{-1}} \pi^* B_1\big) + \sigma_2^* \big(g_1^*(\Omega)\big)$$
$$= \big(Ad_{\sigma_2^*(g_1^{-1})} (\pi\sigma_2)^* B_1\big) + (g_1\sigma_2)^*(\Omega)$$
$$= Ad_{g^{-1}} B_1 + g^* \Omega.$$

And this proves (14.10). ∎

The gauge transformation (14.10) is valid for general Lie groups. In the case of matrix Lie groups, this can also be written as

$$B_2 = g^{-1} B_1 g + g^{-1} dg,$$

where dg is the matrix of 1-forms obtained by taking the exterior derivative d of each entry of $g : M \to G$. And g^{-1} is the matrix inverse of the matrix-valued function g.

Corollary 14.1 *Let $\pi : P \to M$ be a principal bundle with structure group G and Ehresmann connection ω. Let $\{V_\alpha, \phi_\alpha\}$ be a family be local trivializations of P, where $\{V_\alpha\}$ is an open cover of M and $\phi_\alpha : \pi^{-1}(V_\alpha) \to V_\alpha \times G$ is an isomorphism of principal bundles with structure group G covering the identity map of V_α.*

Let $\sigma_\alpha : V_\alpha \to \pi^{-1}(V_\alpha)$ be the canonical section associated to ϕ_α. Define $B_\alpha := \sigma_\alpha^*(\omega)$ for each α. So B_α is a $\mathfrak{g}$-valued 1-form on V_α for each α. Then for all α and β, we have on $V_\alpha \cap V_\beta$ that

$$B_\beta = Ad_{g_{\alpha\beta}^{-1}} B_\alpha + g_{\alpha\beta}^{-1*} \Omega.$$

where $g_{\alpha\beta} : V_\alpha \cap V_\beta \to G$ is the cocycle of the principal bundle with respect to $\{V_\alpha, \phi_\alpha\}$. In particular,

$$\sigma_\beta(x) = \sigma_\alpha(x) g_{\alpha\beta}(x)$$

for all $x \in V_\alpha \cap V_\beta$.

This is basically a restatement of the theorem. The possibly infinite number of trivializations in the family $\{V_\alpha, \phi_\alpha\}$ is not relevant since we are only stating a result about each pair of these trivializations. The real value of this corollary is that it leads us to consider its converse.

Theorem 14.4 *Let $\pi : P \to M$ be a principal bundle with structure Lie group G. Let $\{V_\alpha, \phi_\alpha\}$ be a family be local trivializations of P, as above. Suppose that B_α is a given $\mathfrak{g}$-valued 1-form on V_α for each α and that for all α, β we have*

$$B_\beta = Ad_{g_{\alpha\beta}^{-1}} B_\alpha + g_{\alpha\beta}^{-1^*} \Omega, \tag{14.11}$$

where $g_{\alpha\beta} : V_\alpha \cap V_\beta \to G$ is the cocycle of the principal bundle with respect to $\{V_\alpha, \phi_\alpha\}$.

Then there exists a unique Ehresmann connection $\omega : T(P) \to \mathfrak{g}$ on P such that $B_\alpha = \sigma_\alpha^*(\omega)$ for each α.

The proof of this is typical of the do-it-yourself system of assembling some object from a kit. The kit always consists of two parts: the pieces to be assembled and the instructions for assembling them. Here the pieces are the B_αs and the instructions are basically the "consistency" conditions (14.11) they satisfy. However, the assembly is on P, not on M. So one must construct the local Ehresmann connection ω_α from B_α on each piece $\pi^{-1}(V_\alpha)$ of P and then show that they are *equal* on the overlaps $\pi^{-1}(V_\alpha) \cap \pi^{-1}(V_\beta)$. The details can be found in most introductory texts on gauge theory, but it would be better if the readers did this exercise for themselves.

We will use Theorem 14.4 in the next chapter in order to construct the Ehresmann connection associated with the Dirac magnetic monopole.

The moral of Theorem 14.4 is that we can represent the given family of $\mathfrak{g}$-valued 1-forms on the manifold M as an Ehresmann connection on some principal bundle $P \to M$ with structure Lie group G. Even if we are only given the family $\{V_\alpha, \phi_\alpha\}$ of trivializations and the family B_α satisfying the compatibility conditions (14.11), then we also have the cocycle $g_{\alpha\beta}$ at our disposal. This means we can construct the associated principal bundle with that given cocycle with respect to the family $\{V_\alpha, \phi_\alpha\}$ of trivializations. And then by Theorem 14.4, we can also construct the Ehresmann connection ω on P. So the formula (14.11), which is the transformation rule from physics for gauge fields under local gauge transformations, determines a unique principal bundle and a unique Ehresmann connection on it. Conversely, a principal bundle with an Ehresmann connection on it determines the local gauge fields on the base space, and these satisfy the transformation rule from physics for gauge fields. So the reader should realize that Theorem 14.4 is intimately related to physics. One briefly says that "a gauge field is a connection on a principal bundle."

So we want to study Ehresmann connections on a principal bundle in the geometric approach to gauge theory. There are various structures associated to Ehresmann connections, and each should have its physical interpretation. For now we consider the curvature and the Bianchi identity of an Ehresmann connection. We will discuss the covariant derivative in the next section.

First of all, the curvature $\Theta = D\omega$ of an Ehresmann connection ω is a geometric structure on the total space P of a principal bundle $\pi : P \to M$, while its pullback by the canonical section $\sigma : V \to \pi^{-1}(V)$ to an open subset V of a chart in the base space M has the physics interpretation of the field strength F in the gauge theory in the gauge associated to the explicit trivialization of $\pi^{-1}(V) \to V$. Putting this into an equation:

$$F = \sigma^*(\Theta) := \sigma^*(D\omega).$$

Also, the gauge field A in this trivialization by definition is

$$A := \sigma^*(\omega).$$

Using the Cartan structure equation $\Theta = d\omega + \frac{1}{2}\omega \wedge \omega$, we arrive at

$$F = \sigma^*(\Theta)$$

$$= \sigma^*\left(d\omega + \frac{1}{2}\omega \wedge \omega\right)$$

$$= d(\sigma^*\omega) + \frac{1}{2}(\sigma^*\omega) \wedge (\sigma^*\omega)$$

$$= dA + \frac{1}{2}A \wedge A.$$

This is the correct generalization of (14.6) for this setting. But also notice that in the above we commuted the pullback with the operations of exterior derivative and wedge product. While the reader may feel comfortable doing that for $\mathbb{R}$-valued forms, let's recall that ω is a form with values in the vector space $\mathfrak{g}$. So the reader should verify that this is okay in this setting too. Next, for any form on the base space V which is the pullback of a form θ on the total space $\pi^{-1}(V)$, define its covariant derivative as the pullback of the covariant derivative of θ. The formula for this is

$$D(\sigma^*\theta) := \sigma^*(D\theta).$$

One says that we have pulled D back to the base space, but this is only so for forms that themselves are pullbacks. Just to keep everyone on their toes, we often say that

D is *pulled down* to the base space. So we now use Bianchi's identity $D\Theta = 0$ to get

$$DF = D(\sigma^*(\Theta)) = \sigma^*(D\Theta) = 0.$$

Hence, we end up with the identity $DF = 0$, where $F = dA + (1/2)A \wedge A$ is the field strength of the gauge field A. And this is typically how Bianchi's identity is written in the physics literature.

This is actually an instance of a general phenomenon. When working with the theory of principal bundles, there is a tendency toward a dichotomy since one is usually dealing with what is going on either "upstairs" in P or "downstairs" in M. Typically, mathematicians work upstairs and physicists downstairs. The melodrama consists of seeing how the people of these two worlds relate to each other. One small problem is a difference in terminology due to the historically independent developments in these two worlds. Much more fundamental are the differences in scientific methodology, which also has historical roots. And the biggest difference is that each world excludes from its own discourse basic methodologies from the other world.

Bianchi's identity happens to appear here in a seemingly innocuous way. But now we can consider it in the very first example that we presented of a gauge theory: electromagnetism. In that abelian case, $DF = dF$. So Bianchi's identity in this case just says $dF = 0$, which are the four scalar homogeneous Maxwell's equations.

So we have the dichotomy in electromagnetism that the four homogeneous equations come from the geometry of an associated principal bundle while the four nonhomogeneous (or source) equations are Yang–Mills equations of motion derived from something in physics known as the least action principle. Moreover, not only does this dichotomy persist at the general level of gauge theory, but it is more readily noticeable at the general level than in the particular case.

Again, let us emphasize that *every* Yang–Mills field strength "upstairs" is the curvature of the Ehresmann connection and hence satisfies Bianchi's identity. We can pull Bianchi's identity (or any other equation) down to the base space M locally in a trivialization via the associated section and get the corresponding equation "downstairs." And this was how we understood the identity (14.6).

This section gives the motivation for the following definition, which should be considered a two-way dictionary for translating between mathematics terminology and physics terminology. Also, notice that we no longer require G to be a matrix Lie group.

Definition 14.3 *Suppose that G is a Lie group with Lie algebra $\mathfrak{g}$ and that $\pi : P \to M$ is a principal bundle over the differential manifold M with structure group G.*

1. *A gauge field on P is an Ehresmann connection ω on P with values in the Lie algebra $\mathfrak{g}$ of G. If $\sigma : V \to \pi^{-1}(V)$ is a local section of the principal bundle, then $\sigma^*(\omega)$ is called a (local) gauge field on V in the gauge (i.e., trivialization) associated to σ.*

In this case, a principal bundle $P \to M$ together with a gauge field ω is called a gauge theory.

2. *The curvature $D\omega$ of this connection is called the* field strength on P *of the gauge theory. If $\sigma : V \to P$ is a local section of the principal bundle, then $\sigma^*(D\omega)$ is called a* field strength on V *in the gauge (i.e., trivialization) associated to σ.*

3. *A (local)* gauge transformation *of the principal bundle is defined to be a diffeomorphism $\phi : P \to P$ covering the identity map of M that also commutes with the right action of G on P; that is,*

$$\phi(ph) = \phi(p)h$$

for all $p \in P$ and all $h \in G$.

In short, ϕ is a fiber-preserving G-equivariant diffeomorphism. Putting this even more abstractly, ϕ is an automorphism of P in the category of principal bundles over M with structure Lie group G.

Notice that the definition of gauge transformation does *not* depend on the choice of the Ehresmann connection ω of the gauge theory. Rather, it is a property of a principal bundle. We have already seen this in the examples of electromagnetism and Yang–Mills theories, where the local gauge transformations act on many things, including the gauge fields, and only later the gauge fields are identified as the defining element of a connection.

It is to be hoped that parts 1 and 2 of this definition have been sufficiently motivated. But part 3 needs further attention to see what it is all about. First, let $U : V \to G$ be a local gauge transformation for the case of the principal bundle $p_1 : V \times G \to V$ discussed in the previous section. Then we define $\phi_U : P \to P$ by $\phi_U(x, h) := (x, U(x)h)$, where $x \in V$ and $h \in G$.

Exercise 14.8 *Prove that this ϕ_U satisfies the conditions in part 3 of the previous definition.*

Conversely, suppose that $\phi : P \to P$, where $P = V \times G$, satisfies the conditions in part 3 of the previous definition. Since ϕ covers the identity map of M, it must have the form

$$\phi(x, h) = (x, f(x, h)),$$

where $x \in V, h \in G$, and $f : V \times G \to G$. Since ϕ is G-equivariant, we have

$$f(x, h) = f(x, e)h = U_\phi(x)h,$$

where $e \in G$ is the identity element and $U_\phi : V \to G$ is defined as it must be by $U_\phi(x) := f(x, e)$. So this map U_ϕ is a local gauge transformation that is defined in terms of the given function ϕ.

Exercise 14.9 *We have the two functions $U \to \phi_U$ and $\phi \to U_\phi$. Prove that these functions are inverses of each other.*

Moreover, in a given, fixed local trivialization $V \times G \cong \pi^{-1}(V) \to V$ of the principal bundle $\pi : P \to M$, the reader can check that ϕ restricted to $\pi^{-1}(V)$ is a local gauge transformation. That restriction is a local gauge transformation on $V \times G$, using the given isomorphism $V \times G \cong \pi^{-1}(V)$. But that isomorphism is not unique (except in trivial cases). So this identification depends on the choice of an isomorphism, that is, a local trivialization.

So the moral of the story is that part 3 of Definition 14.3 is a good generalization of local gauge transformation from the setting of trivial principal bundles to that of a principal bundles in general.

Definition 14.4 *Let $P \to M$ be a principal bundle over the manifold M with structure group G. Define the* gauge group *of P to be*

$$\mathcal{G} := \{\phi \mid \phi \text{ is a local gauge transformation of } P\}.$$

Be warned that the phrase "gauge group" of a principal bundle is used in the literature to refer to either G or $\mathcal{G}$.

Note that $\mathcal{G}$ is indeed a group since it is the group of automorphisms of an object in the category of principal bundles.

Continuing in this more general setting, we can define the Lagrangian density, the action integral, and the Yang–Mills equations. This is done in such a way that when we view it in local coordinates in a chart of M, we will have the previously derived theory if the Lie group G is a matrix group. Then the Yang–Mills equations in the setting of an oriented manifold M with nondegenerate metric are compactly written as

$$D(*F) = 0, \tag{14.12}$$

where $*$ is the Hodge star on M. This follows from a standard variational argument that we do not present here. Thinking of W-valued k-forms as sections, we note that the Hodge star is defined in the usual way on the expressions dx_I, where I is a k-index, and then extended by acting trivially on the vector coefficients of the k-form. Equation (14.12) is a second-order, in general nonlinear, partial differential equation in the field strengths.

But a common (though not universal) situation in dimension 4 is that $**$, the composition of the Hodge star with itself, is the identity map on 2-forms. In this case, as we already saw at the level of the fiber spaces, we can uniquely write any $F = F_S + F_A$, where $*F_S = F_S$ is called the *self-dual (SD)* part of F and $*F_A = -F_A$ is called the *anti-self-dual (ASD)* part of F.

For example, suppose that $F = DB$ is a g-valued, anti-self-dual 2-form. Then we see immediately that

$$D(*F) = -DF = 0,$$

where the first equality holds because F is ASD and the second holds because of Bianchi's identity. So instead of solving the Yang–Mills equations, we can solve the equation $*F = -F$ (or really, $*DB = -DB$) for the unknown 1-form B. And this is a first-order, though still in general nonlinear, partial differential equation in the gauge field B. And any such solution will then also immediately give a solution $F = DB$ of the Yang–Mills equation $D(*F) = 0$.

By a very similar argument, SD 2-forms $F = DB$ also solve the Yang–Mills equations. There is no claim that every solution of the Yang–Mills equations must have one of these two forms. By no means!

One rather nice way to think about this aspect of the theory, even though it is only an analogy, is to compare it with complex analysis. There we study the Cauchy–Riemann (CR) equations whose solutions are the holomorphic functions. The solutions of the CR equations are also harmonic functions, that is, complex solutions of the equation $\Delta\psi = 0$, where Δ is the Laplacian, and so is a second-order partial differential equation. However, not all such complex-valued functions ψ will be holomorphic. Far from it! But the holomorphic functions are solutions to the CR equations, which are first-order equations and so more manageable. In this analogy, solutions of the Yang–Mills equations correspond to harmonic functions, while the solutions of the ASD equation correspond to the holomorphic functions.

Moreover, in complex analysis there is a theory that is a trivial "mirror image" of the theory of holomorphic functions. This is the theory of the *antiholomorphic functions*, which are solutions of the *anti-CR equations*. These also are harmonic functions. (Speaking a bit colloquially, the space of "antiharmonic" functions is the space of harmonic functions.) In this analogy, the antiholomorphic functions correspond to the solutions of the SD equation. For whatever historical (but not mathematical) reasons, the study of antiholomorphic functions is rarely even mentioned in texts of complex analysis. Of course, everything in the theory of antiholomorphic functions has a "mirror" in the theory of holomorphic functions. Supposedly, everyone knows this, and so it is seen as not even worth mentioning. Similarly, it seems that the mathematical world has focused on ASD solutions to the Yang–Mills equations and that the "mirror" world of SD solutions is not considered to be worth mentioning.

Unfortunately, these analogies cannot be pressed too far. The partial differential equations of Yang–Mills theory are in general nonlinear, while those of complex analysis are linear. That is one very big difference.

14.3 Matter Fields in Geometry

In this brief section, we introduce an essential ingredient in the construction of any meaningful physical theory: matter. The gauge fields we have been considering are certainly important, but they entered the theory to describe interactions among material particles, such as the nucleons. Of course, in the nonabelian case, we have the added feature that the particles that mediate interactions also interact among themselves. But still how do we include into this geometrical setting the usual fermions that constitute ordinary matter?

The usual answer to this question in quantum theories is that matter is represented by so-called wave functions, which are *complex-valued* functions (actually, equivalence classes of almost-everywhere-equal functions) defined on an appropriate *configuration space M* associated with the physical system. As far as I am concerned, no one has ever explained why quantum theory requires complex vector spaces. This can hardly be described as physically intuitive. The best we can say is that real vector spaces are not adequate, while other structures turn out to be unnecessary generalizations.

We note that the introduction of a nontrivial principal bundle $P \to M$ over the configuration space M in order to represent a gauge field as an Ehresmann connection on P requires a modification of the way to represent the matter fields ψ. Intuitively, P is locally trivial and so ψ should be as described above for each local trivialization of P. Moreover, these local representatives of ψ should be consistent under a change of local trivialization, that is, under local gauge transformations.

The appropriate concept is a section; that is, ψ should be a section of some bundle. This idea is already clearly formulated in 1976 in the paper [53] by Wu and Yang. Since we still wish to form a complex Hilbert space out of these sections, the bundle should be a complex vector bundle so that we can add wave functions as well as multiply them by *complex* scalars. And this vector bundle should be related to the principal bundle P of the gauge theory. But how? Well, intuition only gets me this far and then I for one am forced to make some appeal to what has worked. And this has worked:

Definition 14.5 *Let $\pi : P \to M$ be a principal bundle with structure Lie group G. Suppose that W is a (finite-dimensional) vector space over the complex field $\mathbb{C}$ and that there is a left action of G on W. Form the associated complex vector bundle $E := P \times_G W$ over M. Then we say that a section $\psi : M \to E$ of this vector bundle E is a* matter field.

Of course, this is not yet enough structure to define a complex Hilbert space of matter fields. To do this, one would like to define the *square integrable* matter fields. To achieve this, one needs a measure μ on the base space M and a hermitian inner product $\langle \cdot, \cdot \rangle_E$ on the fibers of E. Then one requires the section ψ to satisfy

$$\int_M d\mu \, \langle \psi, \psi \rangle_E < \infty.$$

The complex vector space of all such (smooth!) sections will then become a *pre-Hilbert space*, which then has a unique *completion* that is a Hilbert space. In examples, one can take W to be a real vector space since we can easily pass to its complexification.

One example of a matter field is obtained by taking $W = \mathfrak{g}$, the Lie algebra of G with the left adjoint representation of G acting on $\mathfrak{g}$. The resulting matter fields are called *Higgs fields*.

In Definition 14.5, there is no reference to an Ehresmann connection ω on P. But this is required to define a gauge theory. In particular, the covariant derivative of a matter field has to be defined so that one can construct gauge-invariant field theories.

In the Yang–Mills theory with gauge field

$$\sum_{\mu=0}^{3} B_\mu \, dx_\mu \quad \text{where} \quad B_\mu : \mathbb{R}^4 \to \mathfrak{su}(2),$$

the components D_μ of its covariant derivative for $\mu = 0, 1, 2, 3$ are given by

$$D_\mu \psi = (\partial_\mu - ig B_\mu)\psi$$

since in this case the bundles are trivial, and so matter fields of nucleons are given by wave functions $\psi : \mathbb{R}^4 \to \mathbb{C}^2$. Here g is the coupling constant of Yang–Mills theory and ∂_μ is a 2×2 diagonal matrix. Similarly, for the electromagnetic gauge field

$$\sum_{\mu=0}^{3} A_\mu \, dx_\mu \quad \text{where} \quad A_\mu : \mathbb{R}^4 \to \mathbb{R},$$

the components of its covariant derivative acting on $\psi : \mathbb{R}^4 \to \mathbb{C}$ are

$$D_\mu \psi = \left(\partial_\mu - ie A_\mu\right)\psi,$$

where e is electric charge of the particle described by ψ. So the minimal coupling criterion in physics corresponds to using the covariant derivative in these two cases.

Chapter 15
The Dirac Monopole

Dirac introduced the theory of the magnetic monopole which bears his name in 1931 in [9], quite a few years before the advent of either the Yang–Mills theory or the theory of fiber bundles. However, by 1975, Wu and Yang had learned enough about fiber bundles to reformulate Dirac's theory in terms of a connection on a principal bundle. See [52].

In this chapter we present this example of a $U(1)$ gauge theory that is not topologically trivial even though the group $U(1)$ is abelian. In fact, the nonzero magnetic charge of the monopole will be related to a nonzero topological invariant of the bundle. For this reason, the magnetic charge of the Dirac monopole is called a *topological charge*.

15.1 The Monopole Magnetic Field

Maxwell's equations in vacuum in the case when there are no electric charges, no electric currents, and time-independent fields are as follows:

$$\nabla \cdot E = 0, \qquad \nabla \cdot B = 0,$$
$$\nabla \times E = 0, \qquad \nabla \times B = 0.$$

We next introduce a source term for magnetic charges, but not for magnetic currents, by changing the equation $\nabla \cdot B = 0$ to

$$\nabla \cdot B = 4\pi \rho_m,$$

where ρ_m is the magnetic charge density. This is in analogy with the case of electric charge. Continuing this analogy, we consider a point source of magnetic charge sitting at the origin of $\mathbb{R}^3$. What this really means is that there exists a magnetic field defined on $\mathbb{R}^3 \setminus \{0\}$ as the analog of the electric field generated by a point electric charge, namely,

© Springer International Publishing Switzerland 2015
S.B. Sontz, *Principal Bundles*, Universitext, DOI 10.1007/978-3-319-14765-9_15

$$B = g\frac{1}{r^2}\,\hat{r} = g\frac{1}{r^3}\,(x, y, z), \tag{15.1}$$

where $r = (x^2 + y^2 + z^2)^{1/2} > 0$ and $\hat{r} = (x, y, z)/\|r\|$ is the unit radial vector field on $\mathbb{R}^3 \setminus \{0\}$. We are using the convention that (x, y, z) represents a point in $\mathbb{R}^3$. Also, $g \in \mathbb{R}$ measures the strength (magnetic charge) of the monopole.

However, the vector field defined in (15.1) is quite different from the analogous electric field. Of course, (15.1) defines a central field, that is, a field in the radial direction. But it is *not* a force field and therein lies a tale. The force associated to any magnetic field B on a point particle is given by the Lorentz law $F = q(v \times B)$, where q is the electric charge of the particle and v is its velocity, which is a vector. So the force associated with the magnetic field in (15.1) is always orthogonal to the radial direction. The actual direction of the force is further determined by the direction of the velocity v and the sign of the electric charge q. Recall also from Newton's law of motion that the direction of the force is the same as the direction of the acceleration, which itself can have absolutely any angle with the direction of the velocity. It is this last direction (namely, that of the velocity) that we "see" (with our eyes and our measuring instruments) as the particle moves through space.

The reader may well object that (15.1) defines a central force field when the test particle itself has a magnetic charge and no electric charge; that is, the test particle is another magnetic monopole. And this is what one would expect if there exist at least two monopoles in the universe. Nonetheless, at least in our local environment, monopoles are certainly scarce. So if we were ever to find one, the way to study it would be via its interaction with readily available electrically charged particles, such as electrons, alpha particles, and so on.

Exercise 15.1 *Let B be the point monopole given in (15.1). On $\mathbb{R}^3 \setminus \{0\}$, show that*

$$\nabla \cdot B = 0 \quad \text{and} \quad \nabla \times B = 0.$$

And for those who know distribution theory, show that in the sense of distributions we have on $\mathbb{R}^3$ that

$$\nabla \cdot B = 4\pi g\,\delta_0,$$

where δ_0 is the Dirac delta at the origin in $\mathbb{R}^3$.

As the reader is now well aware, the divergence operator, $\nabla \cdot$, is not defined on vector fields but on 2-forms. We can see this as passing from the vector field

$$B = B_1\frac{\partial}{\partial x} + B_2\frac{\partial}{\partial y} + B_3\frac{\partial}{\partial z}$$

to the 2-form

$$\mathbf{B} = B_1\,dy \wedge dz + B_2\,dz \wedge dx + B_3\,dx \wedge dy = *(B_1\,dx + B_2\,dy + B_3\,dz),$$

where $*$ is the Hodge star on $\mathbb{R}^3$ with the standard inner product and the orientation determined by $dx \wedge dy \wedge dz$. So the previous exercise shows that $d\mathbf{B} = 0$ on $\mathbb{R}^3 \setminus \{0\}$, or in other words, $\mathbf{B}$ is a closed 2-form on $\mathbb{R}^3 \setminus \{0\}$. However, the de Rham cohomology

$$H^2_{dR}(\mathbb{R}^3 \setminus \{0\}) \cong H^2_{dR}(S^2) \cong \mathbb{Z}$$

is not trivial. We are using some results from topology here. First, we are using that the inclusion map $S^2 \subset \mathbb{R}^3 \setminus \{0\}$ is a homotopy equivalence and so induces isomorphisms in cohomology. Second, we are using the identification of $H^2_{dR}(S^2)$ as $\mathbb{Z}$, which is another standard result in topology.

In short, the cocycle determined by $\mathbf{B}$ need not be exact on $\mathbb{R}^3 \setminus \{0\}$. We shall see later on that indeed it is not exact for $g \neq 0$; that is, there is no 1-form A defined on $\mathbb{R}^3 \setminus \{0\}$ such that $B = dA$ on $\mathbb{R}^3 \setminus \{0\}$.

15.2 The Monopole Vector Potential

But by Poincaré's lemma, we do have $\mathbf{B} = dA$ on any star-shaped subset $S \subset \mathbb{R}^3 \setminus \{0\}$. The 1-form A will not be unique and will depend on the subset S. In vector calculus notation, this says that $B = \nabla \times A$; that is, A is a vector potential for the magnetic field B, but only in the region S.

How do we quickly find such vector potentials? Well, since the magnetic field (15.1) is spherically symmetric, this tells us that things should look simpler in spherical coordinates. Recall that

$$x = r \sin\theta \cos\phi,$$
$$y = r \sin\theta \sin\phi, \qquad (15.2)$$
$$z = r \cos\theta,$$

where $r > 0$, $0 < \theta < \pi$, and $0 < \phi < 2\pi$, give spherical coordinates on a dense, open subset of $\mathbb{R}^3$. Thinking of equations (15.2) as a C^∞-mapping

$$\Psi : \{(r,\theta,\phi) \mid r > 0, 0 < \theta < \pi, 0 < \phi < 2\pi\} \to \mathbb{R}^3,$$

we have that the determinant J of its Jacobian matrix is

$$J = \det(\Psi_*) = r^2 \sin\theta.$$

Exercise 15.2 *Verify this formula for J.*

In particular, this immediately gives the transformation of the unit volume element in the Euclidean coordinates to the spherical coordinates as

$$dx \wedge dy \wedge dz = (r^2 \sin\theta)\, dr \wedge d\theta \wedge d\phi. \qquad (15.3)$$

Since $r^2 \sin \theta > 0$, the orientations of $dx \wedge dy \wedge dz$ and $dr \wedge d\theta \wedge d\phi$ are the same.

In terms of a person standing anywhere on Earth (viewed as S^2), except for the North and South Poles, the three unit tangent directions,

$$\hat{r} = \frac{\partial}{\partial r}, \qquad \hat{\theta} = \frac{1}{r} \frac{\partial}{\partial \theta}, \qquad \hat{\phi} = \frac{1}{r \sin \theta} \frac{\partial}{\partial \phi}, \tag{15.4}$$

correspond to up, south, and east. Notice that the vector field $\partial/\partial\phi$ (east) extends smoothly to every point on Earth except the two poles even though ϕ is discontinuous on that region. So unless you are reading this at either the North or South Pole, stand up and look south. Put your right hand in front of you with the thumb up and the index finger (the one next to the thumb) pointing straight ahead, that is, south. Then straighten the finger next to the index finger; it will now be pointing east. For this reason, we say that the spherical vector fields (15.4) form a *right-handed* coordinate system.

By definition, the vector fields

$$\hat{i} = \frac{\partial}{\partial x}, \qquad \hat{j} = \frac{\partial}{\partial y}, \qquad \hat{k} = \frac{\partial}{\partial z}$$

also form a right-handed coordinate system. These are unit vectors by the very definition of the standard metric on $\mathbb{R}^3$.

Exercise 15.3 *Show that the vector fields in (15.4) have unit length in the tangent space at every point of $\mathbb{R}^3 \setminus Z$, where $Z := \{(0, 0, z) \mid z \in \mathbb{R}\}$ is the z-axis.*

Here is a quick way to get everything into spherical coordinates. We first note that the 1-form corresponding to the vector field B in (15.1) is $B = (g/r^2) \, dr$, and so the associated 2-form is

$$\begin{aligned}
\mathbf{B} &= *B \\[4pt]
&= g\frac{1}{r^2}(* \, dr) \\[4pt]
&= g\frac{1}{r^2}\left((r^2 \sin \theta) \, d\theta \wedge d\phi\right) \\[4pt]
&= (g \sin \theta) \, d\theta \wedge d\phi \\[4pt]
&= d(-g \cos \theta \, d\phi).
\end{aligned}$$

Note that $* \, dr = (r^2 \sin \theta) \, d\theta \wedge d\phi$ follows from (15.3) and the definition of the Hodge star.

Since θ and $d\phi$ are only defined on $\mathbb{R}^3 \setminus Z$, the 2-form $\mathbf{B}$ is only defined on $\mathbb{R}^3 \setminus Z$, while B itself is defined on the larger set $\mathbb{R}^3 \setminus \{0\}$. Nonetheless, this shows that we can take

$$A = -g \cos \theta \, d\phi$$

on $\mathbb{R}^3 \setminus Z$ or, equivalently, for $0 < \theta < \pi$. But, of course, the vector potential A is not uniquely determined by the condition $B = dA$. Far from it! Now comes the clever part. We would like to modify the definition of A so that it can be smoothly extended to at least some subset of the z-axis while still having $B = dA$ on that extension. Of course, $d\phi$ is undefined on the z-axis. And there is no way to smoothly extend it to the z-axis. (Draw some sketches to convince yourself of this.) But by a general meta-principal (which is usually left unexpressed), zero times an undefined object (say, a vector) is the zero object (say, the zero vector). But $-g \cos\theta$ is not zero for either of the values $\theta = 0$ or $\theta = \pi$. But $-g \cos\theta$ is constant for $\theta = 0$; namely, it is equal to $-g$. So we define the 1-form

$$A^N = A^N(\theta) := (g - g\cos\theta)\, d\phi = g(1 - \cos\theta)\, d\phi$$

for $0 \leq \theta < \pi$. The equation $\theta = 0$ corresponds to the open upper part of the z-axis, that is, the set $\{(0, 0, z) \mid z > 0\}$. Therefore, $A^N(\theta = 0) = 0$ since we arranged it to happen this way! So we next define $Z^- := \{(0, 0, z) \mid z \leq 0\}$.

Exercise 15.4 *Prove that $B = dA^N$ on $V_N := \mathbb{R}^3 \setminus Z^-$. Prove that A^N is a C^∞-function on V_N.*

As the reader has already realized, the symbol N in A^N refers to north since the vector potential A^N is defined on the northern part of the z-axis. The result of the previous exercise is not too surprising, given the result of the next exercise.

Exercise 15.5 *Prove that V_N is star-shaped, but it is not convex.*

We can similarly modify A so that we can extend it to $\mathbb{R}^3 \setminus Z^+$, where $Z^+ := \{(0, 0, z) \mid z \geq 0\}$, by defining

$$A^S = A^S(\theta) := (-g - g\cos\theta)\, d\phi = -g(1 + \cos\theta)\, d\phi.$$

Here S stands for south.

We now have enough material available in order to give an elementary argument for a fundamental property of the Dirac monopole.

Proposition 15.1 *The Dirac magnetic monopole has no vector potential defined on $\mathbb{R}^3 \setminus \{0\}$ if $g \neq 0$.*

Proof: For suppose that such a vector potential, say A^0, did exist. That means $B = dA^0$ on $\mathbb{R}^3 \setminus \{0\}$. In particular, on the smaller set $V_N = \mathbb{R}^3 \setminus Z^-$, we would have

$$d(A^0 - A^N) = 0.$$

But V_N is star-shaped [implying $H^1(V_N) = 0$], and so

$$A^0 - A^N = df_N$$

on V_N for some function $f_N \in C^\infty(V_N)$. Repeating this argument for A^S, we see that

$$A^0 - A^S = df_S$$

on V_S for some function $f_S \in C^\infty(V_S)$. These two results combine to give $A^N - A^S = df_S - df_N = d(f_S - f_N)$ on $V_N \cap V_S$.

But we also have $A^N - A^S = 2g\, d\phi$ on $V_N \cap V_S$. Since we are assuming that $g \neq 0$, we arrive at

$$d\phi = \frac{1}{2g}(A^N - A^S) = d\left(\frac{1}{2g}(f_S - f_N)\right)$$

on $V_N \cap V_S$. And this says that the 1-form $d\phi$ is exact on $V_N \cap V_S$.

Now it is well known that $d\phi$ is not exact on $V_N \cap V_S$. This is intuitively clear enough. So we have arrived at a contradiction that arose by assuming the existence of A^0. So no such A^0 exists.

However, let us rely on a proof rather than just pure intuition. Again, we argue by contradiction. So suppose to the contrary that $d\phi$ is exact on $V_N \cap V_S$, which means $d\phi = df$, where $f \in C^\infty(V_N \cap V_S)$. Now we do have the standard azimuthal angle, denoted for now by ϕ_0 with $\phi_0 : U \to (0, 2\pi)$, where

$$U := (V_N \cap V_S) \setminus \{(x, 0, z) \mid x \geq 0,\ z \in \mathbb{R}\}.$$

Then $\phi_0 \in C^\infty(U)$ and $d\phi_0 = d\phi$ on U. The latter implies that $d\phi_0 = df$ on U. But U is a connected set. So $\phi_0 = f + c$ on U for some constant $c \in \mathbb{R}$. But both f and c extend to smooth functions on $V_N \cap V_S$. So ϕ_0 extends to a smooth function on $V_N \cap V_S$. But ϕ_0 does not even have a continuous extension, let alone a smooth one, to $V_N \cap V_S$. So this contradicts the assumption that $d\phi$ is exact on $V_N \cap V_S$. We thus conclude that $d\phi$ is not exact on $V_N \cap V_S$. ∎

Exercise 15.6 *In most introductory texts, another, somewhat more abstract, argument for the result of this proposition is given. Try to find it.*

Exercise 15.7 *Prove $B = dA^S$ on $V_S := \mathbb{R}^3 \setminus Z^+$. Prove that V_S is star-shaped, but it is not convex.*

15.3 The Monopole Principal Bundle

The two subsets V_N and V_S form an open cover of $\mathbb{R}^3 \setminus \{0\}$. The sets Z^- and Z^+ are examples of *Dirac strings*, which in general are the images of simple curves in $\mathbb{R}^3$ connecting the origin to infinity. We have well-defined vector potentials on the complements of the Dirac strings Z^- and Z^+. This can be done in general for any

Dirac string, but it suffices to do it on enough strings so that we get an open cover of $\mathbb{R}^3 \setminus \{0\}$. Clearly, the minimal number of such Dirac strings is two.

So we now have gauge fields A^N and A^S, each defined on an open set of an open cover of the base space $\mathbb{R}^3 \setminus \{0\}$ of a principal bundle. We still have to identify the structure Lie group of this bundle and show that there is an Ehresmann connection ω on it that pulls back via the canonical sections to A^N and A^S. And we have to define the principal bundle!

The structure Lie group of this bundle has to be $U(1)$ since we are still working in the setting of electromagnetism. We can also see this by computing on $\mathbb{R}^3 \setminus Z = V_N \cap V_S$ the difference

$$A^N - A^S = g(1 - \cos\theta)\, d\phi - (-g(1 + \cos\theta)\, d\phi) = 2g\, d\phi.$$

We define the transition function $g_{NS} : V_N \cap V_S \to U(1)$ by

$$g_{NS}(r, \theta, \phi) := e^{(2ig)\phi}$$

so that

$$g_{NS}^{-1}\, dg_{NS} = e^{-(2ig)\phi}(2ig)e^{(2ig)\phi}\, d\phi = (2ig)d\phi.$$

This is a $[\mathfrak{u}(1) = i\mathbb{R}]$-valued 1-form defined on $V_N \cap V_S$ in the mathematics convention for the Lie algebra of $U(1)$. The corresponding $[\mathfrak{u}(1) = \mathbb{R}]$-valued 1-form in the physics convention is obtained by dividing by i. Then we see that

$$A^N = A^S + 2g\, d\phi = A^S + i^{-1} g_{NS}^{-1}\, dg_{NS}.$$

Since this is the local gauge transformation for a gauge field on the overlap $V_N \cap V_S$, we see that we obtain an Ehresmann connection ω on the principal $U(1)$-bundle according to Theorem 14.4. Thus, ω contains the information of the two local gauge fields A^N and A^S as well as for all the other local gauge fields associated to local trivializations of the principal bundle.

But all of this depends on having a *smooth* function $\phi \mapsto e^{(2ig)\phi}$ on $V_N \cap V_S$, that is, for $0 \leq \phi \leq 2\pi$. Since e^z is periodic for $z \in \mathbb{C}$ with period $2\pi i$, we can do this if and only if $g = k\pi$ for some integer k.

Thus, the condition for constructing the cocycle and hence the principal bundle determines the magnetic charge g of the monopole. For each integer k, we denote the associated cocycle as $g_{NS,k}$ defined by

$$g_{NS,k}(r, \theta, \phi) := e^{(2k\pi i)\phi}.$$

Conversely, every principal bundle over $\mathbb{R}^3 \setminus \{0\}$ (or, equivalently, over the unit sphere S^2) with structure group $U(1)$ is determined by an integer k via the associated cocycle $g_{NS,k}$. Moreover, a theorem in algebraic topology asserts that the mapping from the integers $\mathbb{Z}$ to the set of equivalence classes of $U(1)$

principal bundles over S^2 [given by k going to the bundle determined by $g_{NS,k}$] is a bijection. The upshot is that the set of all Dirac monopoles is in bijection with the set $\mathbb{Z}$. Moreover, the magnetic charge $g = k\pi$ of the monopole is encoded in the topological structure of the principal bundle. Again, this is the reason that we say that magnetic charge is a topological charge.

Chapter 16
Instantons

In this chapter we will find certain solutions of the Yang–Mills equations that were given in the now-classic paper [3] by the four authors Belavin, Polyakov, Schwartz, and Tyupkin. Those solutions were called pseudoparticles in that paper, but this is now generally considered to be antiquated terminology. Rather, they are currently called *BPST instantons*.

This chapter is motivated in part by the rather lengthy presentation in Naber's books [38] and [39]. But also see Darling's book [6] for a rather telegraphic exposition.

16.1 Quaternions

We will use the quaternions in the construction of the BPST instanton. So we start with very standard material, in part to establish notation.

We define $\mathbb{H} := \mathbb{R}^4$ to be the *set* of quaternions. This set clearly has the structure of a vector space of dimension 4 over the reals. To introduce a multiplication, we will use special symbols for the elements of the canonical basis of $\mathbb{R}^4$, now thought of as quaternions:

$$1 = (1,0,0,0), \quad i = (0,1,0,0), \quad j = (0,0,1,0), \quad k = (0,0,0,1).$$

Then we first define the multiplication on pairs of basis elements and then extend bilinearly (over $\mathbb{R}$) to get the quaternionic multiplication. So we define

$$ij = k, \, ji = -k, \, i^2 = -1,$$
$$jk = i, \, kj = -i, \, j^2 = -1,$$
$$ki = j, \, ik = -j, \, k^2 = -1,$$

© Springer International Publishing Switzerland 2015

S.B. Sontz, *Principal Bundles*, Universitext, DOI 10.1007/978-3-319-14765-9_16

and we also define $1q = q1 = q$ for all $q \in \{1, i, j, k\}$. Then $\mathbb{H}$ is an associative, noncommutative algebra over $\mathbb{R}$. It has a multiplicative unit, namely, the basis element 1. Notice that multiplication by *real* multiples of 1 (on the right or on the left) can be canonically identified with the scalar multiplication by reals on $\mathbb{R}^4$.

Writing an arbitrary element $q \in \mathbb{H}$ as

$$q = a + bi + cj + dk$$

with $a, b, c, d \in \mathbb{R}$, we define the *real part* of q as

$$Re(q) := a,$$

the *imaginary part* of q as

$$Im(q) := bi + cj + dk,$$

the *conjugate* of q as

$$\overline{q} := a - bi - cj - dk,$$

and the *norm* of q as

$$|q| := (a^2 + b^2 + c^2 + d^2)^{1/2}.$$

Notice that $Re : \mathbb{H} \to \mathbb{H}$ and $Im : \mathbb{H} \to \mathbb{H}$ are projections (that is, idempotents) of rank 1 and 3, respectively. The image of the projection Im is denoted by $Im(\mathbb{H})$ and is called the imaginary subspace. Then it is easy to show that a quaternion q is an element of $Im(\mathbb{H})$ if and only if $\overline{q} = -q$ if and only if $Im(q) = q$.

Next, we have the identity $q\overline{q} = \overline{q}q = |q|^2$. This immediately implies that every nonzero quaternion q has a two-sided multiplicative inverse, which is given by

$$q^{-1} = \frac{1}{|q|^2}\overline{q}.$$

In short, the quaternions satisfy all of the axioms for a field *except* for the commutativity of multiplication.

Here are a few more useful facts about quaternions. The mapping $q \mapsto \overline{q}$ is linear over the reals, is an involution ($\overline{\overline{q}} = q$), is an isometry ($|\overline{q}| = |q|$), and is a multiplicative antiisomorphism ($\overline{q_1 q_2} = \overline{q_2}\,\overline{q_1}$). Moreover, $|q_1 q_2| = |q_1|\,|q_2|$. The multiplicative group $\mathbb{H} \setminus \{0\}$ is a Lie group, and the sphere (or 3-sphere) defined by

$$S^3 := \{q \in \mathbb{H} \mid |q| = 1\}$$

is a compact Lie subgroup of $\mathbb{H} \setminus \{0\}$.

Since $\mathbb{H}$ is not a field, we do not have available the usual theory of vector spaces over $\mathbb{H}$. But we can construct free, finitely generated right $\mathbb{H}$-modules. These are

$$\mathbb{H}^n := \{ (q_1, \ldots, q_n) \mid q_1, \ldots, q_n \in \mathbb{H} \} = \mathbb{H} \times \cdots \times \mathbb{H} \quad (n \text{ factors}),$$

where n is a positive integer. (We can also define $\mathbb{H}^0 := \{0\}$.) These are additive groups in the usual way, while the right $\mathbb{H}$-module structure is given by

$$(q_1, \ldots, q_n)r := (q_1 r, \ldots, q_n r)$$

for all $r, q_1, \ldots, q_n \in \mathbb{H}$. Clearly, we can identify $\mathbb{H}^n$ with $\mathbb{R}^{4n}$.

We will next define the Hopf bundles in the quaternionic setting. This parallels the discussion of the Hopf bundles given in Section 10.3. First, the quaternionic projective space $\mathbb{H}P^n$ for any integer $n \geq 1$ is defined as the space of quaternionic lines in $\mathbb{H}^{n+1}$ passing through the origin. Each such quaternionic line (being actually a copy of the four-dimensional space $\mathbb{H}$) intersects the unit sphere S^{4n+3} in $\mathbb{H}^{n+1}$ in a copy of S^3. And the family of these spheres in S^{4n+3} is in one-to-one correspondence with the points of $\mathbb{H}P^n$. Actually, these spheres are the orbits of the natural action of S^3 on S^{4n+3} given by $x \mapsto x\lambda$ for $x \in S^{4n+3}$ and $\lambda \in S^3$. So we get a principal bundle $S^{4n+3} \to \mathbb{H}P^n$ with structure Lie group S^3. This is called a *(quaternionic) Hopf bundle*. The case $n = 1$ will be considered below. In that case, we use the identification of $\mathbb{H}P^1$ with the one-point compactification of $\mathbb{H} \cong \mathbb{R}^4$, which is S^4. So in that case, we get the principal Hopf bundle $S^7 \to S^4$ with structure group S^3.

16.2 Compact Symplectic Groups $Sp(n)$

We say that a map $A : \mathbb{H}^n \to \mathbb{H}^m$ is $\mathbb{H}$-linear if it is additive and commutes with the right action of $\mathbb{H}$, that is, if it is a right $\mathbb{H}$-module map. We then can define the corresponding general linear group:

$$GL(n, \mathbb{H}) := \{ A : \mathbb{H}^n \to \mathbb{H}^n \mid A \text{ is } \mathbb{H}\text{-linear and bijective} \}.$$

We will be more interested in a certain subgroup of $GL(n, \mathbb{H})$. To introduce this, we define an *inner product* on $\mathbb{H}^n$ by

$$\langle q, r \rangle := \sum_{j=1}^{n} \overline{q}_j r_j,$$

where $q = (q_1, \ldots, q_n)$ and $r = (r_1, \ldots, r_n)$ are in $\mathbb{H}^n$. Note that $\langle q, r \rangle$ is an element in $\mathbb{H}$. Also, $\langle q, q \rangle = \sum_{j=1}^{n} |q_j|^2 = ||q||^2$, where this last expression is the square of the usual Euclidean norm in $\mathbb{R}^{4n}$.

Next, for every integer $n \geq 1$, we define the *compact symplectic group* by

$$Sp(n) := \{\, A \in GL(n, \mathbb{H}) \mid \langle Aq, Ar \rangle = \langle q, r \rangle \ \forall q, r \in \mathbb{H}^n \,\}.$$

The name and notation for this group vary. The name we have chosen to use is standard and underlines the fact that this group is compact. If you have read about the "symplectic group" before, it most likely was the noncompact group of $2n \times 2n$ matrices with real entries that preserve a symplectic form on $\mathbb{R}^{2n}$.

Both $GL(n, \mathbb{H})$ and $Sp(n)$ can be made in a natural way into Lie groups. Their dimensions over the field of real numbers are $\dim GL(n, \mathbb{H}) = 4n^2$ and $\dim Sp(n) = 2n^2 + n$.

Exercise 16.1 *Identify $\mathbb{H}$-linear maps $\mathbb{H}^n \to \mathbb{H}^m$ with $m \times n$ matrices in the usual way, but being careful with the order of multiplication! Show that $A \in Sp(n)$ if and only if $(A^t)^- A = I$, where $(A^t)^-$ means the matrix A transposed and then conjugated. (Be careful with matrix multiplication too!)*

We would now like to identify $GL(1, \mathbb{H})$ and $Sp(1)$. First, let $A : \mathbb{H} \to \mathbb{H}$ be $\mathbb{H}$-linear. Define $a := A(1) \in \mathbb{H}$. Then

$$A(q) = A(1q) = A(1)q = aq$$

for all $q \in \mathbb{H}$. So the element $a \in \mathbb{H}$ completely determines the map A.

Conversely, any $a \in \mathbb{H}$ defines an $\mathbb{H}$-linear map $A : \mathbb{H} \to \mathbb{H}$ given by $A(q) := aq$ since $A(qr) = a(qr) = (aq)r = A(q)r$ for all $r \in \mathbb{H}$ by the associativity of the multiplication. Clearly, A is also additive.

So the set of all $\mathbb{H}$-linear maps $A : \mathbb{H} \to \mathbb{H}$ is in bijective correspondence with the elements in $\mathbb{H}$. (These $\mathbb{H}$-linear maps in turn were also identified with 1×1 matrices in Exercise 16.1.) It follows immediately that $GL(1, \mathbb{H})$ is in bijective correspondence with $\mathbb{H} \setminus \{0\}$. Moreover, this is a Lie group isomorphism.

Next, one can show that the elements $A \in Sp(1)$ correspond to elements $a \in \mathbb{H} \setminus \{0\}$ that satisfy $|a| = 1$. So $Sp(1)$ is in bijective correspondence with S^3. Moreover, this is a Lie group isomorphism. It is also a well-known result in Lie group theory that S^3 and $SU(2)$ are isomorphic Lie groups.

Now we are going to analyze $Sp(2)$. Let

$$q = \begin{pmatrix} \alpha & \gamma \\ \beta & \delta \end{pmatrix}$$

be an element in $Sp(2)$, where $\alpha, \beta, \gamma, \delta \in \mathbb{H}$. Then its transpose conjugated is

$$(q^t)^- = \begin{pmatrix} \overline{\alpha} & \overline{\beta} \\ \overline{\gamma} & \overline{\delta} \end{pmatrix}.$$

So the condition that $q \in Sp(2)$, namely, that $(q^t)^- q = I$, translates into these three conditions:

$$|\alpha|^2 + |\beta|^2 = 1,$$
$$|\gamma|^2 + |\delta|^2 = 1,$$
$$\langle (\alpha, \beta), (\gamma, \delta) \rangle = 0.$$

This says that each column in q is in the sphere S^7 and that these two vectors are orthogonal. In general, $Sp(n)$ acts transitively on the unit sphere S^{4n-1} in $\mathbb{H}^n$ and has a stabilizer that is isomorphic to $Sp(n-1)$ for integers $n \geq 1$. This gives us a principal bundle $Sp(n-1) \hookrightarrow Sp(n) \to S^{4n-1}$.

Here we are interested in the case $n = 2$, so that the bundle becomes $Sp(1) \hookrightarrow Sp(2) \xrightarrow{\pi} S^7$. We already know that the base space S^7 here appears as the total space in the Hopf principal bundle $S^3 \hookrightarrow S^7 \to S^4 \cong \mathbb{H}P^1$. We want to construct a connection on the Hopf bundle, namely, a certain type of 1-form on S^7 with values in the Lie algebra of $S^3 \cong SU(2)$, and then pull it back with a local section to the base space, which is four-dimensional. When we write that 1-form on $\mathbb{H}P^1$ in the canonical local coordinates of the projective space, namely, pulling the form back to $\mathbb{H} \cong \mathbb{R}^4$, we will get the BPST instanton on $\mathbb{R}^4$.

We are going to start this by considering the Maurer–Cartan form on $Sp(2)$. We can write this as $q^{-1} dq$, where $q^{-1} = (q^t)^-$ and

$$dq = \begin{pmatrix} d\alpha & d\gamma \\ d\beta & d\delta \end{pmatrix}.$$

So we have

$$q^{-1} dq = \begin{pmatrix} \bar{\alpha} & \bar{\beta} \\ \bar{\gamma} & \bar{\delta} \end{pmatrix} \begin{pmatrix} d\alpha & d\gamma \\ d\beta & d\delta \end{pmatrix} = \begin{pmatrix} \bar{\alpha} d\alpha + \bar{\beta} d\beta & * \\ * & * \end{pmatrix},$$

where we have only computed one of the four entries in the last matrix. Notice that this entry depends only on two entries of q (and their differentials) and that these entries satisfy $\bar{\alpha}\alpha + \bar{\beta}\beta = 1$. Taking the exterior derivative of this equation then gives us

$$(d\bar{\alpha})\alpha + (d\bar{\beta})\beta + \bar{\alpha} d\alpha + \bar{\beta} d\beta = 0,$$

which implies

$$\bar{\alpha} d\alpha + \bar{\beta} d\beta = -((d\bar{\alpha})\alpha + (d\bar{\beta})\beta) = -(\bar{\alpha}(d\alpha) + \bar{\beta}(d\beta))^-.$$

So we have that $\bar{\alpha} d\alpha + \bar{\beta} d\beta$ is a 1-form taking values in $Im(\mathbb{H})$ and therefore $\bar{\alpha} d\alpha + \bar{\beta} d\beta = Im(\bar{\alpha} d\alpha + \bar{\beta} d\beta)$.

The introduction of the Maurer–Cartan form on $Sp(2)$ was not an essential part of this particular discussion. We just used it to isolate one of its matrix entries for further consideration.

16.3   The Ehresmann Connection on the Hopf Bundle

Summarizing, the 1-form $Im(\overline{\alpha}d\alpha + \overline{\beta}d\beta)$ with values in $Im(\mathbb{H})$ defined on $Sp(2)$ "passes" to the quotient space S^7, where it also gives a 1-form, denoted by ω, with values in $Im(\mathbb{H})$. (More on this in the next remarks.) Now a minor miracle occurs.

Proposition 16.1 *The 1-form $\omega = Im(\overline{\alpha}d\alpha + \overline{\beta}d\beta)$ defined on S^7 is an Ehresmann connection on the principal Hopf bundle $S^3 \hookrightarrow S^7 \to \mathbb{H}P^1 \cong S^4$.*

Remarks: Before proving this proposition, we would like to comment on the definition

$$\omega := \mathrm{Im}(\overline{\alpha}d\alpha + \overline{\beta}d\beta).$$

The apocryphal (though possibly true) story has it that one day while a very famous professor was lecturing on differential geometry, he wrote on the blackboard the definition of a differential form in terms of some variables. Some students were confused and, after whispering among themselves, one of them dared to ask the professor if he could please tell them the space on which the form was defined. The professor answered, "No." The professor's answer is not as unreasonable as the students might have first thought. The point is that such a formula in terms of certain variables will define a differential form in any manifold that has functions that we can identify with the variables in the formula. And this is what happens with the above definition of the form ω. The only thing a bit special about this case is that α, β are quaternionic variables and so must be matched up to quaternionic-valued functions. We can consider ω as a differential form on $Sp(2)$ or on S^7. We started thinking that ω was a form on $Sp(2)$ but are now considering it as a form on S^7. So there you have it: One formula gives us two distinct differential forms!

Proof: We have to show the two properties of an Ehresmann connection, as given in the definition. We first discuss the invariance under the right action of the group S^3. This is expressed in the commutativity of this diagram (which we must prove for every $g \in S^3$):

$$
\begin{array}{ccc}
TS^7 & \xrightarrow{R_{g,*}} & TS^7 \\
\downarrow{\omega} & & \downarrow{\omega} \\
Im(\mathbb{H}) & \xrightarrow{Ad_{g^{-1}}} & Im(\mathbb{H})
\end{array}
\qquad (16.1)
$$

Since $S^7 \subset \mathbb{H}^2$, we have that $R_g : S^7 \to S^7$ can be written as

$$R_g(q, r) = (qg, rg), \qquad (16.2)$$

where $(q, r) \in S^7 \subset \mathbb{H}^2$; that is, $q \in \mathbb{H}$ and $r \in \mathbb{H}$ are the "quaternionic coordinates" on S^7. Of course, the previous formula (16.2) extends to the domain $\mathbb{H}^2$, in which case it is a linear map over $\mathbb{R}$ from $\mathbb{H}^2$ to $\mathbb{H}^2$. Viewed this way, it is clear that its derivative

$$R_{g,*} : T_{(q,r)}(\mathbb{H}^2) \cong \mathbb{H}^2 \to T_{(qg,rg)}(\mathbb{H}^2) \cong \mathbb{H}^2$$

is identified with R_g under the two indicated canonical isomorphisms for the two tangent spaces. We will use these identifications from now on and so will say that $R_{g,*}$ is equal to R_g. So, by restriction, $R_{g,*} : T_{(q,r)}S^7 \to T_{(qg,rg)}S^7$ is also equal to R_g.

Note that for any point $(q, r) \in S^7$ and any vector $(v, w) \in T_{(q,r)}S^7$ tangent at that point, we have that

$$\omega_{(q,r)}(v, w) = \mathrm{Im}(\bar{q}v + \bar{r}w). \qquad (16.3)$$

Exercise 16.2 *This formula is an "obvious" consequence of the definition of the 1-form ω when it is written in this quaternionic notation. The reader is encouraged to internalize this result until it is obvious.*

As a mental check that we are not writing totally absurd formulas, let's note that the formula (16.3) is linear in the tangent vector (v, w) and of class C^∞ in the point $(q, r) \in S^7$.

Now to show the commutativity of the diagram (16.1), we propose to chase any arbitrary element $(v, w) \in T_{(q,r)}S^7$ through it, where $(q, r) \in S^7$ is an arbitrary point in the base space. First, we go across and then down. Going across gives

$$(v, w) \xrightarrow{R_g*} (vg, wg) \in T_{(qg,rg)} \in S^7,$$

and then continuing by going down yields

$$(vg, wg) \xrightarrow{\omega_{(vg,wg)}} \mathrm{Im}((qg)^- vg + (rg)^- wg) = \mathrm{Im}(\bar{g}\,\bar{q}vg + \bar{g}\,\bar{r}wg).$$

On the other hand, going down and then across (in one fell swoop), we have

$$(v, w) \xrightarrow{\omega} \mathrm{Im}(\bar{q}v + \bar{r}w) \xrightarrow{Ad_{g^{-1}}} g^{-1}\mathrm{Im}(\bar{q}v + \bar{r}w)g$$

since $ad_g^{-1} : S^3 \to S^3$ is $q \mapsto g^{-1}qg$, which extends to an $\mathbb{R}$-linear map $\mathbb{H} \to \mathbb{H}$ whose derivative, when restricted to $T_e(S^3) \to T_e(S^3)$, is the "same" linear map $v \mapsto g^{-1}vg$ when we identify $T_e(S^3)$ as a subspace of $\mathbb{H}$ using the canonical isomorphism. Here $e = (1, 0, 0, 0)$ is the identity element of S^3, of course.

Now to complete this part of the proof, we will prove the following.

Lemma 16.1 *For all $g \in \mathbb{H} \setminus \{0\}$ and all $h \in \mathbb{H}$, we have*

$$g^{-1}(\mathrm{Im}\, h)\, g = \mathrm{Im}\,(g^{-1}hg).$$

Since $g^{-1} = \bar{g}/|g|^2$, this formula is equivalent to

$$\bar{g}(\mathrm{Im}\, h)\, g = \mathrm{Im}\,(\bar{g}hg).$$

But

$$\mathrm{Im}\,(\bar{g}hg) = \frac{1}{2}\left(\bar{g}hg - (\bar{g}hg)^-\right)$$

$$= \frac{1}{2}\left(\bar{g}hg - \bar{g}\bar{h}g\right)$$

$$= \bar{g}\,\frac{1}{2}(h - \bar{h})g$$

$$= \bar{g}(\mathrm{Im}\, h)\, g.$$

And this proves the lemma. So now we apply this lemma to get

$$g^{-1}\mathrm{Im}(\bar{q}v + \bar{r}w)g = \mathrm{Im}(g^{-1}\bar{q}vg + g^{-1}\bar{r}wg) = \mathrm{Im}(\bar{g}\,\bar{q}vg + \bar{g}\,\bar{r}wg)$$

since $g^{-1} = \bar{g}$ for $g \in S^3$. And this concludes the proof of the commutativity of the diagram (16.1) and therefore that ω is G-equivariant.

To show the other defining property of an Ehresmann connection, consider

$$S^7 \xrightarrow{A^\sharp} TS^7 \xrightarrow{\omega} \mathrm{Im}(\mathbb{H}),$$

where $A \in \mathrm{Im}(\mathbb{H}) =$ the Lie algebra of S^3. We have to show that this composition is the constant function whose value is always equal to A. Recall that by definition,

$$A^\sharp(p) = \iota_{p,*}(A),$$

where $p \in S^7$ and $\iota_p : S^3 \to S^7$ is given by $\iota_p(g) = p \cdot g$.

Note that this same formula defines an $\mathbb{R}$-linear map $\iota_p : \mathbb{H} \to \mathbb{H}^2$. So its derivative at the identity element $e \in S^3$ is the linear map

$$\iota_{p*} : T_e(S^3) \cong \mathrm{Im}(\mathbb{H}) \to T_p(S^7) \subset \mathbb{H}^2$$

given by $A \mapsto pA \in T_p(S^7)$. Writing $p = (q, r) \in S^7$ by using the quaternionic notation, we have that

$$A \xrightarrow{\iota_{p*}} pA = (qA, rA) \xrightarrow{\omega} \mathrm{Im}(\bar{q}qA + \bar{r}rA) = \mathrm{Im}\big((|q|^2 + |r|^2)\,A\big)$$
$$= \mathrm{Im}(A) = A,$$

where the last two equalities follow from $(q, r) \in S^7$ and from $A \in \mathrm{Im}(\mathbb{H})$, respectively. And this concludes the proof of the second defining property of an Ehresmann connection, and hence the proof of the theorem. ∎

16.4 BPST Instantons

We now show how ω gives us the BPST instantons. First, note that instantons are 1-forms on four-dimensional manifolds. And the point here is that ω is the global object defined on the total space S^7 of the bundle, whereas each of its pullbacks to the base space by a local section is a local gauge potential that turns out to be the BPST instanton for specific local sections (i.e., specific local trivializations of the bundle). So we want to look carefully at coordinates on the base space $\mathbb{H}P^1$.

We define two open subsets of $\mathbb{H}P^1$ as follows:

$$U_1 := \{ [q, r] \in \mathbb{H}P^1 \mid q \neq 0 \} = \mathbb{H}P^1 \setminus \{[0, 1]\},$$
$$U_2 := \{ [q, r] \in \mathbb{H}P^1 \mid r \neq 0 \} = \mathbb{H}P^1 \setminus \{[1, 0]\}.$$

Then we define charts on these sets by

- $\phi_1([q, r]) := rq^{-1}$ for all $[q, r] \in U_1$,
- $\phi_2([q, r]) := qr^{-1}$ for all $[q, r] \in U_2$.

We leave it as an exercise for the reader to check that these functions are well defined, that is, independent of the choice of representative of the equivalence class $[q, r]$. Clearly, $\phi_1 : U_1 \to \mathbb{H}$ and $\phi_2 : U_2 \to \mathbb{H}$ are both surjective maps. Moreover, it is easy to check that

$$U_1 \cap U_2 = \mathbb{H}P^1 \setminus \{[0, 1], [1, 0]\}$$

and that $\phi_1(U_1 \cap U_2) = \phi_2(U_1 \cap U_2) = \mathbb{H} \setminus \{0\}$. The inverses of these functions are clearly given by $\phi_1^{-1}(y) = [1, y]$ and $\phi_2^{-1}(y) = [y, 1]$. So for all $y \in \phi_1(U_1 \cap U_2) = \mathbb{H} \setminus \{0\}$, we have that

$$\phi_2 \circ \phi_1^{-1}(y) = \phi_2([1, y]) = y^{-1}$$

and, similarly,

$$\phi_1 \circ \phi_2^{-1}(y) = \phi_1([y, 1]) = y^{-1}.$$

In particular, this shows that $\{(U_1, \phi_1), (U_2, \phi_2)\}$ is an atlas on $\mathbb{H}P^1$. The resulting differential structure is the usual one for a projective space.

Now, since each U_α for $\alpha = 1, 2$ is homeomorphic to the contractible space $\mathbb{H}$, it follows that the bundle $S^3 \hookrightarrow S^7 \xrightarrow{\pi} \mathbb{H}P^1$ is trivial over U_1 and over U_2. Of course, the trivializations of these restrictions are far from being unique. We will find one trivialization for each of these open sets. We represent points in S^7 as pairs (q, r) of quaternions such that $|q|^2 + |r|^2 = 1$.

Define $\Psi_1 : \pi^{-1}(U_1) \to U_1 \times S^3$ by

$$\Psi_1(q, r) := ([q, r], \frac{1}{|q|} q)$$

and similarly $\Psi_2 : \pi^{-1}(U_2) \to U_2 \times S^3$ by

$$\Psi_2(q, r) := ([q, r], \frac{1}{|r|} r).$$

Exercise 16.3 *Show that Ψ_1 and Ψ_2 are local trivializations of the principal bundle* $S^3 \hookrightarrow S^7 \to \mathbb{H}P^1$.

A bit of drudgery yields formulas for their inverses. First, we obtain

$$\Psi_1^{-1}([q, r], y) = |q|(y, rq^{-1}y)$$

for all $y \in S^3$ and all $[q, r] \in U_1$, provided that we still use the normalization condition $|q|^2 + |r|^2 = 1$. Similarly,

$$\Psi_2^{-1}([q, r], y) = |r|(qr^{-1}y, y)$$

for all $y \in S^3$ and all $[q, r] \in U_2$ with $|q|^2 + |r|^2 = 1$. Just a wee bit more drudgery and the transition functions fall out:

$$\Psi_2 \circ \Psi_1^{-1}([q, r], y) = ([q, r], \frac{rq^{-1}}{|rq^{-1}|} y)$$

provided that $[q, r] \in U_1 \cap U_2$. So $g_{21} : U_1 \cap U_2 \to S^3$ is given by

$$g_{21}([q, r]) = \frac{rq^{-1}}{|rq^{-1}|}.$$

Now we can compute the sections associated with these trivializations. First, $s_1 : U_1 \to \pi^{-1}(U_1)$ is given by

$$s_1([q, r]) = \Psi_1^{-1}([q, r], 1) = |q|(1, rq^{-1}) \in \pi^{-1}(U_1)$$

for $q \neq 0$. Similarly, $s_2 : U_2 \to \pi^{-1}(U_2)$ is given by

$$s_2([q, r]) = \Psi_2^{-1}([q, r], 1) = |r|(qr^{-1}, 1) \in \pi^{-1}(U_2)$$

for $r \neq 0$. In both of the previous equalities, we are taking $|q|^2 + |r|^2 = 1$.

Next, we would like to find the pullbacks of ω by the sections s_1 and s_2 to get 1-forms on U_1 and U_2, respectively. Actually, we will study the equivalent pullbacks by $s_1 \circ \phi_1^{-1}$ and by $s_2 \circ \phi_2^{-1}$ to $\mathbb{H}$. In other words, we will evaluate everything in the local coordinates given by the charts $\{(U_1, \phi_1), (U_2, \phi_2)\}$. Recall that

$$\mathbb{H} \xrightarrow{\phi_1^{-1}} U_1 \xrightarrow{s_1} \pi^{-1}(U_1) \subset S^7 \subset \mathbb{H}^2$$

and similarly for U_2. We start by pulling back by $s := s_1 \circ \phi_1^{-1}$, that is, by evaluating $s^*\omega = (s_1 \circ \phi_1^{-1})^*\omega$, which is a 1-form on $\mathbb{H}$ with values in $Im(\mathbb{H})$.

We will need to compute the derivative of s, so first we derive the formula for s. For $q \in \mathbb{H}$, we have

$$
\begin{aligned}
s(q) &= s_1 \circ \phi_1^{-1}(q) = s_1([1, q]) \\
&= s_1 \left(\frac{1}{(1 + |q|^2)^{1/2}}, \frac{q}{(1 + |q|^2)^{1/2}} \right) \\
&= \frac{1}{(1 + |q|^2)^{1/2}} \left(1, \frac{q}{(1 + |q|^2)^{1/2}}(1 + |q|^2)^{1/2} \right) \\
&= \frac{1}{(1 + |q|^2)^{1/2}} (1, q).
\end{aligned}
$$

Notice that we normalized the representative of the equivalence class $[1, q]$ in order to be able to apply our formula for s_1.

Since $(s^*\omega)_q(v) = \omega_{s(q)}(s_*v)$ for $v \in \mathbb{H} \cong T_q\mathbb{H}$, we want to calculate the derivative s_*v for starters. Let's make explicit how an element $v \in \mathbb{H}$ corresponds to a tangent vector in $T_q\mathbb{H}$ for a given point $q \in \mathbb{H}$. Given this data, we define a curve $\gamma_{q,v} : \mathbb{R} \to \mathbb{H}$ as usual by

$$\gamma_{q,v}(t) := q + tv.$$

Then the tangent vector at $q \in \mathbb{H}$ corresponding to this v is just the derivative of this curve at $t = 0$, that is, the directional derivative in the "direction" given by v. So we have

$$s_* v = \frac{d}{dt}\Big|_{t=0} s(\gamma_{q,v}(t)) = \frac{d}{dt}\Big|_{t=0} s(q+tv)$$

$$= \frac{d}{dt}\Big|_{t=0} \left(\frac{1}{(1+|q+tv|^2)^{1/2}}(1, q+tv) \right).$$

To evaluate this derivative, note that

$$|q+tv|^2 = (q+tv)^-(q+tv) = (\bar{q}+t\bar{v})(q+tv)$$

$$= \bar{q}q + t\bar{v}q + t\bar{q}v + t^2\bar{v}v$$

$$= |q|^2 + 2t\,Re(\bar{q}v) + t^2|v|^2$$

since $(\bar{v}q)^- = \bar{q}v$, implying that $\bar{v}q + \bar{q}v = 2\,Re(\bar{q}v)$. Therefore,

$$\frac{d}{dt}\left(\frac{1}{(1+|q+tv|^2)^{1/2}} \right) = -\frac{1}{2}(1+|q+tv|^2)^{-3/2}(2\,Re(\bar{q}v) + 2t|v|^2),$$

which implies in particular that

$$\frac{d}{dt}\Big|_{t=0}\left(\frac{1}{(1+|q+tv|^2)^{1/2}} \right) = \frac{-Re(\bar{q}v)}{(1+|q|^2)^{3/2}}.$$

Also, we see that

$$\frac{d}{dt}\left(\frac{q+tv}{(1+|q+tv|^2)^{1/2}} \right) =$$

$$\frac{v}{(1+|q+tv|^2)^{1/2}} + (q+tv)\frac{d}{dt}\left(\frac{1}{(1+|q+tv|^2)^{1/2}} \right),$$

which in turn immediately yields

$$\frac{d}{dt}\Big|_{t=0}\left(\frac{q+tv}{(1+|q+tv|^2)^{1/2}} \right) = \frac{v}{(1+|q|^2)^{1/2}} - q\frac{Re(\bar{q}v)}{(1+|q|^2)^{3/2}}.$$

Then, putting things together, we see that

$$s_* v = \left(\frac{-Re(\bar{q}v)}{(1+|q|^2)^{3/2}}, \frac{v}{(1+|q|^2)^{1/2}} - q\frac{Re(\bar{q}v)}{(1+|q|^2)^{3/2}} \right).$$

Again, we are trying to evaluate

$$(s^*\omega)_q(v) = \omega_{s(q)}(s_* v) = Im(\bar{\alpha}d\alpha + \bar{\beta}d\beta)_{s(q)}(s_* v),$$

where $(\alpha, \beta) \in S^7 \subset \mathbb{H}^2$ and $v \in \mathbb{H} \cong T_q\mathbb{H}$. Of course, we are engaging in a common abuse of notation. Actually, α and β are not a pair of quaternions, but rather two smooth functions $S^7 \to \mathbb{H}$, which is why $d\alpha$ and $d\beta$ make sense and are $\mathbb{H}$-valued 1-forms. Also, $\overline{\alpha}$ and $\overline{\beta}$ are smooth functions $S^7 \to \mathbb{H}$. For example,

$$\overline{\alpha}(s(q)) = \left(\frac{1}{(1 + |q|^2)^{1/2}} \right)^{-} = \frac{1}{(1 + |q|^2)^{1/2}}$$

and

$$\overline{\beta}(s(q)) = \left(\frac{q}{(1 + |q|^2)^{1/2}} \right)^{-} = \frac{\overline{q}}{(1 + |q|^2)^{1/2}}.$$

So

$$(\overline{\alpha}d\alpha)_{s(q)}(s_*v) = \frac{1}{(1 + |q|^2)^{1/2}} \cdot \frac{-Re(\overline{q}v)}{(1 + |q|^2)^{3/2}} = \frac{-Re(\overline{q}v)}{(1 + |q|^2)^2}$$

and

$$(\overline{\beta}d\beta)_{s(q)}(s_*v) = \frac{\overline{q}}{(1 + |q|^2)^{1/2}} \left(\frac{v}{(1 + |q|^2)^{1/2}} - q \frac{Re(\overline{q}v)}{(1 + |q|^2)^{3/2}} \right)$$

$$= \frac{\overline{q}v}{(1 + |q|^2)} - \frac{|q|^2 Re(\overline{q}v)}{(1 + |q|^2)^2}.$$

Putting this all together, we have

$$(s^*\omega)_q(v) = \omega_{s(q)}(s_*v) = Im(\overline{\alpha}d\alpha + \overline{\beta}d\beta)_{s(q)}(s_*v)$$

$$= Im \left(\frac{-Re(\overline{q}v)}{(1 + |q|^2)^2} + \frac{\overline{q}v}{(1 + |q|^2)} - \frac{|q|^2 Re(\overline{q}v)}{(1 + |q|^2)^2} \right)$$

$$= Im \left(\frac{\overline{q}v}{(1 + |q|^2)} \right).$$

The last equality holds since the first and third terms on the second line are real. Now we would like to eliminate v from both sides of the preceding formula in order to obtain a formula equating 1-forms with values in $Im(\mathbb{H})$. This gives

$$((s_1 \circ \phi_1^{-1})^*\omega)_q = (s^*\omega)_q = Im \left(\frac{\overline{q}}{1 + |q|^2} dq \right), \tag{16.4}$$

which is a BPST instanton on $\mathbb{H} \cong \mathbb{R}^4$. For $0 \neq q \in \mathbb{H}$, we can rewrite this as

$$(s^*\omega)_q = Im \left(\frac{|q|^2}{1 + |q|^2} q^{-1} dq \right) = \frac{|q|^2}{1 + |q|^2} Im \left(q^{-1} dq \right).$$

Interpreting papers written ages ago using the then-currently-used notations and conventions is as much art as science. However, the previous formula "compares well" with the formula for the Yang–Mills field A_μ given in [3] as the solution of their self-duality equation (13), provided that we take the scale parameter in that paper to be $\lambda = 1$.

The reader should now realize that a similar, lengthy calculation will yield an answer for the section s_2. In the same local coordinates as used in (16.4), except with $0 \neq q \in \mathbb{H}$, it is

$$((s_2 \circ \phi_2^{-1})^* \omega)_q = Im \left(\frac{\overline{q}^{-1}}{1 + |q|^2} \, d\overline{q} \right).$$

This becomes

$$((s_2 \circ \phi_2^{-1})^* \omega)_q = Im \left(\frac{1}{\overline{q}(1 + |q|^2)} \, d\overline{q} \right)$$

$$= Im \left(\frac{q}{|q|^2(1 + |q|^2)} \, d\overline{q} \right)$$

$$= \frac{1}{|q|^2(1 + |q|^2)} \, Im \, (q \, d\overline{q}).$$

Let me remark that we have not shown that the 1-form $A := s^* \omega$ solves the $SU(2)$ Yang–Mills equations.

Exercise 16.4 *So your last exercise is to prove that.*

Chapter 17
What Next?

This book leaves off well before arriving at the frontier where modern research is occurring. For those who wish to go that far, there are many possible directions, both in mathematics and in physics. But any way one eventually goes, I highly recommend reading and studying a lot.

For those who wish to learn more about the multitude of relations of geometry and topology with modern physics as of 2003, I highly recommend Nakahara's book [40]. This is definitely at least one big notch up from this book in what it requires from the reader. It may be a giant step, but it is a step in the right direction and well worth taking. There is also a lot of nice physics and mathematics material in Frankel's book [16].

Others might be interested in *supersymmetric* versions of gauge theories. Or in discrete models of gauge theories, known as *lattice gauge theory*. Or in the ongoing challenges in the quantization of gauge theories in a mathematically rigorous way. Or in the efforts for the description of all interactions in one *unified theory*, which of necessity must have some relation to gauge theory. Or looking at the problem another way, in how to describe gravity as a quantum theory having some necessary relation with the gauge theory known as the *standard model*. Or in the context of noncommutative geometry, say in a setting with quantum principal bundles.

Or maybe in a way to use principal bundles in a setting that no one has yet conceived of. That is the stuff that dreams are made of.

And I do wish you a prosperous and productive scientific future!

© Springer International Publishing Switzerland 2015
S.B. Sontz, *Principal Bundles*, Universitext, DOI 10.1007/978-3-319-14765-9_17

Correction to: Principal Bundles

Correction to:
S.B. Sontz, *Principal Bundles*, Universitext,
https://doi.org/10.1007/978-3-319-14765-9

The original version of this book was inadvertently published without updating the following corrections in Chapters 5, 7, and 11. These are corrected now.

Chapter 5

Pages 71–72: Definition 5.4 has been replaced by all of the following. This has been placed immediately after the definition of $\Omega^k(M)$ at the bottom of page 71:

The construction $T^{\diamond k}(M)$ for every smooth manifold M is the first part of the definition of a covariant functor. The second part is a vector bundle morphism $T^{\diamond k} f : T^{\diamond k}(M) \to T^{\diamond k}(N)$ for every smooth function $f : M \to N$ of smooth manifolds M and N. This is defined by

$$T^{\diamond k} f(v_1, \ldots, v_k) := (T(f)v_1, \ldots, T(f)v_k) \in T^{\diamond k}_{f(x)}(N) \qquad \text{for } k \geq 1,$$

where $(v_1, \ldots, v_k) \in T^{\diamond k}_x(M)$ for some $x \in M$ (which means $v_j \in T_x(M)$ for all $1 \leq j \leq k$) and $T(f) : T(M) \to T(N)$ is the vector bundle morphism induced by $f : M \to N$ on the tangent bundles. The fact that this is a vector bundle morphism says that this diagram commutes:

$$
\begin{array}{ccc}
T^{\diamond k}(M) & \xrightarrow{\;T^{\diamond k}(f)\;} & T^{\diamond k}(N) \\
\downarrow & & \downarrow \\
M & \xrightarrow{\quad f \quad} & N
\end{array}
$$

The updated version of this book can be found at
https://doi.org/10.1007/978-3-319-14765-9_5
https://doi.org/10.1007/978-3-319-14765-9_7
https://doi.org/10.1007/978-3-319-14765-9_11
https://doi.org/10.1007/978-3-319-14765-9

We often write $f_* = T^{\circ k} f$ in order to spare the reader from excessive notation, though at the expense of some slight ambiguity. Note this is consistent with the previously introduced notation $f_* = T(f)$, which is the special case here when $k = 1$.

Exercise: *Prove that $T^{\circ k}$ is a covariant functor.*

Definition 5.4. *Suppose $f : M \longrightarrow N$ is a smooth function of smooth manifolds M and N. Let $\eta : T^{\circ k}(N) \longrightarrow \mathbb{R}$ be a k-form on N. Then the pullback of η by f is defined as*

$$f^*(\eta) := \eta \circ f_* : T^{\circ k}(M) \longrightarrow \mathbb{R}.$$

namely, the composition

$$T^{\circ k}(M) \xrightarrow{f_*} T^{\circ k}(N) \xrightarrow{\eta} \mathbb{R}.$$

Exercise: *For $k \geq 1$ prove that $f^*(\eta)$ is a k-form on M and that $f^* : \Omega^k(N) \longrightarrow \Omega^k(M)$ is linear. Then extend all this to the case $k = 0$.*

Chapter 7
Page 95, Definition 7.5:
"$T(L_g)X = X$" has been replaced with "$T(L_g)X = X L_g$".

And the below sentence has been added to that definition:
In other words, $X : G \to T(G)$ is a G-equivariant map with respect to the left actions of G on G and $T(G)$.

Page 98, near the bottom of the page:
The displayed equation "$(L_{g*})X = X$" has been replaced with "$(L_{g*})X = X L_g$".

Page 101, Section 7.5:
After the second paragraph of this section the following Exercise has been added:

Exercise: *Prove that the Lie bracket of the Lie algebra $\mathfrak{gl}(n, \mathbb{C})$ is given by the commutator of matrices.*

Chapter 11
Near bottom of page 147, "as in the preceding on," has been replaced with "as in the preceding one,"

Appendix A
Discussion of the Exercises

This appendix contains comments and hints for most of the problems. Hints are not given in the text itself. It is highly recommended to ponder a problem long and well before availing yourself of this material, which is not claimed to be complete or crystal clear.

Chapter 2

- Exercise 2.1: The condition in the exercise is necessary and sufficient because of the definition of atlas.
- Exercise 2.2: Throw into the given atlas all of the charts compatible with it. Then one proves that the resulting set of charts is (i) a smooth atlas, (ii) maximal, and (iii) unique with these two properties.
- Exercise 2.3: In all cases one has to exhibit a set of charts and show that it is an atlas. In the first two examples, the set of charts is given. In addition, for S^n, one has to show the two atlases are compatible with each other.

  For the real projective space, use the images in the quotient space of the first atlas on S^n. This gives the standard differential structure on this topological space. And ponder why this does not work for the second atlas on S^n. A nontrivial fact here is that the minimal number of charts for defining this particular differential structure on $\mathbb{R}P^n$ is $n + 1$. You are not expected to prove this now.

  For the complex projective space, imitate the case of the real projective space. Recall that the complex plane $\mathbb{C}$ is homeomorphic to $\mathbb{R}^2$. And so $\mathbb{C}^n$ is homeomorphic to $\mathbb{R}^{2n}$.

  For the last example, you have to know what a Riemann surface is. Of course! If you do, use that knowledge. (I will leave you to do that on your own, since we will not use this result later on in this book.) If not, think about acquiring that knowledge.

- Exercise 2.4: The statement in the exercise is a very minor modification of the definition of smooth function. In fact, the modification is so minor the reader may miss it. The exercise does away with reference to the point $x \in M$. In other words, smoothness is a local structure (i.e., depending on the open sets in the spaces) and not a pointwise structure. That's it. So think about why $x \in M$ is an unnecessary detail.
- Exercise 2.5: This is indeed tricky because the manifold $M = \mathbb{R}$ being discussed here is the same as the model space of the manifold. But as the model space, it is the real line $\mathbb{R}$, as learned in introductory calculus. And M is also the real line as learned in introductory calculus, but the identity map from M to $\mathbb{R}$ is not the diffeomorphism that gives the identification. Just work with the definitions of all of these concepts (carefully, of course) and you will see what is happening.
- Exercise 2.6: First, show that if two manifolds are diffeomorphic, then they have the same dimension and are homeomorphic. To do this, you can reduce the problem to a linear problem by taking the derivative of the diffeomorphism to get an isomorphism of vector spaces. This uses material in the following section. For the first part, I only know of proofs that use some tool from algebraic topology. The idea is that there exist invariants (the simplest being homotopy groups and homology groups) that show that S^n and S^m are not homeomorphic for $n \neq m$.

  For the second part, use that the one-point compactification of $\mathbb{R}^n$ is (or more correctly, is homeomorphic to) S^n. This readily implies that $\mathbb{R}^n$ and $\mathbb{R}^m$ are not homeomorphic for $n \neq m$.

  For the third part, think about compactness of the spaces.
- Exercise 2.7: A vector space of dimension zero has *one* point, namely, the zero vector. For example, if the domain space has dimension zero, then the increment vector h can only be zero. And when the codomain space has dimension zero, then the function is constant and you should know from elementary calculus how to take its derivative.
- Exercise 2.8: Use Definition (2.4) very carefully to calculate $T(g \circ f)$ and $T(1_U)$. If you have never worked with commuting diagrams, take your time and be patient with yourself. The advantage of simple diagrams like $A \xrightarrow{f} B$ and $B \xrightarrow{g} C$ is that you have only to paste them together in the obvious way

$$A \xrightarrow{f} B \xrightarrow{g} C,$$

  and not the other way, to get a more complicated diagram.
- Exercise 2.9: The three facts are labeled R, S, and T to stand for the three properties of an equivalence relation: reflexive, symmetric, and transitive. So use each fact to prove the corresponding property. And the facts themselves are proved by applying the definitions of V_α, $V_{\alpha\beta}$, and $\psi_{\alpha\beta}$. But watch out! Some care must be given to the exact domains of the functions.

- Exercise 2.10: Map X to M by first sending each $x \in V_\alpha$ to

$$\phi_\alpha^{-1}(x) \in U_\alpha \subset M.$$

This defines a mapping from $\bigsqcup_\alpha V_\alpha$ to M. Then show that this passes to the quotient of the domain by the equivalence relation $\sim$ to give a map $X \to M$. Then this map is the desired diffeomorphism.
- Exercise 2.11: This follows in analogy to the solution to Exercise 2.9, the only difference being the definitions of the transition functions and their domains. So work with these new definitions but using the pattern established in Exercise 2.9.
- Exercise 2.12: The first part of the problem parallels Exercise 2.10. However, notice that the natural charts $(\tilde{U}_\alpha, \tilde{\phi}_\alpha)$ have codomains of a special form (open times Euclidean space) and that the transition functions also have a special form, namely, $T\psi_{\beta\alpha}$, which we know is of class C^∞.
The dimension of TM is the dimension of the codomains of its charts.
WARNING: One has to show that TM is a Hausdorff space. At that point, one has to use that the given manifold M is itself a Hausdorff space.
The tangent bundle map is a new structure that is not analogous to something previously done. One defines τ_M above each U_α to make the square on the right of the diagram commute; that is, locally it is projection onto the first factor. One shows that this definition is compatible on the intersections $U_\alpha \cap U_\beta$; this justifies the "pasting" together and gives the commutativity of the left square. As with any map, τ_M is smooth if and only if it is locally smooth near every point in its domain. But locally it is projection on the first factor.
- Exercise 2.13: Pick one natural chart $(\tilde{U}_\alpha, \tilde{\phi}_\alpha)$ above a coordinate chart (U_α, ϕ_α) of M. Take a point $x \in U_\alpha$ and consider the fiber $T_x M = \tau_M^{-1}(x)$ over it. Then $\tilde{\phi}_\alpha^{-1} : T_x M \to \{\phi_\alpha(x)\} \times \mathbb{R}^m$ is a bijection. Since the codomain here is a vector space of dimension m over the reals, we can use this bijection to make $T_x M$ into a vector space of dimension m over the reals. If $x \in U_\beta$, then we get a possibly different vector space structure on $T_x M$. Then one shows that these two vector space structures are actually identical. However, the isomorphisms $\tilde{\phi}_\alpha^{-1}$ and $\tilde{\phi}_\beta^{-1}$ are in general not equal.
- Exercise 2.14: You are on your own for this one.
- Exercise 2.15: The essential point of this problem is to see whether the resulting space $(V_1 \sqcup V_2)/\sim$ is Hausdorff. In part (i), the points $0 \in V_1$ and $0 \in V_2$ are not identified under the equivalence relation $\sim$, but any neighborhood of one of them intersects all neighborhoods of the other. In part (ii), these points $0 \in V_1$ and $0 \in V_2$ also give distinct points in $(V_1 \sqcup V_2)/\sim$, but these points can be separated with open sets, which is the Hausdorff property. The space $(V_1 \sqcup V_2)/\sim$ is then a well-known differential manifold. If you have studied stereographic coordinates, then this should be something you already know.
- Exercise 2.16: This is the usual list of wishes for gifts from Santa Claus. But you have to be Santa's helper and make the gifts yourself. The first gift comes from calculating Tf in two overlapping charts and showing that the results agree in the overlap. To check that Tf is smooth, it suffices to check that it is smooth locally.

The commutative diagram holds locally and so must hold globally as well. If $T_x f$ is linear in one choice of charts U and V with $f(U) \subset V$, then it is linear in all such pairs of charts. Why? Because the linear structure of the fiber does not depend on the choice of chart, as shown in Exercise 2.13.

- Exercise 2.17: More playing with definitions and knowing that these properties hold locally. See Exercise 2.8.
- Exercise 2.18: The atlas on $\mathbb{R}^{k+1}$ consisting of one global chart will not work. You need something more refined to take into consideration the sphere. Think of the case $k = 2$ and how it relates to coordinates on the Earth. Yes, the ancient idea is locally correct. The surface of the Earth locally is like a plane sitting as a slice in three-dimensional space.

  The curve $t \mapsto (t, rt) \in \mathbb{R}^2$, where $t \in \mathbb{R}$ passes down to give a curve in the quotient space $\mathbb{T}^2$. The image of this curve in $\mathbb{T}^2$ is closed if and only if r is rational. If r is irrational, then the image of this curve is a dense subset of $\mathbb{T}^2$.
- Exercise 2.19: For every $k \geq 0$, evaluate $g^{(k)}(x)$ for $x \neq 0$ by using elementary calculus. You do not need an exact formula, just enough to be able to calculate $g^{(k)}(0)$ from its definition as a limit. (Warning: Do not use that the kth derivative is continuous at 0 before proving that this is so.) Prove that this limit exists and evaluate it. You can use everything you know from calculus.
- Exercise 2.20: Calculate the derivatives $f^{(k)}(a)$ for all k and construct the Taylor series of f centered at a. Then show that the defining property of a real analytic function does not hold near $x = a$. The same method works for a'.

Chapter 3

- Exercise 3.1: With two charts, there are four functions in its cocycle. Using the notation (U_j, ϕ_j) for the charts with $j = 1, 2$, the functions in the cocycle are $g_{11}, g_{12}, g_{21}, g_{22}$. But g_{11}, g_{22} are identity functions, while $g_{12} = g_{21}^{-1}$. So the cocycle is determined completely by the function g_{21}, say.

  With one chart, there is only one function in the cocycle, and it is the identity function on its domain.
- Exercise 3.2: This commutative diagram is the key to this construction:

$$
\begin{array}{ccc}
E \supset \pi^{-1}(U_\alpha) & \xrightarrow{\tilde{\phi}_\alpha} & \phi_\alpha(U_\alpha) \times \mathbb{R}^l \\
\downarrow{\scriptstyle \pi} \quad\quad \downarrow{\scriptstyle \pi} & & \downarrow{\scriptstyle \pi_1} \\
M \supset \quad U_\alpha & \xrightarrow{\phi_\alpha} & \phi_\alpha(U_\alpha)
\end{array} \quad .
$$

The elements of the atlas $\{(\pi^{-1}(U_\alpha), \tilde{\phi}_\alpha)\}_{\alpha \in A}$ are the natural charts on the total space E.

The construction of the tangent bundle is both a special case and a motivating example for this exercise.

- Exercise 3.3: The manifold M may have billions and billions of charts in a given atlas, and the cocycle is defined for each pair of these charts. But the map $g_{\alpha\beta}$: $U_\alpha \cap U_\beta \to G$ is the constant function $g_{\alpha\beta}(x) = e$ for all $x \in U_\alpha \cap U_\beta$. Of course, this *is* a cocycle, as the reader should check.
- Exercise 3.4: Show that the composition $\rho \circ t_{\beta\alpha}$ satisfies the cocycle condition (3.3), given that $t_{\beta\alpha}$ satisfies that condition.
- Exercise 3.5: Careful here! The linear maps A and $A \otimes A$ are smooth, since all linear maps are smooth. But that is not the issue here, but rather (a) the proof that the mapping $A \mapsto A \otimes A$ maps *invertible* linear maps A to *invertible* linear maps and (b) that it is a smooth mapping. For part (a), one uses that $A^{-1} \otimes A^{-1}$ is the inverse of $A \otimes A$, provided that A is invertible.

  For part (b), it seems to be easier to show that $T \mapsto T \otimes T$ is a smooth map from $End(V)$ to $End(V \otimes V)$, where $T \in End(V)$ or equivalently, for $T : V \to V$ linear. Of course, the mapping $T \mapsto T \otimes T$ is not linear. And to do that, one can take a basis $\{x_i\}$ of V and write T in that basis and write $T \otimes T$ in the basis $\{x_i \otimes x_j\}$. The latter matrix is called the *Kronecker product* of the matrix of T with itself. When it is written this way, one can see that all the entries of the matrix for $T \otimes T$ are very elementary (and smooth!) functions of the entries of the matrix for T.
- Exercise 3.6: This is the generalization of Exercise 3.5 to k factors instead of 2 factors. One way to prove this is to generalize the proof of Exercise 3.5. Another way is to use induction on $k \geq 2$. Note that the cases $k = 0$ and $k = 1$ are trivial.
- Exercise 3.7: It is a question of showing that the matrix entries of $(A^{-1})^*$ are smooth functions of the matrix entries of A. This is a two-step proof. First, one identifies the matrix elements of A^{-1} in terms of those of A. But there is a formula from linear algebra that does exactly that! It is known as Cramer's rule. One has to show that this implies that entries in A^{-1} are smooth functions of the entries of A. Then one identifies the matrix entries of $(A^{-1})^*$ in terms of those of A^{-1} and ultimately in terms of those of A. But this second step is a very easy application of linear algebra.
- Exercise 3.8: To say that "M is something that transforms as M does" seems to be the height of what is known as a *circular definition*. It is not, because it is not a definition but rather a description. As an example, a contravariant vector transforms in a very certain way under coordinate changes, a way that is given by a formula. That is a property of contravariant vectors. Moreover, that certain way of transforming is then known as the way that a contravariant vector transforms. It is a (trivial) consequence of the definitions that a contravariant vector is something that transforms as a contravariant vector does. In fact, in this case this property is even characteristic; that is, it identifies exactly the class of contravariant vectors.

  Keeping all this in mind, it is not so difficult to write out rigorously the *description* of a covariant vector in terms of its transformation property. The reader might also read again the remarks after (2.12) about the similar expression that "a vector is something that transforms as a vector."

- Exercise 3.9: This is a rather personal exercise in creating your own comprehension of this material.
- Exercise 3.10: The calculations are completely straightforward to show that ρ_s and ρ_{ps} are representations. Also, the fact that they are not equivalent is rather transparent. The difficult bit is understanding that ρ_{ps} is important in physics. One might think, as some Nobel laureates thought, that only ρ_s enters in the theory of physically measurable quantities. That is not a mathematical result, of course. The basis for using any of these representations has to come from experiment. When the dust settled in the 1950s, the conclusion was that ρ_{ps} describes certain (but far from all!) measurable quantities associated with the *weak interaction*.
  The upshot is that whatever your sources of intuition might be and whatever conclusions they have led you to, one has to get used to thinking about pseudoscalars. Because there they are in nature! So again this is a challenge for creating your own understanding of this matter.
- Exercise 3.11: Let s_1 and s_2 be smooth sections. So $s_1(x)$ as well as $s_2(x)$ lie in the same fiber E_x for every $x \in M$. But we already know that E_x is a vector space. So we define the sum and scalar multiplication in terms of these vector space operations on E_x. For example, $s_1 + s_2$ is defined pointwise by $(s_1 + s_2)(x) := s_1(x) + s_2(x)$. The tricky bit is to show that these vector space operations on smooth sections then yield *smooth* sections. (That they give sections is easy enough.) As usual, this is shown locally in a system of coordinates. Then it is straightforward to show that the set of smooth sections is a vector space over $\mathbb{R}$.
  To show that the space of sections is in general infinite dimensional, one can consider first the case of a trivial vector bundle. Every vector bundle is locally trivial, and sections with support inside of a coordinate neighborhood, where the bundle is trivial, are easy enough to construct using a smooth approximate characteristic function.
  Notice that in the text when we say "section," we automatically mean a smooth map. Here we include the adjective "smooth" to make things more explicit.
- Exercise 3.12:
  For the first statement, we take two vector bundle maps, which are commutative square diagrams by definition, with the codomain of one vector bundle map being the domain of the second vector bundle map. Then we paste together these two squares on their common vertical edge. Composing the horizontal arrows (above and below) converts this 1×2 rectangle into a 1×1 square. And that square (which does commute) is the composition sought for.
  For the second statement, one is given a collection of objects (vector bundles) and morphisms (vector bundle maps). So one only has to show that the axioms of a category are satisfied.
  Behind the third statement are simply two basic facts from calculus:

  1. The chain rule.
  2. The derivative of the identity is the identity.

However, the definition of the functor on smooth maps $f : M \to N$ was not given in the statement of the problem. So you have to give the definition of the vector bundle map from the bundle $\tau_M : T(M) \to M$ to the vector bundle $\tau_N : T(N) \to N$. The hint is that you have already seen this vector bundle map. Then, when all has been defined, one proves that the defining properties of a functor are satisfied.

The fourth statement is similar to the second statement. This exercise consists of showing that the given objects and morphisms satisfy the axioms of a category.

- Exercise 3.13: Think about the set of equivalences of the explicitly trivial bundle with itself.
- Exercise 3.14: Warning: The approach of this exercise is generally considered not to be elegant, even though it is rigorous. What "elegant" is supposed to mean is anyone's guess.

  However, "does not depend on the choice of the two bases" means that if we choose other bases $\{e'_\alpha \mid \alpha \in A\}$ of V and $\{f'_\beta \mid \beta \in B\}$ of W, then the space constructed with basis $e'_\alpha \otimes f'_\beta$ for $\alpha \in A$, $\beta \in B$, say $(V \otimes W)'$, is isomorphic to the space $V \otimes W$ by a *uniquely* determined isomorphism. The construction of this unique isomorphism is the heart of the problem.

- Exercise 3.15: The first statement can be proved from the construction in Exercise 3.14. Or you can prove it directly.

  For the second statement, given a basis $\{e_\alpha \mid \alpha \in A\}$ of V, one defines the *dual basis* $\{e^*_\alpha : V \to \mathbb{R} \mid \alpha \in A\}$ of V^* by $e^*_\alpha(e_\beta) := \delta_{\alpha\beta}$ (the Kronecker delta) on the basis and then extending linearly to V. Here the number of elements in A is n. The thing to prove here is that the so-called dual basis is in fact a basis for the dual space V^*.

  The space $\mathrm{Hom}(V, W)$ is isomorphic to the vector space of all $m \times n$ matrices with real entries. This latter vector space has a well-known basis whose elements E_{ij} are the matrices with all but the (i, j) entry equal to zero and the (i, j) entry equal to 1. Here $1 \le i \le m$ and $1 \le j \le n$. Since there are nm such matrices E_{ij}, this establishes the dimension of $\mathrm{Hom}(V, W)$.

  The above remarks show that the dimension of $V^* \otimes W$ is nm. So it is isomorphic to $\mathrm{Hom}(V, W)$ by a dimension argument. But we want something more explicit, and that is what $\eta_{V,W}$ is all about. One way to define $\eta_{V,W}$ is to map the basis element $e^*_i \otimes f_j$ of $V^* \otimes W$ to the basis element E_{ij} of $\mathrm{Hom}(V, W)$. If we look at it this way, we do not see the underlying structure of $\eta_{V,W}$. Hence, for any decomposable element $e^* \otimes f \in V^* \otimes W$, meaning that $e^* \in V^*$ and $f \in W$, we define

$$\eta_{V,W}(e^* \otimes f) := E_{e^* f} : V \to W,$$

where $E_{e^* f}(x) := e^*(x) f$ for $x \in V$. For physicists who are used to Dirac notation, we note that if V has an inner product (and therefore a corresponding isomorphism $V^* \cong V$), then $E_{e^* f} = |e\rangle\langle f|$, where $e \in V$ is the element corresponding to $e^* \in V^*$ under the isomorphism.

The naturality of the isomorphism $\eta_{V,W}$ follows by checking that the defining property of naturality holds. See any text containing the basics of category theory (e.g., the classical text [35]) for the exact definition. Since the concept of naturality is not emphasized in this book, we leave this detail to the interested reader.

Recall that the rank of a linear map is the dimension of its range. Given this viewpoint, one realizes immediately that the rank of E_{e^*f} is 1 if and only if both e^* and f are nonzero. Otherwise, its rank is 0. Clearly, there are linear maps in $\mathrm{Hom}(V, W)$ with rank ≥ 2 if both V and W have dimension ≥ 2.

- Exercise 3.16: Just use Definition (3.7).

Chapter 4

- Exercise 4.1: This is less than solving a differential equation. We can consider the equations $\gamma(0) = x$ and $\gamma'(0) = v$ as initial conditions for the unknown curve γ. But there is no differential equation that $\gamma'(t)$ has to satisfy. Thinking of writing $\gamma(t)$ as a Taylor series

$$\gamma(t) = \gamma(0) + \gamma'(0)\,t + \sum_{k=2}^{\infty} \frac{1}{k!}\gamma^{(k)}(0)\,t^k$$

$$= x + vt + \sum_{k=2}^{\infty} \frac{1}{k!}\gamma^{(k)}(0)\,t^k,$$

we see that even for such real analytic functions, there is an infinite-dimensional space of solutions γ. The simplest function of this form is $\gamma(t) = x + vt$.

There are even more C^∞-functions γ satisfying these two conditions.

But what about the definition of the domain J of γ?

- Exercise 4.2: For part (a), by hypothesis the matrix $(A^{-1})^t$ exists. Now multiply it on the left and on the right by the matrix A^t to see whether it really is the inverse of A^t.

For part (c), the matrices under consideration satisfy $A^t A = A A^t = I$, the identity matrix. These are the $n \times n$ *orthogonal matrices*. The set of all of these matrices is typically denoted by $O(n)$. Moreover, $O(n)$ is a group under matrix multiplication. It is also a differential manifold embedded in $\mathbb{R}^{n^2}$. It has a dimension equal to n^2 minus the number of independent constraints in the relations $A^t A = A A^t = I$. Finally, these two structures on $O(n)$ are compatible, which means that the matrix multiplication operation and the inverse operation $(A \mapsto A^{-1})$ are smooth maps. For the time being, we leave these details for the further consideration of the interested reader.

- Exercise 4.3: The condition $df(y) = w$ expands to

$$\frac{\partial f}{\partial x_j}(y) = w_j$$

for all $j = 1, \ldots, n$, where each $w_j \in \mathbb{R}$ is a component of w. One solution (far from being unique) is given for $x \in U$ by

$$f(x) = w_1 x_1 + w_2 x_2 + \cdots + w_n x_n.$$

Can you find some of the many, many more solutions?

- Exercise 4.4:

  (a) Define df locally in a chart by the usual formula from multivariable calculus and show that what you get does not depend on the choice of the chart.

  (b) $d\omega = 0$ is true if and only if it is true locally in each chart. Then use the previous part to calculate $d\omega$ locally.

  (c) First, let's consider the question of uniqueness. But clearly, if $\omega = df$, then we also have $\omega = d(f + c)$ for any constant $c \in \mathbb{R}$. Since M is connected, this gives all possible functions $g : M \to \mathbb{R}$ such that $\omega = dg$.
  The problem of existence reduces to solving the differential equation $df = \omega$ for the unknown f in terms of the given ω. Note that $d\omega = 0$ is a necessary condition for solving this equation. One says that $d\omega = 0$ are the *integrability conditions*. The meat of the matter is that these are also sufficient for solving $df = \omega$. How do we see this? By integration! We define for any $y \in M$

$$f(y) = \int_0^y \omega = \int_0^y \big(\omega_1(x) \, dx_1 + \cdots + \omega_n(x) \, dx_n \big),$$

  where we integrate over any simple curve in M that starts at the origin $0 \in M$ and ends at $y \in M$. The condition $d\omega = 0$ implies that this integral does not depend on the choice of the particular simple curve. Then by a standard argument as given in a course on multivariable calculus, one proves that $df = \omega$.

  (d) A convex set C is star-shaped with respect to every point $p \in C$. To get a star-shaped set that is not convex, think about the usual iconic image for a star in Western culture, be it a starfish, a star fruit, or a piñata representing a star.

  (e) This generalizes part (c), but the proof is essentially the same except that now the integral starts at any star center $p \in S$ instead of at 0. The function f so obtained is again unique up to an additive constant.

  (f) The classical example in calculus of several variables starts by taking $U = \mathbb{R}^2 \setminus \{0\}$. Then one can define $d\theta$ at every point in this open set, where θ is the polar angular coordinate, even though the function θ can only be smoothly defined on a cut plane, such as $\mathbb{R}^2$ minus the closed positive x semiaxis. One way to convince yourself that the domain of this 1-form is U is to write it in Cartesian coordinates.

Chapter 5

- Exercise 5.1: If they are linearly independent, find a basis $v_1, v_2, \ldots, v_n$ with $v_1 = v$ and $v_2 = w$.

  The converse is to show if they are linearly dependent (some linear combination of them is 0), then the wedge product is 0.
- Exercise 5.2: Much as in Exercise 5.1. Linearly dependent means one of the vectors is a linear combination of the rest.
- Exercise 5.3: Definition (5.4) can be written as

$$\left(e_{\alpha_1} \wedge \cdots \wedge e_{\alpha_j}\right) \wedge \left(e_{\beta_1} \wedge \cdots \wedge e_{\beta_k}\right) := e_{\alpha_1} \wedge \cdots \wedge e_{\alpha_j} \wedge e_{\beta_1} \wedge \cdots \wedge e_{\beta_k},$$

where the right side here is the previously defined wedge product of $j + k$ vectors. So the point is whether we can "extract" the j vectors $e_{\alpha_1}, \cdots, e_{\alpha_j}$ from w_1 when we only know that $w_1 = e_{\alpha_1} \wedge \cdots \wedge e_{\alpha_j}$; that is, w_1 is some wedge product of j basis vectors in strictly increasing order.

  And the answer is yes, we can. Now that you understand what the problem is and what the solution is, prove that it is so!
- Exercise 5.4: Expand each vector v_j in the given basis and compute away.
- Exercise 5.5: The preservation of products is an immediate consequence of Exercise 5.4 in the case $k = 2$. So it only remains to show that $\Lambda^* T$ maps the identity to the identity.
- Exercise 5.6: Λ^k is a functor from the category of vector spaces to itself. Λ^* is a functor from the category of vector spaces to the category of associative, graded algebras with identity.
- Exercise 5.7: $r = \det T$.
- Exercise 5.8: For $k = 0$ and $k = n$, the dimension of $\Lambda^k V$ is 1 and so every element in that space is trivially decomposable. For $k = 1$, we have $\Lambda^k V = V$, so again every vector in it is decomposable. The case $k = n - 1$ might not be so obvious, though there seems to be some sort of duality going on here that reduces this case to the $k = 1$ case.

  For the case $n = 4$ and $k = 2$, play around with 2-forms in four variables. The example you find will also work for any $n \geq 4$ and $k = 2$. Your example must use all four variables since any expression for a 2-form in fewer variables is decomposable by the above remarks.
- Exercise 5.9: I don't think this will be used later on. So you are on your own.
- Exercise 5.10: The cases $k = 0$ and $k = 1$ are rather straightforward. So let's consider the case $k \geq 2$. Then what happens to the decomposable element $v_1 \otimes v_2 \otimes \cdots \otimes v_k$ under the interchange $v_1 \leftrightarrow v_2$? Could it be equal to $-v_1 \otimes v_2 \otimes \cdots \otimes v_k$? That is to say, can we have the following?

$$v_2 \otimes v_1 \otimes \cdots \otimes v_k = -v_1 \otimes v_2 \otimes \cdots \otimes v_k$$

Well, this does hold if some $v_j = 0$, and so $v_1 \otimes v_2 \otimes \cdots \otimes v_k = 0$.

Now show that it cannot happen if all $v_j \neq 0$.

- Exercise 5.11: Each eigenvalue has multiplicity 3.
- Exercise 5.12: Just calculate using the definitions.
- Exercise 5.13: I would rather that you think long and hard about this important result. If necessary (but only after really thinking long and hard about it), you may consult the literature.
- Exercise 5.14: More than half the battle is to realize that (5.7) says something that can be proved. And therefore needs to be proved! To prove it, just use the definition of d to calculate $d(x_k)$.
- Exercise 5.15: Use Exercise 5.14.
- Exercise 5.16: This formula holds if and only if it holds locally. So prove it in a chart. When you do that, the natural thing to do is prove the formula for the choices $X = \partial/\partial x_j$ and $Y = \partial/\partial x_k$. And then ones thinks that both sides of the formula are bilinear and so the formula holds for all X and Y. But then one begins to wonder why the third term on the right side of the formula (5.9) is there. After all, $[\partial/\partial x_j, \partial/\partial x_k] = 0$, which you used when checking (5.9). So it seems that (5.9) is true without the third term.

  But there is a slight, but essential, misunderstanding in this reasoning. If $X = \sum_j a_j \partial/\partial x_j$ and $Y = \sum_k b_k \partial/\partial x_k$, then using bilinearity over the reals, we see that (5.9) holds and that the third term is zero, *provided* that the a_j and b_k are real numbers. But general vector fields have the above form when a_j and b_k are smooth functions. So we need to check that (5.9) is bilinear over the ring $C^\infty(V)$, where $V \subset M$ is the open set of the chart. Actually, (5.9) is bilinear over $C^\infty(M)$. The upshot is that you now have a new exercise to show this more general bilinearity. And when you do that exercise, you will see why that third term on the right side has to be there.

  A recurring theme in Helgason's book [21] is that tensor fields are multilinear over the ring of C^∞-functions.

Chapter 6

The exercises in this chapter are meant to be its most difficult part. However, the underlying idea behind them is that they are (or come directly from) local properties of smooth functions, which is to say, calculus of several variables. Do not worry if you cannot do all of them during a first reading of this material. But during that first pass, one should understand the meaning and possible importance of each exercise.

- Exercise 6.1: If necessary, review the theory of the existence, uniqueness, and smoothness of solutions of ordinary differential equations. There you will find the proofs of these two theorems for the case when $M = U$, an open subset of $\mathbb{R}^n$. That is really all that is behind these two theorems. The extension to an arbitrary manifold M is a bother, but it is one of those chores in life that must

260 A Discussion of the Exercises

be attended to. Yes, this is a hard problem, but the idea behind it is simply to use the previously available theory of ordinary differential equations. The idea, of course, is to use that theory in coordinate charts and "paste" the results together.

- Exercise 6.2:
  C^∞ is a local property and the theory of ordinary differential equations gives us that σ_t is locally C^∞.
  The three properties of the family of maps σ_t are condensed into this brief phrase: σ_t is a *group of diffeomorphisms* of M. The first property follows from the initial condition imposed on γ_p, while the second property follows from the uniqueness of the integral curve with a given initial condition. The third property is a consequence of the previous two properties.
- Exercise 6.3: Write the definition of $\mathcal{L}_X$ in local coordinates.
- Exercise 6.4: Again, local coordinates do the trick. This and the previous exercise are actually solved in the subsequent text by using local coordinates. However, the equation

$$f\big(p + tY(p) + o(t)\big) = f(p) + tDf(p) \cdot Y(p) + o(t)$$

is blithely used there without justification. So you should justify it.
- Exercise 6.5: Yes, Z is given by a vector field Y. The "tricky" bits are to find that vector field Y and then to prove that $Z = Y$. For the first step, note that Taylor's theorem says that for any given point $y \in U$, we can write

$$f(x) = f(y) + \sum_{j=1}^{n} \frac{\partial f}{\partial x_j}(y)(x_j - y_j) + r(x - y),$$

where the remainder term $r(x-y)$ is $o(x-y)$. Applying the given linear operator Z to the right side, remembering that $y \in U$ is constant while $x \in U$ is the variable, gives

$$\sum_{j=1}^{n} \frac{\partial f}{\partial x_j}(y)Z(x_j) + Z(r(\cdot - y)).$$

Note that Z sends constants to zero because it satisfies Leibniz's rule. So $Y_j := Z(x_j - y_j) = Z(x_j) \in C^\infty(U)$ and $Z(f(y)) = 0$. This leads one to suspect that the vector field

$$Y := \sum_{j=1}^{n} Y_j \frac{\partial}{\partial x_j}$$

on U is what we are looking for. Of course, we already know that $Y : C^\infty(U) \to C^\infty(U)$ is a linear map that satisfies Leibniz's rule. This quickly leads to

$Zx_k = Yx_k$ for each coordinate function x_k and then to any polynomial in the coordinate functions. Is that enough to show $Zf = Yf$ for all $f \in C^\infty(U)$?

- Exercise 6.6: The point is that the action of a vector field Y on a function f does not depend on the coordinate system used. So the same is true of the operator $\mathcal{L}_Y$ acting on f.
- Exercise 6.7: It is a question of identifying the vector space in which $(T\sigma_t)^{-1}[X(\gamma_p(t))] - X(p)$ lies. By definition, $X(p) \in T_p(M)$, a finite-dimensional Euclidean space. Check that $(T\sigma_t)^{-1}[X(\gamma_p(t))]$ also lies in $T_p(M)$. [It must! Otherwise, the difference $(T\sigma_t)^{-1}[X(\gamma_p(t))] - X(p)$ makes no sense.] To study the limit, one does what one often does: Look at the expression in local coordinates.
- Exercise 6.8: We know that $\mathcal{L}_Y X(p) \in T_p(M)$ by Exercise 6.7. It remains to show that $p \mapsto \mathcal{L}_Y X(p)$ is a smooth map $M \to T(M)$. How? Local coordinates.
- Exercise 6.9: The point is that σ_t always makes sense for t in some small neighborhood of 0.
- Exercise 6.10: First off, what could go wrong with a vector field on an open subset $U \subset \mathbb{R}^n$? In that case, an integral curve could arrive in finite time at the boundary of U and then there would be no way to extend it. Or, if U is unbounded, it could go to infinity in finite time, and then there is no way to extend it to later times. But these pathologies cannot occur in a compact manifold M. Why not? If the domain of the integral curve is (a, b) for some $b \in \mathbb{R}$, then let $c_n \in (a, b)$ be any sequence with $c_n \to b$. Then the images of this sequence must have a converging subsequence in M by compactness. And this allows one to extend the integral curve at least a little bit beyond b by solving the ODE with new initial condition at $x = b$. And that is a contradiction. So we must have $b = +\infty$. Similarly, $a = -\infty$. That's the basic idea, but without all of the details.
- Exercise 6.11: Straightforward calculations.
- Exercise 6.12: The definition of $\mathcal{L}_Y \omega$ is the tricky bit even though the idea is simple enough. One wants to define $\mathcal{L}_Y \omega(p)$ by forming the difference of two vectors in $T_p^*(M)$, dividing by t and taking the limit as $t \to 0$. This is in analogy with the definition of $\mathcal{L}_Y X(p)$. Then the identity becomes an easy calculation using the definitions. This identity is a sort of acid test for verifying that you defined $\mathcal{L}_Y \omega(p)$ correctly, because if that is not defined correctly, there is no way you will be able to prove the identity!

By the way, the reader should realize that this identity is a form of Leibniz's rule.

Chapter 7

- Exercise 7.1: This problem lists some of the basic properties that follow immediately from the definitions.
- Exercise 7.2: This boils down to showing two things: (a) $T(L_e) = id_{T(G)}$ and (b) $T(L_{gh}) = T(L_g)T(L_h)$.

- Exercise 7.3: A rather easy argument.
- Exercise 7.4: Prove that $(T(L_g))^{-1} = T(L_{g^{-1}})$. The rest of the problem is quite immediate.
- Exercise 7.5: This problem looks easier than it actually is. While $T(L_g)$ acts on vector fields, such as X, Y, and $[X, Y]$, it has not been defined so far on expressions such as XY and YX, which in general are second-order differential operators.
- Exercise 7.6: Immediate from definitions.
- Exercise 7.7: The verifications are rather straightforward, though the Jacobi identity may give you pause. There is a triple-cross product identity that helps out. Or just do all the gory details.
  Play around with the standard basis of $\mathbb{R}^3$ to see how associativity fails.
- Exercise 7.8: We have to show that the multiplication map and the inverse map are smooth functions for the group $GL(n, \mathbb{R})$. Now the differential structure on $GL(n, \mathbb{R})$ comes from its being identified with an open subset of the Euclidean space $\mathbb{R}^{n^2}$. So the coordinate functions $A \mapsto A_{ij}$ are smooth on $GL(n, \mathbb{R})$, where A_{ij} is the (i, j) matrix entry of A. Here $1 \leq i, j \leq n$.
  For $A, B \in GL(n, \mathbb{R})$, we note that the matrix product AB has entries that are bilinear, and hence smooth, in the coordinates of A and B. For A^{-1}, we note that its matrix entries are given by Cramer's rule, which explicitly exhibits them as smooth functions of the matrix entries of A.
- Exercise 7.9: I will let you play with this without any help from me. If you get stuck, see any standard text on Lie groups.
- Exercise 7.10: Suppose that $b < +\infty$ and argue by contradiction. Let $g = \phi(b/2)$. Show that $\psi(t) := g\,\phi(t)$ defines an integral curve of X for $t \in (a, b)$ by using a calculation similar to (7.1). But $\psi(0) = g$. So $\psi(t) = \phi(t + b/2)$ for $t \in (a, b/2)$ by the uniqueness of integral curves passing through g. Pasting ψ onto the "right end" of ϕ gives an integral curve of X passing through e with domain $(a, 3b/2)$. This contradicts the maximality of b. So we must have $b = +\infty$. Similarly, one shows that $a = -\infty$.
- Exercise 7.11: This is shown using the chain rule.
- Exercise 7.12: This is a standard exercise in analysis. We actually show that the series converges absolutely, which by a standard result shows that the series itself converges. To converge absolutely means that the corresponding series with norms on each term converges as a series of (nonnegative) real numbers; that is,

$$\sum_{j=0}^{\infty} ||\frac{1}{j!} X^j||_{op} \leq 1 + ||X||_{op} + \frac{1}{2!}||X||_{op}^2 + \frac{1}{3!}||X||_{op}^3 + \cdots$$

$$= \exp(||X||_{op}) < \infty.$$

Here, for the sake of convenience, we used the operator norm $|| \cdot ||_{op}$ since $||M^k||_{op} \leq ||M||_{op}^k$ holds for every matrix M and every integer $k \geq 0$.

• Exercise 7.13: This is the derivative of a power series. Another standard result in analysis says that a power series defines a C^∞-function within its (open) disk of convergence and that its first derivative is given by differentiating the original power series term by term.
To prove that $\exp(tX)$ is a one-parameter subgroup, you need to know how to multiply power series.

Chapter 8

• Exercise 8.1: No comments.
• Exercise 8.2: That Σ is a subgroup and that it is a closed subset of $GL(n)$ are easy enough.
You may wish to write $T \in \Sigma$ in block matrix form with respect to a basis of $\mathbb{R}^n$ that contains a subset that is a basis of W_0. (Such bases of $\mathbb{R}^n$ do exist. Can you prove that?) The number of nonzero elements in such a block matrix will be the dimension of Σ.
• Exercise 8.3: $\dim G(n,k) = \dim GL(n) - \dim \Sigma = n^2 - \dim \Sigma$.
• Exercise 8.4: No comments.
• Exercise 8.5: Put the standard Euclidean inner product and norm on the vector space $\mathbb{R}^n$. Then find the stabilizer of the action of $O(n)$ on $G(n,k)$ in terms of block matrices.
Specifically, the group $O(k) \times O(n-k)$ is realized as a closed subgroup of $O(n)$ by considering it as the $n \times n$ matrices in block form with two nonzero blocks along the diagonal, the first block being a $k \times k$ orthogonal matrix and the second block being an $(n-k) \times (n-k)$ orthogonal matrix.
The dimension calculation now is

$$\dim G(n,k) = \dim O(n) - \big(\dim O(k) + \dim O(n-k)\big).$$

The conclusion that $G(n,k)$ is compact is now trivial since it is being exhibited as a quotient of the compact space $O(n)$. Before when it was seen as the quotient of the noncompact space $GL(n)$, this was not obvious at all.

Chapter 9

• Exercise 9.1: Yes, this is a hard problem. But the idea behind the proof should by now be a familiar technique for the reader. One is given a "kit" plus the "instructions" for putting the kit together. And this is the essence of the proof of Theorem 9.1. The point is that it is a good exercise for you to take the idea and implement it as a detailed proof.

- Exercise 9.2: Yes, they are actually charts according to the definition. But no, it does not really matter. The point is that they are standard model spaces for describing certain spaces locally. The fact that the model spaces are open subsets in a Euclidean space is not so important. Just as long as the model spaces are well-known objects, we can easily use them to describe the assembled "kit space" in familiar terms.
- Exercise 9.3: This is another example of the "kit plus instructions" technique. Here are a few more comments:
Replace the model fiber space G in the construction of a principal bundle by F.
When the space F has some extra structure (say, it is a ring), then it comes down to showing that the natural definition of that structure on the fibers of the bundle in terms of a given local trivialization does not depend on the choice of local trivialization.
Also think of the special case, which we have seen with vector bundles, where $G = GL(n, \mathbb{R})$ acts on $F = \mathbb{R}^n$ from the left by the standard ("canonical") action. Here the extra structure on the fiber is that of a vector space over $\mathbb{R}$.
Another special case was given in Chapter 8 in the construction of the Grassmannian k-planes $G(TM, k)$ over M.
These two special cases should give you the pattern for doing this more general problem where the fiber is modeled on an abstract space F.
- Exercise 9.4:

    Part (a) is more or less immediate from definitions.
    Part (b) is decidedly trickier. Yes, one can use a local trivialization to put a group structure on fibers. But no, this structure depends on the local trivialization and so is not a part of the intrinsic structure of a principal bundle. Part (c) also follows from definitions. We will come back to this fact in the next exercise.

- Exercise 9.5: The element g in part (c) of Exercise 9.4 is given its own notation and consequently realized as being a particular case of a general concept: an affine operation. The proofs of parts (a)–(f) are algebraic in nature and rather easy. The technique for the proof that the affine operation is smooth is hardly a surprise; one proves that it is locally smooth using local coordinates.
- Exercise 9.6: In general, the right multiplication does not preserve the affine operation.
- Exercise 9.7: It is the trivial principal bundle $\pi_1 : M \times G \to M$, where π_1 is the projection onto the first factor. Of course, you should check that the given definition for $g_{\beta\alpha}$ does actually define a cocycle. Then it is a question of constructing the associated principal bundle.
- Exercise 9.8:
For Aut(A), the product is the composition of morphisms ("arrows") in the category and the identity element is the identity morphism, usually denoted as 1_A. The objects in the category of all principal bundles with structure Lie group G is obvious enough. However, the definition of the set of "arrows" or "morphisms" between two such objects could be less than obvious, at least to some

readers. If E_1 and E_2 are the total spaces for two principal bundles over the manifolds M_1 and M_2, respectively, but with the same structure Lie group G, then one reasonable definition for a morphism would be a G-invariant map $F : E_1 \rightarrow E_2$. Then F induces a smooth map $f : M_1 \rightarrow M_2$, making the obvious diagram commute. The product of two such G-invariant maps is obvious enough. Specifically, if $F : E_1 \rightarrow E_2$ and $H : E_2 \rightarrow E_3$ are morphisms of principal bundles, all of which have the same structure group G, then their composition is the usual composition of maps $H \circ F : E_1 \rightarrow E_3$.

Showing that every principal bundle is locally trivial is a question of understanding well all of the definitions involved.

- Exercise 9.9: The commutativity of the diagrams follows directly from the definitions of the various arrows in them.

  $\Psi(s)$ is smooth if and only if its projections onto the two factors of $M \times G$ are smooth.

  $\Psi(s)$ is easily checked to be G-invariant. It is a trivialization because it has an inverse map $M \times G \rightarrow P$.

- Exercise 9.10: Just use all the definitions.

- Exercise 9.11: A *sequence* in a category is a linear diagram of objects and arrows, such as

$$\cdots \rightarrow A_{-2} \rightarrow A_{-1} \rightarrow A_0 \rightarrow A_1 \rightarrow \cdots . \qquad (A.1)$$

A sequence can have either a finite or an infinite number of objects, denoted here by A_j. The index set for j is any interval of integers, whether finite or infinite. Furthermore, suppose that each arrow f in the category has a kernel object, denoted by $\mathrm{Ker}\, f$, and an image object, denoted by $\mathrm{Im}\, f$. This is certainly true for the category of vector spaces with arrows being linear maps. Then any such sequence (A.1) is said to be *exact* if, for every object A_j in the sequence that is both the domain of one arrow and the codomain of another arrow, say

$$\cdots \rightarrow A_{j-1} \xrightarrow{f_{j-1}} A_j \xrightarrow{f_j} A_{j+1} \rightarrow \cdots ,$$

satisfies $\mathrm{Ker}\, f_j = \mathrm{Im}\, f_{j-1}$.

In a category with a zero object, denoted 0, a *short sequence* is a sequence with five objects, the first and last of which are the zero object. So a short sequence in the category of vector spaces has the form

$$0 \rightarrow V_1 \xrightarrow{i} V_2 \xrightarrow{j} V_3 \rightarrow 0,$$

where V_1, V_2, and V_3 are vector spaces and 0 denotes the zero vector space. This sequence is also exact if the kernel of the outgoing arrow is equal to the image of the incoming arrow at each of the objects V_1, V_2, and V_3. In the notation for the above short sequence, this means that i is a monomorphism (one-to-one), j is an epimorphism (onto), and $\mathrm{Ker}\, j = \mathrm{Im}\, i$.

In practice, it is often fairly easy to show that i is a monomorphism, that j is an epimorphism, and that the composition $ji = 0$, which is equivalent to $\operatorname{Im} i \subset \operatorname{Ker} j$. But often to prove the opposite inclusion $\operatorname{Ker} j \subset \operatorname{Im} i$ requires much more work.

- Exercise 9.12: The smoothness of $A^\sharp : P \to TP$ is proved locally.
  The linearity of $A \to A^\sharp$ is not that difficult, but proving that this is a Lie algebra morphism requires that one dig deeper into the definitions.

Chapter 10

- Exercise 10.1: What is the kernel of $T\pi$? See Exercise 9.11.
- Exercise 10.2: Most of this follows quickly enough from definitions. For example,

$$\operatorname{Ker} \omega_p = \operatorname{Ker} (T_e \iota_p)^{-1} \circ \operatorname{pr}_p = \operatorname{Ker} \operatorname{pr}_p$$

since $(T_e \iota_p)^{-1}$ is an isomorphism. Finally, $\operatorname{Ker} \operatorname{pr}_p = \operatorname{Hor}(T_p P)$ by the definition of the projection.

The smoothness of ω is proved, as expected, locally.

The definition of ω is that it is the linear map ω_p on the fiber $T_p P$.

- Exercise 10.3: The map $\zeta : TN \to \mathbb{R}$ determines the map $N \to T^*N$ that sends $x \in N$ to the linear map $\zeta_x : T_x N \to \mathbb{R}$, that is, $\zeta_x \in T_x^* N$. So $\sigma : x \mapsto \zeta_x$ is the section of T^*N corresponding to ζ.
  Conversely, given a section $\sigma : N \to T^*N$, we have that $\sigma_x \in T_x^* N$ for all $x \in N$. But this means that $\sigma_x : T_x N \to \mathbb{R}$ linearly. Then we define $\zeta : TN \to \mathbb{R}$ by letting its restriction to each fiber $T_x N$ be given by σ_x.
  So far, this is all algebraic. One also has to show that ζ is smooth if and only if σ is smooth.
- Exercise 10.4: Since $\Omega^1(N; W)$ consists of maps into the vector space W, one uses the sum and scalar multiplication in W to define the sum and scalar multiplication for the maps in $\Omega^1(N; W)$. The only subtle bit (but not too subtle) is that one has to show closure of these newly defined operations; that is, the resulting map actually lies in $\Omega^1(N; W)$.
  If either N or W is a one-point set, there is not much inside $\Omega^1(N; W)$. Otherwise, look at what happens locally in N to study the dimension of $\Omega^1(N; W)$.
- Exercise 10.5: It is illegal in a mathematically sophisticated nation to call something in mathematics a pullback if it is not a contravariant functor. It is a question of defining pullback as a composition on the right. Here we have defined $f^*(\zeta) = \zeta \circ f_*$. So there is no suspense. This will be a contravariant functor. It really has nothing to do with the particular structures in play in this problem.

The only particular property here is that f^* is linear, something not too difficult to prove.

- Exercise 10.6: Hint: Construct the inverse map.
- Exercise 10.7: The diagrams are proved to be commutative as usual by chasing an arbitrary element through them. Then it all comes down to the definitions of the arrows.
- Exercise 10.8: No comment.
- Exercise 10.9: First, show that $g^{-1}dg$ is the Maurer–Cartan form of $GL(n)$. Then use the first part of Exercise 10.7 for the other groups.
- Exercise 10.10: This is a straightforward application of linear algebra techniques.

Chapter 11

- Exercise 11.1: This follows from the fact that $d\omega$ is a $\mathfrak{g}$-valued 2-form.
- Exercise 11.2: Take a chart and define a horizontal vector field X there with $X(p) = w$. Then multiply X by any compactly supported smooth approximate characteristic function defined in the local coordinates of the chart. Then extend by zero to the rest of M to define W. This construction shows that W is not unique.

  For the second statement, take the derivative of the right side of (11.3).

Chapter 12

- Exercise 12.1: Nothing fancy here. Just show that gauge equivalence is a reflexive, symmetric, and transitive relation.
- Exercise 12.2: No further comments.
- Exercise 12.3: A straightforward check.
- Exercise 12.4: You just have to check that the two transformations do the same thing to the Schrödinger equation.
- Exercise 12.5: No further comments are needed.

Chapter 13

- Exercise 13.1: This depends on your particular needs.
- Exercise 13.2: This is the generalization of Exercise 12.1 to this context. But the proof is a bit trickier since more identities have to be used. Maybe the proof of symmetry will be the trickiest part for you.
- Exercise 13.3: These elementary results have elementary proofs. Just use the definitions.

- Exercise 13.4: This is a slight modification of Part 2 of Exercise 13.3. The hypothesis here is a bit different and so is the conclusion.
- Exercise 13.5: In this context, "formal" means take the derivative of $e^{-iHt/\hbar}$ with respect to t as if H were a real number. Continuing to pretend that H is a real number, one "sees" that the operators H and $e^{-iHt/\hbar}$ commute.

Chapter 14

- Exercise 14.1: This is a quick interlude because the proof can be written on one line.
- Exercise 14.2: Either use Exercise 14.1 or prove it in its own one-line proof.
- Exercise 14.3: As promised, no comments.
- Exercise 14.4: Use the definitions of $A^{\sharp P}$ and of $A^{\sharp G}$.
- Exercise 14.5: Work on this now to see if you can understand what is going on here. That is the best strategy. But if you can't figure it out, don't worry. The continuation of the text will make this point amply clear.
- Exercise 14.6: Reflect away! If you cannot recollect this fact, consult a (good!) introductory text on differential equations.
- Exercise 14.7: We have enough information to apply the "kit-plus-instructions" method. The notation here is also different from that used in Theorem 9.1.
- Exercises 14.8 and 14.9: These are straightforward verifications using the definitions.

Chapter 15

- Exercise 15.1: This is exactly what happens in the more well-known theory of electrostatics. Since the equations for E in electrostatics are the same as the equations for B in magnetostatics, the same arguments apply.
- Exercise 15.2: This is simply a matter of writing down the 3×3 matrix Ψ_* and then calculating its determinant.
- Exercise 15.3: Use the chain rule to write the vector fields

$$\frac{\partial}{\partial r}, \qquad \frac{\partial}{\partial \theta}, \qquad \frac{\partial}{\partial \phi} \tag{A.2}$$

in terms of the vector fields

$$\frac{\partial}{\partial x}, \qquad \frac{\partial}{\partial y}, \qquad \frac{\partial}{\partial z}.$$

Using those identities and the fact that, by definition, these last three vector fields form an orthonormal basis in the standard metric on $\mathbb{R}^3$, one can then calculate the lengths of the vector fields in (A.2).

As an extra exercise, one can calculate the inner product among all pairs of vector fields in (A.2). And this calculation should help you understand why one says that (A.2) is an *orthogonal basis*, but not an orthonormal basis.

- Exercise 15.4: The first part is a straightforward computation. The only delicate part is establishing the domain on which that computation makes sense.
  For the second part, it suffices to show that A^N is differentiable on V_N. Why? But you already showed that when you proved the first part, right?
- Exercise 15.5: So it is a question of finding a point in V_N that can be connected to every other point in V_N by a line segment lying entirely in V_N. Such a point (and there are many) can be found by drawing a picture of V_N, or merely a two-dimensional slice of V_N, such as the plane containing both the x-axis and the z-axis.
  Generically, for a pair of points $p_1, p_2 \in V_N$, the straight-line segment between them in $\mathbb{R}^3$ will lie in V_N, that is, will not intersect the removed semiaxis Z^-. But for some pairs that line segment does intersect the removed semiaxis Z^-. Consequently, V_N in not convex. Again, drawing some sketches may help you see that this is true.
- Exercise 15.6: Use Stokes' theorem.
- Exercise 15.7: This exercise is a combination of Exercises 15.4 and 15.5 though now for V_S instead of V_N.

Chapter 16

- Exercise 16.1: This is a check of your understanding of matrix algebra for matrices whose entries come from a noncommutative ring. We assume that you have studied the case where the matrix entries come from a field. The results in this problem are straightforward analogs of the field case. So are their proofs.
- Exercise 16.2: Obvious is in the mind of the beholder.
- Exercise 16.3: Grind away using the definitions.
- Exercise 16.4: Show that it is SD.

Bibliography

1. I. Agricola and T. Friedrich, Global Analysis, Am. Math. Soc., Providence, 2002.
2. V.I. Arnold, V.V. Kozlov, and A.I. Neishtadt, Mathematical Aspects of Classical and Celestial Mechanics, 2nd edition, Springer, 1997.
3. A.A. Belavin, A.M. Polyakov, A.S. Schwartz, and Y.S. Tyupkin, Pseudoparticle solutions of the Yang–Mills equations, Phys. Lett. **59B**, no. 1, (1975) 85–87.
4. Y. Choquet-Bruhat, C. DeWitt-Morette, and M. Dillard-Bleick, Analysis, Manifolds and Physics, North-Holland Pub. Co., revised edition, 1982.
5. R. Courant and H. Robbins, What Is Mathematics?, Oxford University Press, 1941.
6. R.W.R. Darling, Differential Forms and Connections, Cambridge University Press, 1994.
7. E.B. Davies, Spectral Theory and Differential Operators, Cambridge University Press, 1995.
8. J. Dieudonné, Foundations of Modern Analysis, Academic Press, 1960.
9. P.A.M. Dirac, Quantised Singularities in the Electromagnetic Field, Proc. R. Soc. Lond. A **133** (1931) 60–72. doi: 10.1098/rspa.1931.0130
10. W. Drechsler and M.E. Mayer, Fiber Bundle Techniques in Gauge Theory, Lecture Notes in Physics, Vol. **67**, Springer, 1977.
11. C. Ehresmann, Les connexiones infinitésimales dans un espace fibré différentiable, in: Colloque de topologie, Bruxelles (1950), pp. 29–55, Masson, Paris, 1951.
12. L. Faddeev, Advent of the Yang–Mills field, in: Highlights of Mathematical Physics, Eds. A. Fokas et al., Am. Math. Soc., 2002.
13. R.P. Feynman, Quantum theory of gravitation, Acta Phys. Polonica **24** (1963) 697–722.
14. R.P. Feynman, Feynman's office: The last blackboards, Phys. Today, **42**, no. 2, (1989) 88.
15. H. Flanders, Differential Forms with Applications to the Physical Sciences, Academic Press, 1963. (Reprinted by Dover, 1989.)
16. T. Frankel, The Geometry of Physics, 2nd edition, Cambridge, 2003.
17. W. Fulton and J. Harris, Representation Theory: A First Course, Graduate Texts in Mathematics, vol. **129**, Springer, 1991.
18. I.M. Gelfand and S.V. Fomin, Calculus of Variations, Prentice-Hall, 1963.
19. S.J. Gustafson and I.M. Sigal, Mathematical Concepts of Quantum Mechanics, Springer, 2003.
20. B.C. Hall, Lie Groups, Lie Algebras, and Representations, Springer, 2003.
21. S. Helgason, Differential Geometry and Symmetric Spaces, Academic Press, 1962.
22. P. Higgs, Spontaneous symmetry breakdown without massless bosons, Phys. Rev. **145** (1966) 1156–1163.
23. F. Hirzebruch, Topological Methods in Algebraic Geometry, Springer, 1978 (reprinted 1995).
24. G. 't Hooft, Gauge theories of the forces between elementary particles, Sci. Am. **243** (1980) 104–138.
25. G. 't Hooft editor, 50 Years of Yang–Mills Theory, World Scientific, 2005.

© Springer International Publishing Switzerland 2015
S.B. Sontz, *Principal Bundles*, Universitext, DOI 10.1007/978-3-319-14765-9

26. L. Hörmander, The Analysis of Linear Partial Operators I, Springer-Verlag, 1983.
27. D. Husemoller, Fibre Bundles, Graduate Texts in Mathematics, Vol. 20, Springer.
28. J.D. Jackson, Classical Electrodynamics, John Wiley & Sons, 2nd edition, 1975.
29. M. Kervaire, A manifold which does not admit any differentiable structure, Commun. Math. Helv. **34** (1960) 304–312.
30. S. Kobayashi and K. Nozimu, Foundations of Differential Geometry, John Wiley & Sons, Vol. I, 1963 and Vol. II 1969.
31. S. Lang, Algebra, Addison-Wesley, 1965.
32. S. Lang, Fundamentals of Differential Geometry, Graduate Texts in Mathematics, Vol. **191**, Springer, 1999.
33. J. Lee, Introduction to Smooth Manifolds, Graduate Texts in Mathematics, Vol. **218**, Springer, 2003.
34. E. Lubkin, Ann. Phys. (N.Y.) **23** (1963) 233.
35. S. Mac Lane, Categories for the Working Mathematician, Springer, 1971.
36. W. Miller, Jr., Symmetry Groups and Their Applications, Academic Press, 1972.
37. J. Milnor, On manifolds homeomorphic to the 7-sphere, Ann. Math. **64** (1956) 399–405.
38. G.L. Naber, Topology, Geometry, and Gauge Fields: Foundations, Springer-Verlag, 1997.
39. G.L. Naber, Topology, Geometry, and Gauge Fields: Interactions, Springer-Verlag, 2000.
40. M. Nakahara, Geometry, Topology and Physics, 2nd edition, Institute of Physics Publishing, 2003.
41. L. O'Raifeartaigh, The Dawning of Gauge Theory, Princeton University Press, 1997.
42. R. Penrose, The Road to Reality, Knopf, 2005.
43. S. Smale, Generalized Poincaré's conjecture in dimensions greater than four, Ann. Math. **74** (1961) 391–406.
44. S.B. Sontz, Principal Bundles: The Quantum Case, Springer, 2015.
45. M. Spivak, Calculus on Manifolds, W.A. Benjamin, 1965.
46. M. Spivak, A Comprehensive Introduction to Differential Geometry, Vol. 2, 3rd edition, Publish or Perish, 1999.
47. R.S. Strichartz, The Way of Analysis, Jones and Bartlett, 2000.
48. R. Utiyama, Invariant theoretical interpretation of interaction, Phys. Rev. **101** (1956) 1597–1607.
49. V.S. Varadarajan, Lie Groups, Lie Algebras, and Their Representations, Graduate Texts in Mathematics, Vol. **102**, Springer, 1984.
50. S. Weinberg, The Quantum Theory of Fields, Vol. II, Modern Applications, Cambridge University Press, 1996.
51. H. Weyl, Gravitation und Elektrizität, Sitz. König. Preuss. Akad. Wiss. **26** (1918) 465–480.
52. T.T. Wu and C.N. Yang, Concept of non-integrable phase factors and global formulation of gauge fields, Phys. Rev. D **12** (1975) 3845.
53. T.T. Wu and C.N. Yang, Dirac monopole without strings: monopole harmonics, Nucl. Phys. B **107** (1976) 365–380.
54. C.N. Yang and R.L. Mills, Conservation of isotopic spin and isotopic gauge invariance, Phys. Rev. **96** (1954) 191–195.
55. D.Z. Zhang, C.N. Yang and contemporary mathematics, Math. Intelligencer **15**, no. 4, (1993) 13–21.

Index

A

Abstract nonsense, 41, 116
Action (associated to a Lagrangian density), 187
Action, free, 94
Action, left, 64, 93–95, 98, 111, 114, 128, 222
Action, right, 93, 94, 112, 123, 128, 129, 137, 219, 235, 238
Action, transitive, 107
Action functional, 199
Action integral, 187–189, 191, 220
Action principle, least, 187, 218
Adjoint map, 94
Adjoint representation, 94, 223
Affine operation, 2, 114, 115, 117, 210
Alternating k-vector, 65
Ampère, André–Marie, 156
Ampère–Maxwell law, 154, 158, 164, 165
Ampère's law, 152, 154
Angular momentum, 174–176, 187
Angular momentum quantum number, orbital, 175
Anti-CR-equations, 221
Anti-hermitian, 102, 196, 197, 201
Anti-holomorphic function, 221
Anti-self-dual (ASD), 70, 220, 221
Anti-symmetric, 59, 60, 64, 65, 71, 96, 150, 184, 201, 203
Anti-symmetric tensor, 64, 65
Arrow, 11, 14–17, 35, 113, 117, 118, 123, 126–128, 132, 206
ASD. See Anti-self-dual (ASD)
Associated bundle, 36
Atlas, 7–10, 12, 17, 18, 20, 25, 32–35, 107, 111, 113, 114, 133, 202, 242

B

Automorphism, 116, 212, 219, 220
Automorphism group, 212

Base space, 15, 21, 23, 31, 35, 74, 113, 121, 135, 137, 205, 214, 217, 218, 222, 231, 237, 239, 241
Bianchi's identity, 149–150, 218, 221
Bilinear, 40, 42–45, 47, 59–61, 63, 67, 68, 73, 96, 140, 141, 203
BPST instanton, 233, 237, 241–246
b quanta, 192
Broken symmetry, 174
Bundel, fiber, 21, 114, 225
Bundle, principal, 2, 4, 5, 21, 111–119, 121–146, 149, 150, 195, 204, 205, 207, 209–220, 222, 230–232, 235, 237, 247
Bundle, vector, 4, 21, 24, 31–48, 71, 72, 81, 83, 108, 111, 113, 116, 117, 138–141, 222
Bundle map, 15, 21, 35, 41, 113, 145, 211
Bundle of frames, 116

C

C^{∞}, 8, 12, 14, 26, 89
Calculus of several real variables, 49
Calculus of variations, 187, 189, 199
Cartan, Élie, 195
Cartan structure equation, 146, 217
Categorical imperative, 2, 35
Category theory, 2, 3, 10, 11, 24, 41, 116, 132, 211, 212
C^{∞}-atlas, 9

© Springer International Publishing Switzerland 2015
S.B. Sontz, *Principal Bundles*, Universitext, DOI 10.1007/978-3-319-14765-9

Cauchy problem, 167, 178
Cauchy–Riemann equations; CR equations, 221
C^∞-function, 4, 11, 14, 23, 26–29, 49, 50, 73, 78, 140, 147, 179
Change of coordinates; change of variables, 56
Characteristic class, 200
Charge, electric, 85, 86, 151–155, 170, 171, 178, 189, 223, 225, 226
Charge, magnetic, 86, 151, 154, 155, 158, 160, 225, 226, 231, 232
Charge, topological, 225, 232
Chart, 1, 7–12, 16, 19–25, 32–35, 38, 39, 49, 73–75, 82, 111, 113, 114, 133, 135, 137, 139, 198, 199, 201, 202, 204, 217, 220, 241, 243
Circular definition, 22
C^k, 8, 9, 14, 26
C^k-atlas, 8, 9
Closed form, 70
Cochain complex, 75
Cocycle, 2, 33–39, 107, 111, 114, 116, 139, 213, 214, 216, 227, 231
Cocycle condition, 33
Codimension, 25, 26
Color, 192
Commutative diagram, 14, 17, 18, 21, 23, 117, 123, 131, 213
Commutator, 91, 95, 101, 185, 186, 191, 196, 197, 203
Compact symplectic group, 235–238
Compatible, 9, 32, 189
Complementary subspace, 122, 124, 129, 130
Complete vector field, 91
Completion (of a pre-Hilbert space), 222
Configuration space, 87, 189, 222
Conjugate (of a quaternion), 234
Connection, 1, 2, 4, 5, 81, 121–150, 195, 196, 204–211, 214–219, 222, 223, 225, 231, 237–241
Conservative electric force, 151
Conserved quantity, locally, 182
Continuity equation, 153, 154
Contravariant vector, 37, 38
Coordinates, 1, 7, 8, 13, 20, 49–53, 56, 67, 73–76, 82, 87–89, 133, 139, 159, 165, 168, 177, 187, 193, 198–203, 205, 220, 227, 228, 237, 239, 241, 243, 246
Coordinates, generalized, 187
Cotangent bundle, 3, 38, 40, 56, 72, 74, 125
Coulomb's law, 154, 158, 162, 166
Covariant, 24, 38, 126, 139–143, 145, 150, 162, 181, 185, 196, 217, 223

Covariant derivative, 139, 141–143, 145, 150, 185, 217, 223
Covariant derivative, exterior, 142
Covariant vector, 38
Covector, 38, 49–57, 124, 143–144
Covering space, 174, 175, 182
Covering space, universal, 174, 175, 182
Covers, covering, 41, 108, 112, 114, 115, 128, 136, 174, 175, 182, 212, 215, 219, 230, 231
Cramer's rule, 32
Cross product, 70, 86
Curl, 78–79, 157
Curvature, 5, 145–150, 196, 204, 205, 217–219

D

Decomposable element, 46, 63–65, 67
Decomposable tensor, 45, 46
Density 1-form, 162
de Rham cohomology, 76
de Rham theory, 3, 73
Derivative
    covariant, 139, 141–143, 145, 150, 185, 217, 223
    variational, 187, 188
Deterministic, 167
Diagram, 1, 11, 14–18, 21, 23, 35, 41, 43, 81, 82, 112, 113, 115, 117, 118, 122–124, 126–128, 131, 132, 189, 190, 206, 213, 238–240
Diagram chase, 15, 127
Diffeomorphic, 12, 18, 35, 108, 114, 133, 134
Diffeomorphism, 12, 18, 49, 84, 90, 93, 94, 101, 107, 108, 112, 122, 219
Differentiable, 13, 24, 73, 161, 198
Differential manifolds, 9–12, 18–25, 32, 33, 36, 40, 56, 76, 91, 201, 218
Differential structure, 9, 10, 12, 18, 22, 31, 32, 56, 96, 97, 242
Dimension/dim, 9–11, 13, 16, 22, 38, 41, 46, 49, 56, 59, 63, 65, 66, 86, 87, 106–108, 116, 144, 159, 174–176, 182, 197, 220, 233
Dirac delta, 226
Dirac magnetic monopole; Dirac monopole, 216, 225–232
Dirac matter field, 191
Dirac, Paul, 176
Dirac string, 230, 231
Directional derivative, 87, 89, 140, 243
Disjoint union, 17, 21, 22, 34, 111
Displacement current, 87, 154, 186

Distribution, 2, 4, 105–110, 122, 146, 226
Distribution of k-planes, 106, 107, 109, 110
Divergence; div, 79, 153, 158, 226
Divergence theorem, 153
Dual basis, 55
Dual connection, 143–144

E

Ehresmann connection; connection form;
 gauge fields, 122
Eigenvalue, 70, 174, 197
Electric charge, 85, 86, 151–155, 170, 171,
 178, 189, 223, 225, 226
Electric charge density, 151
Electric field lines, 85, 86
Electric fields, 85, 86, 151, 152, 154, 157, 160,
 163, 172, 225, 226
Electric force, 85, 86, 151
Electric potential; voltage, 171
Electromagnetic field, 87, 151, 155, 162, 163,
 171, 172, 185–187, 189
Electromagnetism, 151–172, 177, 178, 180,
 186, 189, 195, 218, 219, 231
Electroweak theory, 192, 193
Equivalence of differential structures, 10
Equivalence of vector bundles, 41
Essentially self-adjoint, 171
Euclidean norm, 10, 100, 133, 235
Euler–Lagrange equations (E–L equations),
 187–189, 199
Exact form, 57
Exact sequence, 119
Explicitly trivial bundle, 41, 209, 210
Exponential map, 100–101, 197
Exterior algebra, 59–79, 140
Exterior covariant derivative, 142
Exterior derivative, 73–76, 78, 79, 142,
 160–162, 202, 215, 217, 237
Exterior differential calculus, 73–76
Exterior power, 60, 64
Exterior product, 65

F

Factorization, 43, 64, 176
Faraday, Michael, 86
Faraday's law, 86, 152, 157, 160
Fermion, 222
Feynman diagram, 189, 190
Feynman path integral, 189
Fiber
 bundle, 21, 114, 225
 product, 2, 40, 71, 114

Field strength, 159, 162, 163, 182–186, 188,
 199, 204, 205, 217–220
First order contact, 55
FitzGerald, George Francis, 165
Flow, 4, 84, 90, 91, 148
Force as a 1-form, 156
Force field, 85, 226
Formal approach, 45
4-vector, 159, 162, 172
Free action, 94
Frobenius integrability condition, 109, 110
Frobenius theorem; Frobenius integrability
 theorem, 4, 81, 105–110, 122, 146
Functional analysis, 171, 178, 179
Functor
 contravariant, 73, 126
 covariant, 24, 71, 72, 126
Functorial, 16, 36, 37, 66, 82, 126
Fundamental vector field, 119, 132, 148

G

Galilean group, 156
Gauge equivalent, 163, 181
Gauge field, local/trivialization, 218, 231
Gauge field; potential, 2, 5, 137, 163, 164, 166,
 169, 171, 172, 179–182, 186, 188, 195,
 198–223, 227–231, 241
Gauge group, 220
Gauge invariance, 179, 183, 186, 187, 192,
 195
Gauge theory
 lattice, 247
 non-abelian, 195
 supersymmetric, 247
Gauge transformation
 global, 177, 178
 local, 172, 177–182, 185, 186, 188, 198,
 212–214, 216, 219, 220, 222, 231
Gaussian units, 151
Generalized coordinates, 187
General linear group, 36, 96, 235
General relativity (GR), 193, 195
G-equivariance condition, 127, 128, 144
G-equivariant map; G-morphism, 94, 112, 113,
 118, 129, 137
Global gauge transformation, 177, 178
Global section, 117
Global solution, 136
Gluon, 192
Graded algebra, 61–63, 73
Graded commutativity, 62, 63
Graded Leibniz rule, 75
Graded vector space, 61, 62, 73

Gradient/grad, 52–54, 56, 77, 79
Grassmannian, 106, 108
Grassmannian bundle of k-planes, 108
Group action
  left, 94
  right, 94
Group of diffeomorphisms, 122
Group, structure, 2, 113–117, 215, 218, 220, 231, 235

H
Hausdorff, 7, 9, 12, 25, 107
Heaviside, Oliver, 155
Heisenberg, Werner, 173, 174, 176, 177
Hermitian matrix; self-adjoint matrix, 179, 181, 182, 191, 197
Higgs boson, 192
Higgs fields, 223
Higgs, Peter, 192
Hilbert space, 174, 176, 177, 190, 198, 222
$\mathbb{H}$-linear, 235, 236
Hodge star, 65–71, 78, 79, 158, 160–162, 220, 227, 228
Holomorphic function, 221
Homeomorphism, 1, 7, 8, 18, 20, 25
Homogeneous of degree k, 61
Hopf bundle, 133, 134, 235, 237–239
Horizontal co-vector, 143
Horizontal k-form, 142
Horizontal lift, 134–138
Horizontal subspace, 127, 134, 137, 149
Horizontal vector, 121–124, 142, 143, 146, 148–150
Hydrogen atom, 174
Hypersurface, 26

I
Imaginary part (of a quaternion), 234
Imaginary subspace (of the quaternions), 234
Implicit Function Theorem, 25, 97
Infinitesimal, 3, 36, 54, 55, 72, 74, 98, 106, 139, 157, 159, 160, 162, 175
Inner product, 54, 55, 66–68, 70, 77–79, 122, 156, 157, 160–162, 175, 187, 227, 235
Instanton, 233–246
Integrability condition, 109–110
Integral curve, 3, 4, 55, 81–88, 90, 98, 99
Invariant, 3, 39, 67, 94, 95, 98, 134, 137, 148, 161, 165, 166, 175, 178, 181–183, 185, 186, 225
Irreducible representation, 174–176, 182
Isomorphic, 16, 21, 34, 35, 38, 39, 44, 47, 66, 106, 107, 114–116, 124, 130, 134, 172, 211, 236, 237

Isomorphism, 12, 16, 35, 41, 44, 46, 47, 65–68, 77–79, 82, 90, 95, 114–116, 119, 122–124, 127, 132, 139, 140, 175, 211, 212, 214, 215, 220, 227, 234, 236, 239, 240
Isospin; isotopic spin, 173, 174, 177, 178, 181, 182, 185, 186, 189, 192

J
Jacobi identity, 96
Jacobi matrix, 13, 14, 49–53

K
Kepler problem; two body problem, 174, 175
Kepler's law, 175
k-form, differential, 59, 71–73, 201
k-index, 202, 220
k-multi-linear, 71, 201
Koszul connection, 129, 138–141, 143
k-plane, 24, 106–110
Kronecker delta, 55, 96
Kronecker product, 55, 96
k-slice, 24, 25
kth exterior power, 60, 64
kth order contravariant vector, 37, 38

L
Lagrangian, 187, 188, 191, 192, 199, 220
  density, 187, 188, 191, 192, 199, 220
  formalism, 187
  mechanics, 187
Laplacian; Laplacian operator, 165, 167–170, 221
Least action principle, 187, 218
Lebesgue measure, 167
Left action, 64, 93–95, 98, 111, 114, 128, 222
Left group action, 94
Left invariant vector field; livf, 95, 98, 100, 148
Leibniz, Gottfried Wilhelm, 75, 89, 91, 141, 155, 260, 261
Leibniz rule, 75, 89, 91, 141, 155, 260, 261
Levi-Civita connection, 129, 140
Levi-Civita, Tulio, 195
Lie algebra, 94, 96, 98, 100–103, 119, 124, 129–131, 136, 138, 142, 174–176, 182, 196–199, 201, 203, 209, 210, 214, 218, 223, 231, 237, 240
Lie algebra valued 1-form, 201
Lie algebra valued k-form, 201
Lie bracket, 95, 96, 101, 109, 110, 119, 146, 148, 196, 197, 203
Lie derivative, 4, 55, 81–91, 138, 140

Lie group, 2, 4, 32–34, 67, 93–103, 107, 108, 111, 114–116, 128–134, 138, 148, 149, 172–176, 182, 192, 193, 195, 196, 198, 201, 204, 205, 207, 208, 214–216, 218–220, 222, 231, 234–236
Lie subgroup, 25, 97, 98, 101, 133, 234
Light, speed of, 151, 201
Linear action, 33, 93, 95, 108
Linear map, 13, 16, 23, 31, 34, 36, 38, 41, 43–48, 50, 53, 54, 63, 69, 73, 75, 83, 98, 107, 124, 137, 200, 201, 236, 239, 240
Local gauge transformation, 172, 177–182, 185, 186, 188, 198, 212–214, 216, 219, 220, 222, 231
Locally conserved quantity, 153
Locally trivial, 116, 118, 212, 222
Local section, 117, 118, 218, 219, 241
Local trivialization, 34, 112–114, 137, 139, 215, 216, 220, 222, 241, 242
Lorentz covariance, 163
Lorentz–FitzGerald contraction, 165
Lorentz's force law, 155, 163
Lorentz group, 67, 156
Lorentz, Hendrik, 165
Lorenz condition, 165
Lorenz gauge, 165
Lorenz, Ludvig, 165
lth order covariant vector, 38

M
Magnetic charge, 86, 151, 154, 155, 158, 160, 225, 226, 231, 232
Magnetic charge density, 225
Magnetic current, 225
Magnetic field, 4, 39, 86, 87, 151, 152, 154, 155, 157–159, 162, 163, 166, 171, 172, 225–227
Magnetic field lines, 86
Magnetic force, 86
Massive particle, 85, 87
Matrix
    anti-hermitian, 197
    hermitian, 179, 181, 182, 191, 197
    orthogonal, 108
    traceless, 102, 191
    unitary, 172, 201
Matter, 2, 3, 5, 12, 64, 86, 113, 151, 156, 160, 174, 179, 191, 192, 198, 200, 205
Matter field, 191, 198, 221–223
Maurer–Cartan form, 2, 130–133, 135, 149, 208–210, 214, 237, 238

Maxwell, James Clerk, 86, 87, 152, 154, 156, 186
Maxwell's equations, 86, 87, 151–156, 158, 160, 162–166, 187, 188, 191, 218, 225
    homogeneous, 164, 218
    non-homogeneous, 160, 165, 188, 191
Meson, 191–193
Method of stationary phase, 190
Metric, 40, 187, 188, 199, 220, 228
Minimal coupling, 138, 172, 223
Minkowski inner product, 162
Minkowski space, 67, 158–161
Model fiber space, 33, 35, 72, 111, 139
Model space, 1, 7, 9, 106
Monopole, magnetic, 216, 225, 226, 229
Monopole principal bundle, 230–232
Morphism, 12, 63, 64, 94, 113–115, 119, 128, 138, 175
Morphism of principal bundles, 114, 115
Multi-linear, 47, 60, 62

N
Natural chart, 20, 21, 34, 35, 113
Natural transformation, 132
Negatively oriented; negative orientation, 68
Neutron, 173, 176
Newtonian mechanics, 156
Newton, Isaac, 85, 156, 163, 166, 226
Nobel prize, 176, 192, 196
Non-abelian gauge theory, 195
Non-commutative geometry, 4, 247
Norm (of a quaternion), 234
Normalization condition, 167, 168, 176
n-sphere, 10
n-th remainder term, 29
Nucleon, 173, 176, 221, 223

O
1-form, 2, 3, 40, 56, 57, 74–78, 91, 124–128, 130, 137, 141, 143, 146, 150, 157–160, 162, 163, 179, 186, 200–210, 214–216, 221, 227–231, 237–239, 241, 243, 245, 246
One-parameter subgroup, 98–101, 197
Orbit, 107, 114, 133, 134, 235
Orbital angular momentum quantum number, 175
Ordinary differential equation; ODE, 4, 83, 84, 134
Orientation, 67, 68, 70, 78, 86, 158, 227, 228
Orthogonal, 66–68, 97, 105, 106, 108, 122, 172, 226, 237

Orthogonal basis, 68, 269
Orthonormal basis, 66–69, 77, 160

P

Parallel transport, 137, 139, 140
Partial difierential equation, 167, 178, 191,
     199, 221
Partial order, 9
Particle, massive, 85, 87
Path-ordered integral, 136
Pauli, Wolfgang, 176
Pauli matrices, 176
Perturbation theory, 190
Phase factor, 168, 177
Phase space, 87
Photon, 175, 177, 189, 192
Pion, 192
Poisson's equation, 166
Position, 85, 86
Positively oriented; positive orientation, 68,
     69, 78, 158
Potential, 5, 163, 164, 166, 169, 171, 172, 179,
     180, 186, 188, 227–230, 241
Potential, scalar, 166
Potential theory, 166
Potential, vector, 164, 172, 227–230
pre-Hilbert space, 222
Principal bundle, 2, 4, 5, 7, 21, 111–119,
     121–146, 149, 150, 195, 204, 205, 207,
     209–220, 222, 230–232, 235, 237, 247
   of frames, 116
   morphism, 114, 115, 128
Probabilistic, 167
Probability density, 167
Projective space
   complex, 10, 133
   quaternionic, 235
   real, 10, 107
Propositional logic, 142
Proton, 173, 174, 176, 192
Pseudoparticle, 233
Pseudoscalar, 39
Pseudovector, 39
Pullback, 71, 72, 76, 125, 126, 208, 210, 217,
     241, 243
Pure gauge, 163, 181, 182, 199

Q

Quantization, 190, 192, 247
Quantum chromodynamics (QCD), 192, 193
Quantum electrodynamics, 166

Quantum mechanics, 156, 166–172, 174–176,
     178, 197
Quantum number
   orbital angular momentum, 175
   spin, 175
Quantum principal bundle, 247
Quaternionic Hopf bundle, 235
Quaternionic projective space, 235
Quaternions, 133, 233–235, 242, 245
Qubit; quantum bit, 174
Quotient space, 3, 25, 45, 65, 76, 107, 133, 238

R

Real analytic, 29
Real part (of a quaternion), 234
Reducible representation, 39
Regular value, 25, 26
Renormalization, 192
Representation, 33–39, 55, 69, 72, 74, 94, 95,
     111, 122, 124, 157, 174–176, 182, 198,
     202, 211, 223
Riemannian metric, 40
Riemann surface, 10
Riesz representation theorem, 69
Right action, 93, 94, 112, 123, 128, 129, 137,
     219, 235, 238
Right action map, 93, 94
Right G-space, 93, 95
Right handed coordinate system, 228
Right H-module, 235

S

Scalar, 39, 40, 79, 83, 124, 126, 152, 155, 157,
     159, 164–166, 179, 181, 186, 189, 203,
     218, 222, 234
Scalar potential, 166
Second exterior power, 60
Second order contravariant vector, 37
Section
   global, 117
   local, 117, 118, 218, 219, 241
Section valued k-form, 141
Self-adjoint, 171, 190
Self-adjoint, essentially, 171
Self-dual (SD), 70, 220, 221, 246
semi-Riemannian manifold, 129, 140
Sequence (in a category), 265
Short exact sequence, 119
Short sequence, 117
Smooth
   atlas, 9, 10, 12, 33

function, 11–12, 14, 17, 19, 21, 23, 24, 26,
    32, 33, 39, 40, 56, 71, 72, 73, 75, 76,
    81, 82, 87–90, 94, 111, 117, 125, 142,
    159, 168, 198, 200–202, 208, 212, 230,
    231, 245
manifold, 2, 9, 11, 12, 21, 25, 31, 34,
    35, 41, 71, 72, 73, 87, 90, 93, 94, 97,
    107–111, 116, 125, 129, 130
section, 2, 56, 81, 106, 108, 109, 140, 222
vector bundle, 34–35, 40, 41
Smooth approximate characteristic function,
    29, 141
Spacetime, 67, 158, 168, 179, 186, 187, 193
Spectral theory, 190
Spectrum, 197
Speed of light, 151, 201
Sphere, 12, 25, 105–107, 133, 134, 152, 231,
    234, 235, 237
Spin
    angular momentum, 176
    quantum number, 175
Spontaneous symmetry breaking, 192
Stabilizer, 107, 237
Standard model, 192, 193, 247
Star center, 57
Star-shaped, 57, 227, 229, 230
Static, 166
Stationary phase, method of, 190
Steady state, 85, 151, 152, 166
Stokes' theorem, 153
Strong (nuclear) interaction, 177
Structure equation of a Lie group, 149
Structure group, 2, 113–117, 215, 218, 220,
    231, 235
Sub-bundle, 110, 122, 123, 196
Sub-manifold, 24–26, 97, 105, 108, 109
Sylvester's Law of Inertia, 68
Symmetric second order tensor, 39
Symmetric tensor, 39, 140
Symmetry breaking, spontaneous, 192
Symmetry, broken, 174
Système International (SI) d'Unités, 171

T
Tangent bundle, 2–4, 15–23, 31–33, 35–41, 56,
    71, 72, 74, 81, 90, 107, 108, 116, 125,
    129, 131, 132, 135, 140, 141, 144, 145,
    196
Tangent bundle map, 15, 21, 145
Tangent space, 21, 22, 36, 50, 55, 71, 90, 105,
    106, 108, 118, 141–143, 145, 196, 239
Tangent vector, 3, 20, 38–40, 54–56, 71, 118,
    135, 137, 138, 147, 189, 239, 243

Taylor polynomial, 27
Taylor series, 27–29
Taylor's theorem, 26, 28, 29
Tensor, 4, 36–40, 42–48, 59, 64–65, 91, 140,
    159
Tensor field, 40, 91
3-vector, 159
Time, 2, 3, 12, 22, 32, 62, 67, 84–87, 89,
    90, 97, 98, 136, 151–155, 159, 163,
    165–169, 172, 173, 175, 177–179, 185,
    189, 191, 192, 195, 200, 201, 207, 208,
    225
Time evolution, 87, 151, 163, 178, 189
Topological charge, 225, 232
Topological space, 7–10, 12, 17, 18, 25, 98,
    112
Total space, 2, 15, 17, 21, 23, 35, 108, 113,
    114, 125, 129, 130, 132, 204, 205, 213,
    214, 217, 237, 241
Trace (Tr), 102, 181, 189
Traceless, 102, 191
Trajectory, 85, 86
Transition function, 8, 16–20, 31–34, 108, 112,
    213, 231, 242
Transitive action, 107
Translation function, 2, 115
Triple product, 71
Trivial bundle, 41, 76, 115–117, 209–213
Trivial bundle, explicitly, 41, 209, 210
Trivializable, 41
Trivialization
    global, 177, 178, 181
    local, 34, 112–114, 137, 139, 215, 216,
        220, 222, 241, 242
Trivial, locally, 116, 118, 212, 222
Truth table, 143
Two body problem; Kepler problem, 174, 175

U
$U(1)$ gauge theory, 172, 225
Unified theory, 247
Unitary, 97, 102, 172, 176, 177, 190, 201
Universal covering space, 174, 175, 182
Universal property, 42–45, 47, 64

V
Vacuously true, 142
Variational derivative, 187, 188
Vector bundle, 4, 21, 24, 31–48, 71, 72, 81, 83,
    108, 111, 113, 116, 117, 138–141, 222
Vector bundle map, 35, 41

Vector calculus, 70, 77, 79, 96, 151–156, 164
Vector field
  complete, 91
  fundamental, 119, 132, 148
Vector potential, 164, 172, 227–230
Vector product, 70, 96
Vector valued 1-form, 124
Vector valued k-form, 201
Velocity vector field, 85
Vertical co-vector, 143–144
Vertical subspace, 122, 124, 130
Vertical vector, 118–119, 121, 123, 124, 128,
    130, 143, 205, 207
Volt, 171
Voltage, 171
Volume element, 68, 227

W
Wave function, 167, 168, 172, 176–179, 190,
    198, 222, 223
Weak (nuclear) interaction, 192
Wedge product, 59, 61–63, 65, 72, 150, 217
Weyl, Hermann, 195

Y
Yang, Frank, 173, 176–178, 182, 184–188,
    191, 192, 195, 196, 201
Yang–Mills action integral, 189
Yang–Mills equations, 187–191, 199, 218,
    220, 221, 233, 246
Yang–Mills theory, 173–193, 195, 197–200,
    221, 223, 225
Yukawa, Hideki, 177

Printed in the United States
By Bookmasters